Coordination Compounds of Porphyrins and Phthalocyanines

Coordination Compounds of Porphyrins and Phthalocyanines

B. D. Berezin
Head of Organic Chemistry Department,
Ivanovo Chemical Technology Institute,
Moscow

Translation by
V. G. Vopian

A Wiley–Interscience Publication

JOHN WILEY & SONS
Chichester · New York · Brisbane · Toronto

First published Moscow 1978 as
Koordinatsionnye soedineniya porfirinov
i ftalotsiyanina

British Library Cataloguing in Publication Data:

Berezin, B. D.
Coordination compounds of porphyrins and phthalocyanines.
1. Porphyrin and porphyrin compounds
2. Coordination compounds
3. Phthalocyanines
I. Title
547'.593 QD401 80-40958

ISBN 0 471 27857 2

Typeset in VIP Times by Preface Ltd, Salisbury, Wilts
Printed by The Pitman Press, Bath, Avon.

Preface

Porphyrins are compounds widely occurring in nature, and include the well known chlorophyll *a*, chlorophyll *b*, and haem of the blood, which have been the object of numerous studies over the past 100 years.

Of particular interest is the group of structural analogues of natural porphyrins which includes phthalocyanine and its various metal derivatives. The investigation of these compounds has been especially productive on account of the work by Academician A. N. Terenin, who proposed their use as convenient molecular models for the study of the behaviour of chlorophyll in various photophysical and photochemical processes of photosynthesis. Other factors contributing to successful studies of metal phthalocyanines are their prominent colour-related properties as well as their capacity to behave as semiconductors and catalysts. The close structural affinity of porphyrins and phthalocyanine is the reason why they are considered together in most reviews, adequately complementing each other.

The chemistry of these compounds is covered in specialized publications, such as the well known works by Fischer and Orth, Fischer and Stern, Rabinovich, Lemberg, and Legge. However, the physical aspects of these molecules as a specific and extremely interesting group of coordination compounds are not covered in the above works. In his monograph (*Porphyrins and Metalloporphyrins*, Amsterdam, Elsevier, 1964 and 1975) J. Falk deals with the chemistry of porphyrins and metalloporphyrins with primary emphasis on the preparative methods of isolation and purification of these compounds. Less attention is given to the kinetics and thermodynamics of the processes involving porphyrins and metalloporphyrins. Phthalocyanine as a model compound is mentioned but briefly.

The only monograph concerned with phthalocyanine compounds, by F. Moser and A. Thomas, deals mainly with the synthesis and production of phthalocyanine and, to a much lesser extent, with the physicochemical processes that determine the conditions of their formation and stability, forms in which they exist in solutions, the mechanisms of formation and dissociation, and conditions for further coordination.

Interesting data on the occurrence of ionic forms of porphyrins and phthalocyanine, having to do with the coordination chemistry of these compounds, can be found in a monograph entitled *The Spectroscopy of Chlorophyll and Allied Compounds* [*Spektroskopiya khlorofilla i rodstvennykh soedineniy*] (Minsk, Nauka i Tekhnika Publishers, 1968) by G. P. Gurinovich, A. N. Sevchenko, and K. N. Solovyev.

Until the present a great number of studies into the physicochemical properties of natural and synthetic porphyrins, phthalocyanine and its polymers, and their numerous complexes exhibiting a great variety of properties have been conducted by many investigators, including members of the present author's laboratory. In this monograph an attempt is made to analyse the specific features of the processes of their formation and dissociation, the mechanisms of these processes, and conditions for further coordination on metalloporphyrins. A great deal of attention is given to studies into their ionization and dissociation, structural aspects, electronic effects of coordination and their manifestation in the optical, coordination, acid–base, and oxidation–reduction properties.

I am fully aware of the complexity of the objects and phenomena considered in this book and shall gratefully accept all critical comments.

I should like to thank R. P. Yevstigneeva, Corresponding Member of the USSR Academy of Sciences, and Professor G. A. Krestov, Distinguished Scientist of the RSFSR, for their having reviewed the manuscript and their constructive criticism, as well as G. V. Antonycheva and V. G. Andrianov for their generous help in preparing the manuscript. Special thanks are due to R. A. Berezina for the unflagging support and assistance I have received from her in working on the monograph.

B.D.B.

Preface to the English Edition

I profoundly appreciate this opportunity to address readers in Western Europe and North America who are interested in coordination chemistry in general and in the chemistry of porphyrin complexes and their structural analogues in particular. This opportunity has a special significance to me in view of the fact that the chemistry of synthetic analogues of porphyrins, known as phthalocyanines, saw dramatic progress in Great Britain as far back as the thirties and forties, as a result of the brilliant work by R. Linstead and his followers. Because of his efforts, phthalocyanines rank among the most valuable dyes of various classes.

In the forties and fifties phthalocyanines started being widely used as stable and adequate models of chlorophyll. Academician A. N. Terenin followed by Academician A. A. Krasnovsky and his collaborators carried out extensive studies into the primary photophysical processes of photosynthesis, photsensitization of chemical reactions, photochemical destruction of dyes and chlorophyll, and many other processes involving phthalocyanines. Similar studies were under way in many countries.

I began my physicochemical studies of phthalocyanine, its numerous complexes, and porphyrins in the mid fifties.

Porphyrins and phthalocyanines have been investigated extensively in many countries, including in particular the Soviet Union, the United States, Great Britain, Japan, and France. The main reason is that these molecules exhibit great structural variety, a capacity for coordination, a diversity of complex forms, and unusual structure and properties. Those who work in this field have no doubt that porphyrin complexes are highly promising as catalysts for many major processes in the chemical and other industries, as enzyme models, as photosensitizers, as medicinal preparations, and even as blood substitutes. The possibilities of using them as natural pigments and dyes are extensive.

This is what makes it necessary to study such molecules theoretically on a broader scale and has prompted me to write the present book covering the fundamentals of coordination chemistry of porphyrins and phthalocyanine.

It summarizes the results of studying various forms in which porphyrins and phthalocyanines are present in solutions, their acid and basic ionization, the mechanism of formation of metalloporphyrins, and further coordination of simple and complex ligands on metalloporphyrins, that is, formation of axial complexes. Model phthalocyanine molecules are used as examples illustrating the manifestation of electronic effects occurring in the coordinated ligand as a consequence of different types of chemical bonding. The electronic effects of coordination, together with the steric factor, are responsible for the stability of complexes, their spectral, acid–base, oxidation–reduction, catalytic and other properties.

Particular attention is given to the effect of the solvent on the solvation of porphyrins, metalloporphyrins, and coordinating salts. The nature of the solvent determines the formation rates, stability and many other properties of porphyrins. The kinetic stability of porphyrin and phthalocyanine complexes is also treated at length in view of the great practical interest displayed at present in the synthesis of catalysts derived from porphyrins, which are stable in highly aggressive media. Almost for the first time, data on the thermodynamic stability of metalloporphyrins are presented. It should be emphasized that most of the studies carried out in our laboratory and reported in this book have involved chlorophyll and its metal analogues, structural and metal analogues of haemin, phthalocyanines and its polymers.

Over the past five years, a wealth of new facts has been accumulated in our laboratories (at the Ivanovo Chemical Technology Institute and in the Department of Chemistry of Non-Aqueous Solutions of the USSR Academy of Sciences), and in those of other investigators, concerning tetraphenylporphin, tetrabenzoporphin, together with many aza derivatives of porphin and phthalocyanine. These findings are fully consistent with the basic concepts set forth in the monograph and may serve as a sound basis for further theoretical studies in the physical, coordination, and applied chemistry of porphyrins.

I shall consider my task accomplished if this book contributes in some way or another to advances in experimental and theoretical studies into these vital compounds.

B. D. Berezin
Head of the Organic Chemistry Department
Ivanovo Chemical Technology Institute

Honoured Scientist of the RSFSR

List of abbreviations

MX_2	salt of ion of metal M with anion X
H_2P	porphyrins and unsubstituted porphin
H_2OETP	octaethylporphin
H_2TPP	tetraphenylporphin
H_2TMOPP	tetramethoxyphenylporphin
H_2TBP	tetrabenzoporphin
H_2EP	etioporphyrin
H_2PP	protoporphyrin
H_2MP	mesoporphyrin
H_2DP	deuteroporphyrin
H_2HP	haematoporphyrin
H_2PyP	pyrroporphyrin
H_2PhP	phylloporphyrin
H_2CP	coproporphyrin
H_2Chln	synthetic chlorin
H_2Chlne_6	chlorin e_6
H_2Rh	rhodine g_7
H_2EChln	etiochlorin
H_2Chla	chlorophyllic acid *a* or phaeophytin *a*
H_2Chlb	chlorophyllic acid *b* or phaeophytin *b*
H_2Phe	phaeophorbide
H_2TPyP	tetrapyridylporphin
H_2TAP	tetrazaporphin
$H_2TCHTAP$	tetracyclohexenetetrazaporphin
H_2Pc	phthalocyanine
$(H_2Pc)_i$	polymeric phthalocyanine from pyromellitic acid
H_2PcTS	tetrasulphophthalocyanine tetraanion
MP	metalloporphyrin
MPc	phthalocyanine complex
DME	dimethyl ether
solv	solvent

Contents

IONIZATION PROPERTIES

COORDINATION PROPERTIES

Chapter 1

General concepts of porphyrin and phthalocyanine complexes

1.1 Specific Features of Natural and Synthetic Porphyrin Complexes

Metalloporphyrins, which include such important natural complexes as chlorophyll, haem of the blood, and others, represent a vast and unique group of intercomplex compounds. The central metal atom displaces two hydrogen ions from the porphyrin ligand and practically finds itself in a symmetrical electrostatic field of four nitrogen atoms with which it may form four equivalent, or almost equivalent, coordinate donor–acceptor bonds. If the interaction between the metal and the porphyrin anion is primarily electrostatic, labile ion complexes are formed. These include complexes of Na^{+}, K^{+}, Rb^{+}, Cs^{+}, Be^{2+}, Sr^{2+}, Ba^{2+}, Ca^{2+}, and some other ions. But if the electrostatic interaction involves filling of the vacant orbitals of the central atom by the electrons of the donor N atoms of the ligand, stable porphyrin complexes of the covalent or predominantly covalent type are formed. In this case we have complexes of Fe^{2+}, Fe^{3+}, Co^{2+}, Ni^{2+}, Cu^{2+}, Zn^{2+}, Mn^{3+}, Cr^{3+}, Al^{3+}, Ga^{3+}, Si^{4+}, Ge^{4+}, Sc^{3+}, Ti^{4+}, VO^{2+}, Pt^{4+}, Od^{4+}, and other cations. Most of the above complexes are formed by biometals. As will be seen from what follows, complexes of porphyrins with Mg^{2+}, that is chlorophyll and its structural analogues, do not belong either to purely ionic complexes or to the stable group of predominantly covalent complexes. This peculiarity of chlorophyll, stemming from its intermediate position in the complex stability series, is to some extent responsible for its unique role in nature.

In contrast to chlorophyll, the complex of protoporphyrin with Fe^{2+} (Fe^{3+}) is one of the most stable metalloporphyrins of the covalent type. This property is fully consistent with both the biological composition of the medium in which haem acts in a complex with protein and with its biological functions.

The specific features of metalloporphyrins as intercomplex compounds are due not only to the polydentate (tetradentate) nature of the ligand but also to its rigidity. The latter is determined by the planar structure of the large ring of the porphyrin molecule, by the unique conjugation in it which is due to the strong π-electron interaction over the entire large ring (the so-called macroring), and by the particularly favourable structure of the coordination centre, consisting of four nitrogen atoms, on account of which most metal ions easily enter this space, coming into close contact with the nitrogen atoms. Because of its high rigidity the porphyrin ligand imposes specific requirements on the geometric parameters of the metal ion, whereby they form two clearly defined groups of stable or labile complexes.

Characteristically, the metal ion that has entered into coordination with a porphyrin is in fact a partner of the conjugated porphyrin system and may either stabilize or destabilize it. Owing to this direct contact with the atoms of the conjugated system, it influences all, even the most remote parts, of the large molecule and alters the oxidation–reduction, acid–base, electron-optical, and all other properties of the porphyrin. In this connection it is of particular interest for a coordination chemist to use metalloporphyrins in examining the effect of the nature of the metal involved, the type of chemical bonding, and its strength on the major properties of porphyrins.

A specific feature of metalloporphyrins is their insolubility in water and solubility in organic solvents in cases where no significant solvation centres are present among the functional substituents of the ligand. This is the reason why most studies of porphyrins and metalloporphyrins have been conducted in non-aqueous media.

Another salient feature of porphyrins is their structural diversity, which can be seen from structural formula **I**.

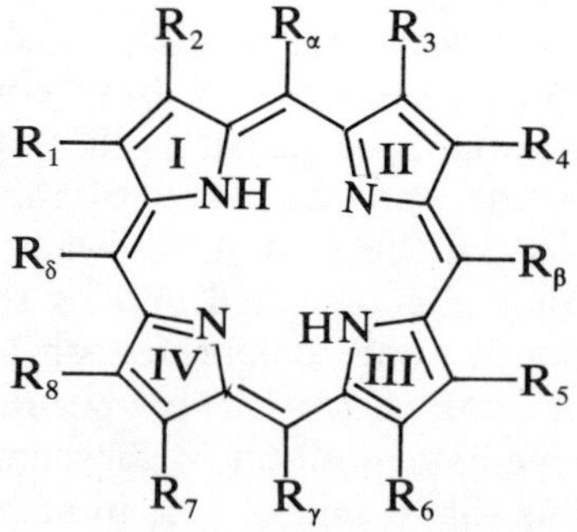

I

Porphyrins differ in the nature of the bridging groups occupying meso positions in the porphyrin molecule. The bridging groups may only be —CH=, —(X)C=, —N=, or their combinations. The structural diversity of porphyrins is further enhanced by the presence of various pyrrole substituents (R_1–R_8) which may be H, CH_3, C_2H_5, $CH{=}CH_2$, $CH(OH)CH_3$, C(O)H, COOH, CH_2COOH, and CH_2CH_2COOH. If it is also taken into consideration that one or more pyrrole double bonds may be hydrogenated and that adjacent substituents R_1–R_8 and R_α–R_δ may form closed cycles, then the diversity of molecular structures of porphyrins, chlorins, azaporphyrins, phthalocyanines, phorbins, and so on becomes more evident.[1] It should be emphasized that a porphyrin molecule ceases to exist as soon as at least one double bond in the macroring is hydrogenated.

The specific features and great diversity of porphyrins, particularly metalloporphyrins, are the factors responsible for their importance and wide distribution in nature, their extensive use in various physical, physicochemical, quantum-chemical, and biological studies, as well as in the production of dyes, semiconductors, and catalysts. The structural diversity of porphyrins and metalloporphyrins opens up enormous possibilities for studying the effect of functional substituents and central atoms on the properties of porphyrins. Such studies will make it possible in the future to provide an answer to the question why Nature has chosen, as the basic component of the photosynthetic and respiratory molecular system, porphyrins with a particular combination of functional substituents such as can be found in chlorophyllic acids (phaeophytins) and protoporphyrin.

An important characteristic of porphyrin complexes is their extremely high stability to dissociation. This is one of the reasons why nobody has been able so far to measure quantitatively the equilibria of their formation in a direct manner and to characterize this stability in terms of a definite constant. The kinetic method has been widely used in investigating the stability of porphyrin and phthalocyanine complexes.

The capacity of complexes for additional coordination is to some or other degree their common property. However, one can hardly find another instance where the coordination of additional ligands is so manifest as in the case of planarly coordinated metalloporphyrins. It is referred to as extra coordination for the ligands are attached along the z axis, passing through the metal atom at a right angle to the plane of the porphyrin molecule. Extra coordination is an important property of such metalloporphyrins as chlorophyll, haem of the blood, catalase, and others for it results in attachment of the active metal complex to the molecule of the carrier protein and is essentially a transport reaction as far as oxygen, hydrogen peroxide, and the electron are concerned.

Finally, an important and unique feature of porphyrins and their metal derivatives is their electronic absorption and emission spectra.[1] The diversity of the latter, corresponding to the great variety of porphyrins, the typical arrangement of spectral bands, the sensitivity to certain effects on the molecule, and their complexity, make them not only a most reliable means for identifying porphyrins but also a rich source of information on the structure of porphyrins, the type of their coordinations, the interaction between substituents and ionizing media, and the intramolecular energies.

1.2 Metalloporphyrins in Biological Systems. Industrial Uses of Phthalocyanine Complexes

A large number of biological systems are known in which metalloporphyrins act as initiators of a particular biological process. Most investigations have involved chlorophyll[1–21] which, being part of the protein–lipid system, initiates photosynthesis in green plants and in photosynthetic bacteria,† that is it initiates the stage of accumulation and transfer of the energy of a light quantum, the stage of electron transfer reaction through the complex photosynthetic system, all the way from the reducing to the oxidizing agents. At this initial stage, chlorophylls *a* and *b* (**IIa**, **IIb**) behave as supracomplex compounds because their basic ligand, known as phaeophytin or chlorophyllic acid, is coordinated with the magnesium ion (primary coordination).

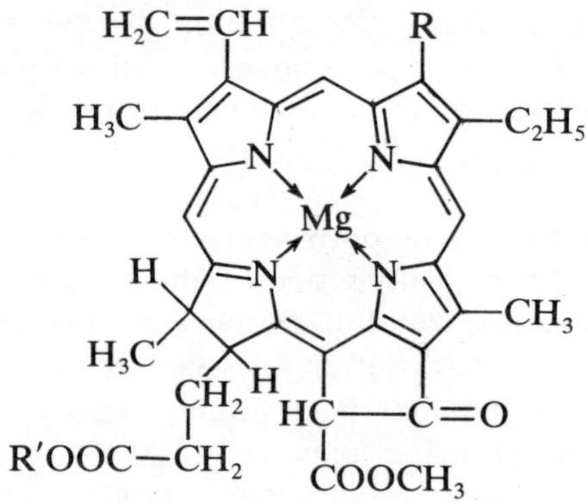

IIa: $R' = C_{20}H_{39}$; $R = CH_3$
IIb: $R' = C_{20}H_{39}$; $R = CHO$

The central magnesium atom in chlorophyll is coordinatively unsaturated and readily attaches two more ligands (so-called extra ligands) along the *z* axis perpendicular to the molecular plane (secondary coordination). It is not yet known what kind of ligands these are. One of them may be a protein molecule weakly coordinated to magnesium through the C—O or NH centres of the carbamide bond. It is also possible that the chlorophyll–protein complex is the result of forces other than the donor–acceptor interactions between the central atom and protein. Such forces may be donor–acceptor interactions between the protein molecule and oxygen atoms of the functional groups of chlorophyll. These are weak acid–base interactions. Van der Waals' forces may also be involved. Taking into consideration the large size of both molecules, one may assume that these forces are sufficiently strong to retain the molecules within the chlorophyll–protein complex.[22] Similar interactions often occur in clathrate

†In photosynthetic bacteria the function of chlorophyll is performed by its structural analogue—bacteriochlorophyll.

(inclusion) compounds. At least one of the coordination positions of magnesium in chlorophyll must be occupied by a water molecule with which its coordination is sufficiently stable.[23] Such a coordination of water is highly favourable for its oxidation after one or two electrons have left the triplet levels of the chlorophyll molecule.[15,21]

One can hardly imagine another situation as favourable for oxidation of the water participating in the oxidation–reduction chain of photosynthesis. Complex formation is known[24] to be the initial stage of oxidation–reduction when the central metal atom acts as the oxidizing agent and the ligand serves as the reducing agent. In the case of metalloporphyrins the process of electron transfer from the extra ligand (e.g. water molecule) to the central atom is particularly favourable in view of the possibility of the charge being distributed over the entire conjugated system of the macroring. Thus coordination phenomena may play a predominant role as early as the first, initial stage of photosynthesis.

It may be assumed that the particularly favourable conditions for coordination of extra ligands at the Mg atom determine the specific role of chlorophyll (rather than another one of its metal analogues) in the process of photosynthesis. Metal analogues of chlorophyll such as complexes of Cu^{2+}, Co^{2+}, Zn^{2+}, Fe^{2+}, and other metals with chlorophyllic acid may also occur in small amounts in green plants and play a major role in bioprocesses. Metal analogues of chlorophyll with iron and copper in chloroplast are assumed to act as electron traps (see p. 177 of reference 6).

Many theoretical and experimental studies involve another important metalloporphyrin—iron protoporphyrin.[25–27] Being part of the protein complex of haemoglobin, it is known as haem of the blood (**III**).

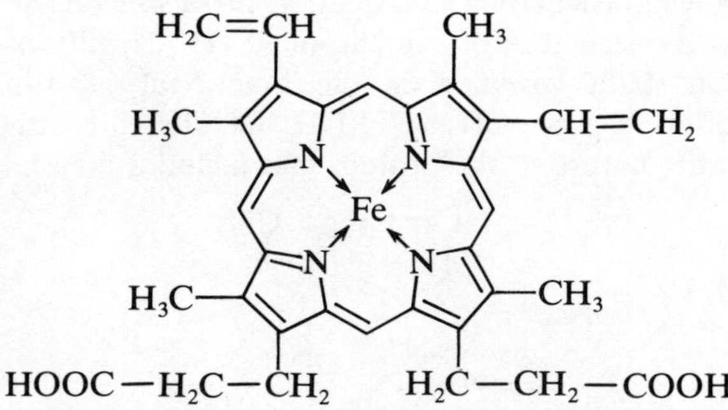

III

When separated from protein, Fe^{2+} protoporphyrin (haem) immediately oxidizes to Fe^{3+} complexes. In this case Fe^{3+} coordinates to extra ligands in addition to the protoporphyrin anion. If Cl^- is such a ligand, the complex is called haemin, (Cl)FePP, and if the extra ligand is OH^-, the compound is known as haematin. The function of the second extra ligand occupying the sixth coordinate place is performed by a molecule of water or another solvent, for example (HO)FePP(H_2O). The phenomenon of complexing plays a decisive role in the formation of haem from protoporphyrin and iron salts in living organisms, determines the stability of haem in a biological medium, and ensures functioning of haem in a complex with protein as an oxygen carrier, that is as a respiratory pigment. In recent years an intensive search has been under way for blood substitutes,[28] in which studies into the nature of chemical bonding in metalloporphyrins and the relationship between primary and secondary coordination are given primary consideration. No less important is the effect of the functional substituents of porphyrin on its primary and secondary complex-forming capacity.

Other metalloporphyrins cannot participate in the process of reversible attachment of oxygen as readily as haem in haemoglobin. In the course of O_2 being absorbed by the iron atom, coordination of imidazole of the histidine present in the protein molecule takes place. The protein molecule itself is attached to the molecule of the iron–protoporphyrin complex due to the acid–base interaction between the nitrogen atoms of the amino acid residues of protein and residues of the propionic acid of porphyrin.[26] The fact that haem contains a central iron atom is significant. The bonding of metalloporphyrin to an O_2 molecule calls for participation of open *d* orbitals of the metal and coordinative unsaturation of the metal atom in the metalloporphyrin. No other central atom (Mn^{2+}, Zn^{2+}, Ni^{2+}, Cu^{2+}) apart from Co^{2+} exhibits both properties at any one time.

The best known enzyme systems containing metalloporphyrins include cytochromes, catalase, and peroxidase.[23,25,29] Cytochromes are essentially porphyrin complexes of iron which form, within a protein complex, oxidizing enzymes participating in oxidation processes on the cellular level. Given below is the structure of cytochrome *c* (**IV**). It differs from haem of the blood by a slight variation in the functional substituents (cysteine molecule residues, Y = S—CH_2—CH(NH_2)COOH, are attached via vinyl groups) and in the nature of the protein that includes a cysteine molecule.

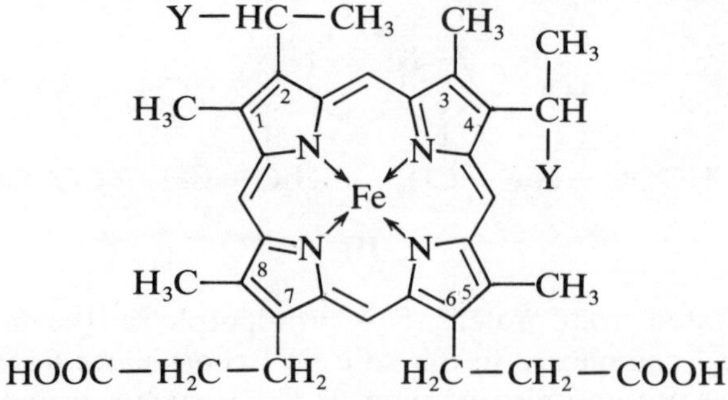

IV

These variations impart new properties to the metalloporphyrin, namely easy reversibility of the process

$$\text{ferrocytochrome} \underset{+e}{\overset{-e}{\rightleftharpoons}} \text{ferricytochrome}$$

with normal oxidation–reduction potential $E_0 = +0.26$ at pH ≈ 7. The central Fe^{2+} atom in ferrocytochrome is coordinated to histidine 18 and methionine 80 of the protein molecule with the result that it is not oxidized by the molecular oxygen within the enzyme. A large number of cytochromes are known,[23] differing in the nature of the protein associated with porphyrin and in side substituents. It is assumed that cytochrome *a* contains 1,3,5,8-tetramethyl-2-formyl-4-ethylporphin-6-acryl-7-propionic acid. Cytochrome a_3, which forms part of cytochrome oxidase, is assumed to be 1,5,8-trimethyl-3-formyl-2,4-divinylporphin-6,7-dipropionic acid.†[29] It is susceptible to oxidation by air and plays an important role as the final link in the respiratory chain.[29]

Cytochrome *b*, catalase, and peroxidase contain, along with a protein molecule protoporphyrin (1,3,5,8-tetramethyl-2,4-divinylporphin-6,7-dipropionic acid‡).

The functioning of catalase (HO)FePP, which exchanges ligand OH^- for OOH^- then reacts bimolecularly with the second hydrogen peroxide molecule, is accompanied by liberation of H_2O and O_2 and regeneration of catalase. Thus catalase protects the living cell against the destructive action of hydrogen peroxide.

Peroxidase, which performs the function of an enzyme in oxidation reactions, also has the composition of extra complex (HO)FePP. The sixth coordinate place is occupied by a protein molecule which is additionally bonded to the skeleton of the porphyrin molecule. Iron–porphyrin enzymes provide a good example of the subtle relationship between their functions and the extra-coordination phenomena occurring at the central atom. The substitution of OH^-, OOH^-, or protein molecules of different nature for the solvent molecule in the coordination sphere of Fe^{3+} radically affects the biological function of the complex, which clearly suggests that comprehensive experimental and theoretical studies into the coordination properties of porphyrin molecules are required with due account for their structure, the central metal atom, and environmental factors. This is the primary objective pursued by the present monograph.

The diversity of useful properties and structures of metalloporphyrins provides favourable conditions for varying the electronic states of the central metal atom which determines the coordination and catalytic properties, as well as searching for new effective catalytic systems that may be compatible with natural biocatalysts and even surpass them. We are speaking of synthesis, on the basis of porphyrin systems, of new effective catalysts, blood substituents, thermo- and photostabilizers, medicinal substances, organic semiconductors, pigments and dyes, solar cells, and

†There exist other hypotheses as to the structures of cytochromes *a* and a_3.
‡The author uses Fischer's nomenclature.

inhibitors of homolytic degradation of protein molecules in living organisms exposed to X and gamma rays.

At present, natural porphyrins and their synthetic analogues are finding a host of industrial uses. Phthalocyanine (**V**), which is a structural synthetic analogue of porphyrins, and its numerous intercomplex salts and functional derivatives are especially well known pigments.[30–38]

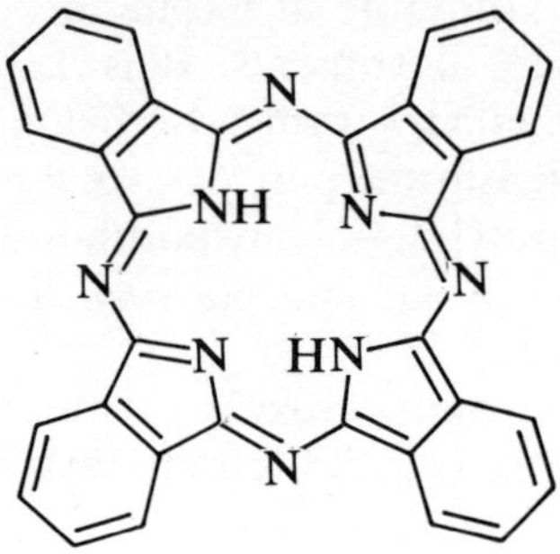

V

Their extremely high thermal stability, fastness to light, inertness to acids and alkalis, insolubility in most solvents, high dyeing power, purity, and colour intensity have ensured their reputation and wide application in the painting, printing, textile, and paper industries, as well as in chemical fibre and plastic dyeing processes, as superior quality blue, blue-green, and green pigments. Also used as pigments are high-chlorine derivatives of phthalocyanine and its complexes (e.g. 'phthalocyanine green', which contains up to 15 chlorine atoms in the copper–phthalocyanine nucleus). Sulpho derivatives of copper phthalocyanine are employed as valuable direct dyes for cotton, known as 'Sirius turquoise' and 'direct light-fast turquoise'. In addition to the most widely used copper complex, complexes of phthalocyanine with Co, Ni, Al, and other metals are also used industrially. The Co^{2+} complex applied on fabrics in a vat dyeing process[34] is known as 'indanthrene brilliant blue 4G'.

The introduction of various functional substituents into the benzene nuclei of phthalocyanine, whereby it acquires solubility, new chromophore properties, and the ability to be fixed on fibres from solution, permits numerous dyes of different classes to be obtained on its basis.[32,33,36]

A popular process in the textile industry is phthalogenic dyeing of fabrics, based on synthesis of phthalocyanines from intermediate products (so-called phthalogens) directly on fibres.[34,35]

Broad possibilities for using porphyrin complexes are being opened up in the technology of semiconductors[39–42] and lasers,[43–45] as well as in quantum-electronics research,[44–48] because of their exhibition of semiconductor and other valuable properties.

Another promising area of their application as oxidation–reduction catalysts includes industrial synthesis,[48–52], enzyme catalysis,[53–56] and electrochemical processes involved in the catalytic reduction of oxygen in fuel cells.[57,58] Other uses of phthalocyanines include, for example, processes for producing heavy isotopes of some elements[59] and oxidation–reduction indicators.[60]

1.3 The Role of Complexing in the Formation, Dissociation, and Functioning of Metalloporphyrin Systems

Porphyrin molecules perform their biological and catalytic functions only within metalloporphyrins. These functions are determined by the coordination (donor–acceptor, complexing) properties of the porphyrin molecule and metal. Since they are bound in the metalloporphyrin molecule by a very strong chemical interaction which ensures the integrity of the entire molecule, we may speak of coordination properties of metalloporphyrins, meaning their capacity for further coordination to extra ligands and for staying together, that is coexisting in proton-donor media.

Thus the coordination properties of porphyrins determine the conditions for and rates of formation of metalloporphyrins and their ability to survive in various media (including proton-donor, oxidation–reduction, and photoexciting ones) without undergoing protolytic dissociation and oxidative or photo-oxidative degradation. They also determine the capacity of porphyrins for acting as photosynthetic, respiratory pigments and enzyme (catalytic) systems.

Investigations of coordination phenomena involving porphyrin molecules are, thus, of paramount importance in the chemistry of porphyrins. Work in this field has been under way over the past two or three decades[18] and is rapidly progressing with studies being conducted in many countries. However, the total volume of research into coordination phenomena involving metalloporphyrins is still far from being adequate to the scientific magnitude of the problem. This lag of the coordination chemistry of porphyrins is primarily due to the complexity and unavailability of most porphyrin ligands and to the long established concept of poor stability and ephemeral nature of porphyrin molecules isolated from natural sources under conditions normally prevalent in complex formation studies. This is probably why in many physicochemical studies use is made of stable synthetic porphyrins such as tetramethyltetraethylporphin (etioporphyrin), α,β,γ,δ-tetraphenylporphin, and α,β,γ,δ-tetrapyridylporphin.

Naturally, of greatest practical interest are the coordination properties of two most important natural porphyrins—chlorophyllic acid (phaeophytin) and protoporphyrin—as well as their complexes. In this work they are given primary attention. At the same time it should be borne in mind that no solution to problems concerning the effect of functional substituents and the structure of a porphyrin molecule as a whole on its activity in biological and other processes can be found without investigating as many porphyrins of different structure as possible, including synthetic ones. This aspect of the problem is also treated at length in the present monograph.

The question as to the role of the central metal atom in molecules of biologically active metalloporphyrins is yet to be answered. Magnesium is present in chlorophyll, while the haem of haemoglobin, cytochromes, and catalase contains an iron atom. To answer the question why these two and no other metals are involved in biological evolution one must examine the effect of the species of the metal atom on all physicochemical phenomena occurring with the participation of porphyrins in photosynthesis, oxygen transport, intracellular oxidation, and other processes. The most interesting in this respect are the electron-optical, oxidation–reduction, and coordination properties. Systematic studies into the kinetic stability of metal analogues of chlorophyll and haemin have given insight into the specific role the magnesium atom plays in the chlorophyll molecule from the standpoint of coordination properties of metalloporphyrins.

As far as the iron atom is concerned, the picture is much clearer. Its particular role stems from the abundance of iron in the Earth's crust rather than from the specific features of its electronic structure since the function of reversible addition and oxygen transport can be performed by other metals as well, for example, cobalt,[28] which may form complexes with porphyrins, possessing a sufficiently high thermodynamic stability and capable of withstanding oxidative degradation in the presence of active haemoglobin.

Chapter 2

Synthesis, isolation, and purification of porphyrin and phthalocyanine complexes

Complex compounds of chlorophyllic acid, protoporphyrin, and most of their structural analogues are synthesized from available ligand molecules of porphyrins and metal salts. Therefore the problem of synthesis of complexes is determined entirely by the availability of respective porphyrins. Only in the case of phthalocyanine or its polymers are complexes easier to synthesize in sizable amounts from semiproducts, that is from structural units making up a phthalocyanine ligand, in the presence of a metal.

The synthesis and isolation of various porphyrin ligands have been described in detail.[23,61,62] In their monograph,[8] Sevchenko, Gurinovich, and Solovyev critically review and summarize the basic principles of synthesizing porphyrins without metals (i.e. porphyrin ligands) from simpler organic molecules and present proven methods for synthesis of about 20 porphyrins belonging, as a rule, to different structural types (porphin, chlorin, tetraphenylporphin, tetrabenzoporphin, tetrazaporphin, phthalocyanine, and some of their functional derivatives). Therefore there is no need to dwell on this problem. This chapter deals primarily with methods for obtaining porphyin complexes and their specific features.

2.1 Isolation of Porphyrin Complexes (Chlorophyll, Haemin) and the Corresponding Porphyrins from Natural Sources

The most convenient way to isolate chlorophyll is from nettles by the method described in reference 68.

Dried nettle leaves (100 g) are shredded in a ball mill with the addition of a small amount of $CaCO_3$, and 400–500 ml of acetone are added as an extractant for several minutes. The mixture is filtered through a Schott funnel, and the filtrate is mixed with 2 l of acetone–petroleum (1:1) solvent. Then a saturated aqueous solution of $CaCO_3$ is added until two layers are formed. The aqueous layer will contain acetone and water-soluble impurities. The solution of chlorophyll $(a + b)$ in petroleum ether is separated from the aqueous layer and washed with water as it is poured into the latter through a narrow tube immersed to a depth several centimetres below the water level. The ether layer is then washed off several times with 80% methanol (in 200 ml amounts) saturated with petroleum ether and containing 10 mg of oxalic acid to prevent chlorophyll from allomerization.[2] The last wash liquid must be light (greenish-yellow). The chlorophyll solution is added dropwise into distilled water, and chlorophyll starts precipitating. The suspension of chlorophyll ($a + b$) is filtered, after having been dried in a desiccator over anhydrous Na_2SO_4, through a 3 cm layer of confectioner's sugar containing 3% of starch and placed in a Büchner funnel (d = 14 cm). The sugar layer is flushed in advance with petroleum ether and pressed until a smooth surface is formed. The pigment precipitated on confectioner's sugar is washed off with petroleum ether to remove carotinoids, separated mechanically, and extracted with not more than 60 ml of petroleum ether having a boiling point of 35–45 °C. Ethyl ether (30% by volume) is added to the solution, whereafter the solution is fed into a chromatographic column (5 cm × 35 cm) with confectioner's sugar after having been dried briefly at 90 °C in a vacuum. The separation normally reveals two chromatographic zones—a green zone 4–10 cm long containing chlorophyll *b* and a lower grey-blue zone 2–6 cm long containing chlorophyll *a*.

The upper zone is mechanically withdrawn, and the column walls are wiped clean with cotton soaked in ether. This zone is used for isolation of chlorophyll *b*. Its central portion is dissolved in a mixture of petroleum ether (70%) and ethyl ether (30%), then chromatographed again on confectioner's sugar. This is followed by a third chromatographic analysis. Note that two chromatographic separations are sufficient in kinetic and thermodynamic studies of chlorophyll *b*. The central part of the chromatographic zone is extracted with ethyl ether; the extract is washed with water, dried over Na_2SO_4, and filtered through a glass Schott funnel No. 3; then the filtrate is evaporated to dryness at room temperature in a vacuum.

The ether solution of chlorophyll *b* exhibits bands at the following wavelengths in the electronic absorption spectrum: 642.5 (I), 594.0 (II), 550.0 (III), 453.0 (IV), and 430.0 (V) nm. The band numbers are given in brackets.

The central portion of the lower grey-blue zone of the chromatogram is separated from the upper and lower portions and extracted with ethyl ether. To isolate solid chlorophyll *a* the procedure is the same as in the case of chlorophyll *b*. In the ether solution chlorophyll *a* is characterized by absorption bands at the following wavelengths: 660 (I), 614 (II), 576 (III), 532 (IV), 429 (V), and 410 (VI) nm.

Experience shows that for chromatographic separation of chlorophylls use can be made of the initial acetone (or ethanol) extract of chlorophyll ($a + b$) passed into a benzene–petroleum ether mixture.[3]

The chlorophyllic acids themselves (known in biochemistry as phaeophytins) are more easily isolated directly from an extract of green leaves rather than separated chlorophylls *a* and *b*. To this end, chlorophyllic acid ($a + b$) is extracted from nettle leaves by Willstätter's method.[61]

Finely shredded moderately dry nettle leaves are drenched in 96% ethanol, allowed to stand for a day, then 6 N HCl is added to the extract. The resulting suspension of chlorophyllic acid ($a + b$) is immediately passed through a paper filter, washed with ethanol to remove HCl and carotene, and dried in darkness. The ($a + b$) mixture is separated chromatographically into components.[69] A saturated benzene solution of the mixture is chromatographed on a column with confectioner's sugar and talc taken in a ratio of 6:1.

The chromatogram is eluted using a petrol–benzene (14:1) mixture and separated mechanically into two zones. The upper, light green zone contains chlorophyllic acid *b*, and the lower, grey-blue zone contains chlorophyllic acid *a*. To isolate both acids the central portion of each zone is taken, the pigment is extracted with an ether–acetone (1:1) mixture, and the solution is evaporated to dryness in a vacuum at room temperature or with slight heating.

Chlorophyllic acids are waxy green-black substances. In most physicochemical studies H_2Chla may be used without additional purification, whereas H_2Chlb requires another chromatographic separation. The absorption bands for pure H_2Chla in ethanol occur at the following wavelengths: 668 (I), 610 (II), 537 (III), 507 (IV), and 409 (V) nm. In the case of H_2Chlb in glacial acetic acid we have 656 (I), 602 (II), 558 (III), 530 (IV), and 438 (V) nm bands.

The mixture of chlorophyllic acids *a* and *b* may also be separated by the extraction method,[8] but the chromatographic technique is more reliable.

Protoporphyrin dimethyl ether can be easily obtained by Grinstein's method.[70]

Defibrinated blood is mixed with two or three volumes of acetone; the residue is separated on a Büchner funnel and washed with acetone. The resulting haemoglobin precipitate is suspended in a 10% methanol solution of oxalic acid. In this case haem passes into solution as haematin, while globin remains in the residue. After filtration ferrous sulphate is added to the haematin solution (3–5 g of $FeSO_4$ per 100 g of the solution), and the latter is saturated with HCl until the red-brown solution turns red-purple. Then the solution is mixed with $CHCl_3$ and a large amount of water. It takes 25 min for protoporphyrin to concentrate in the $CHCl_3$ layer, which is washed with water three times, then with 10% ammonia, and again with water. For preliminary purification H_2PP DME is crystallized from a chloroform–methanol mixture. The final purification is carried out by way of chromatographic separation on alumina with chloroform being used as the eluent. Porphyrin stored for a long period of time must be chromatographically separated without fail. Protoporphyrin DME in ethanol is characterized by absorption bands at the following wavelengths: 631, 603, 576, 532, 504, and 406 nm.

The chlorophyllic acids *a* and *b* as well as protoporphyrin dimethyl ester obtained in this manner from readily available natural sources may subsequently be used for synthesis of their numerous complexes and a wide variety of natural porphyrins[8,23,62] such as phaeophorbids, chlorin e_6, rhodine g_7, dihydroxychlorin e_6, chlorin e_4, phyllochlorin, pyrrochlorin, phaeoporphyrin a_7, phylloerythrin, desoxophylloerythrin, chloroporphyrins e_6 and e_4, mesoporphyrin, pyrroporphyrin, phylloporphyrin, haematoporphyrin, and many others.

2.2 Synthesis of Complexes of Various Porphyrins

In view of the specific biological role of magnesium and iron porphyrins the introduction of these metals into porphyrins is of particular interest despite the serious difficulties involved. It has long been considered that substituting Mg^{2+} for the hydrogen ions of imino groups by treating porphyrins with magnesium salts is impossible because the solvation sheath of magnesium ion $Mg(solv)_6^{2+}$ is highly stable and is similar in this respect to those of transition metal ions,[24] while the tendency of Mg^{2+} to form complexes with nitrogen-containing ligands is weak.

It is for this reason that until recently magnesium had been introduced into porphyrins with the aid of the Grignard reagent.[66]

First, ethyl magnesium bromide is produced by a conventional method, then passed together with *n*-propyl alcohol into C_3H_7OMgBr. Porphyrin is added to the reaction mixture, and the contents are heated at 75 °C until the absorption bands of free porphyrin disappear. The magnesium complex is isolated in a sequence of simple steps; however, this sequence calls for certain precautions to be taken.[23]

As was established later,[67] Mg^{2+} may be introduced into porphyrin with the aid of viologens. The use of magnesium viologen facilitates the introduction of Mg^{2+} into all porphyrins, chlorins, and phthalocyanine.

The most effective agent was found to be magnesium viologen prepared by refluxing Mg and I_2 taken in a molar ratio of 10:1, in absolutely dry pyridine.

Baum and Plane[63–65] made a most amazing discovery when they developed a method of direct introduction of magnesium from a solution of its perchlorate in acetone, methanol, and other organic solvents into dimethyl ether of deuteroporphyrin in the presence of nitrogen-containing bases as catalysts. Without catalysts the rate of the coordination reaction is infinitesimally low ($k = 4.7 \times 10^{-4}\ \mathrm{l\ mol^{-1}\ h^{-1}}$):

$$Mg(solv)_6^{2+}\cdots(ClO_4)_2 + H_2P \rightarrow MgP + 2H^+_{solv} + 2ClO_4^- \quad (2.1)$$

Reaction (2.1) catalysed with pyridine proceeds at high rates ($k'_v = 3.6\ \mathrm{l\ mol^{-1}\ h^{-1}}$) typical of the incorporation of Co^{2+} in porphyrin systems. Practically, the synthesis of magnesium complexes, that is chlorophyll and its structural analogues, is carried out by boiling an acetone solution of porphyrin in the presence of pyridine with an excess amount of $Mg(ClO_4)_2$.

The reagents are thoroughly dried and heated in a sealed ampoule at 100 °C for 6 h. The reaction is considered to be complete when the absorption bands of porphyrin disappear from the electronic spectra. The magnesium complex is extracted with ether, the ether solution being carefully washed off with water to remove magnesium salts and other impurities, dried with the aid of Na_2SO_4, and ether is evaporated in darkness in a flow of nitrogen.

In this fashion magnesium protoporphyrin[69] and magnesium mesoporphyrin[70] were obtained, their ethanol solutions featuring absorption bands at the following wavelengths: 590, 551, 419 and 582, 543, 422 nm, respectively.

Iron is easily introduced into porphyrins from solutions of $FeCl_2$ and $FeAc_2$ in polar organic solvents, best of all from solutions in glacial HAc with moderate heating. Thus metal analogues of chlorophyll with iron are easily produced.[71] Bivalent iron is already oxidized to Fe^{3+} in the course of synthesis of the complex. The final result of introducing Fe^{2+} is XFeP, where X is Cl, Ac.

Other metal analogues of chlorophyllic acid can be readily synthesized from its acetic acid solution and acetates of such metals as Zn^{2+}, Co^{2+}, Ni^{2+}, Cu^{2+}, Ag^{2+}, Pd^{2+}, and Pt^{2+}, which are taken in a 50:1 ratio with respect to porphyrin. Normally moderate heating (50–60 °C) for 1 h is sufficient. This method has been proposed by Zalesky[72] and is most suitable for synthesis of stable porphyrin complexes. Metalloporphyrin is extracted from the acetic acid solution with benzene. The extract is washed with water to remove excess salt and HAc, dried over Na_2SO_4, and benzene is evaporated in a flow of warm dry air.[71] Glacial HAc is not suitable as a solvent in synthesis of labile porphyrin complexes (e.g. with Hg^{2+} and Cd^{2+}). In this case alcohols, acetone, ethyl acetate, and other polar solvents are recommended.[73] Studies of infrared spectra of metal analogues of chlorophyll, synthesized in the above manner, indicate that no perceptible changes in the side substituents of H_2Chl take place.[74]

Unlike Fe^{2+}, the Fe^{3+} ion coordinates reluctantly to porphyrins. Studies of the reaction between $FeCl_3$ and protoporphyrin and its water-soluble derivatives[75] have shown that the coordination reaction takes place in weakly polar (CCl_4, $CHCl_3$) or aprotic (acetone) solvents. First, a primary product of coordination, or intermediate $H_2PP{\cdot}FeCl_3$, is formed which yields, after prolonged heating, haemin Cl·FePP. The method has no preparative value since haemin is more readily obtained from blood,[76] while complexes of Fe^{3+} with chlorophyllic acid and other porphyrins are more easily produced from $FeAc_2$. In the same manner it is more convenient to produce chlorophyll directly from natural sources than to introduce Mg into chlorophyllic acid.

Because of the instability of chlorophyllic acids *a* and *b* in alkali solutions it is impossible to isolate those of their complexes that are obtainable only in alkaline media, namely complexes of alkali and alkali earth metals (except for Mg). The influence of highly alkaline media, which becomes manifest in attempts to introduce these metals from solutions of their hydroxides, results in opening of the cyclopentanone ring[77] with formation of chlorin e_6 or rhodine g_7.

At present, in the case of stable porphyrin molecules such as 1,2,3,4,5,6,7,8-octaalkylporphyrins, tetraphenylporphyrin, and some others, complexes with practically all metals of the periodic table (with the exception of actinoids and bismuth) are obtained. Also obtained are complexes with such non-metals as B and Si. Complexes of Rb and Cs are extremely unstable. Complex salts of alkali (Li–Cs) and alkali earth (Ca–Ba) metals are normally prepared in solutions in the following manner.

Added to a porphyrin or chlorin solution having a concentration of 10^{-5}–10^{-3} M in absolutely dry pyridine is 1/40th part by volume of a 1 M solution of hydroxide of Li, Na, or another alkali metal in absolute ethanol. In the case of Ca–Ba the concentration of the alkaline solution may be lower. Green salts form immediately, which are stable for many hours in the absence of moisture. For isolation of solid salts see reference 78.

For investigation purposes one can prepare alkali and alkali earth salts of porphyrins and chlorins in a solution of alcohols, ethers, and hydrocarbons, in nitrous heterocycles, in dimethylsulphoxide, in sulpholane, and in other alkali-resistant media by introducing a 100- to 1000-fold excess of alcoholate of the respective metal in an alcohol solution. Complex alkali and alkali earth salts dissociate as soon as water[79] and other prononced proton-donor solvents are introduced into the system. Because of the weak alkali metal–porphyrin coordination interaction on the one hand and the high stability of the solvation sheaths of their cations on the other, no solvent has been found so far which is suitable for direct introduction of Li^+, Na^+, K^+, Sr^{2+}, and Ba^{2+} from solutions of their salts. Cations Mg^{2+} and Ca^{2+} can be introduced from a solution of their iodides or perchlorates in pyridine or in an acetone–pyridine mixture. A beryllium complex is produced under similar conditions.[67]

Most cations of heavy and transition metals, such as Mn^{2+}, Fe^{2+}, Co^{2+}, Ni^{2+}, Cu^{2+}, Zn^{2+}, Ag^{2+}, Pd^{2+}, and Pt^{2+}, form complexes with all porphyrins and chlorins from solutions of their acetates or chlorides in most organic solvents. The highest rate of complex formation is observed in glacial acetic acid, methanol, and acetone. The conditions for complete complex formation and methods for isolating solid intercomplex salts from solution are the same as in the case of chlorophyllic acid and protoporphyrin.

Complexes of Cd^{2+}, Hg^{2+}, and Pb^{2+} cannot be obtained in such proton-donor media as carboxylic acids. The best solvent for their synthesis is pyridine with acetates or chlorides taken as salts.

Complexes of three-, four-, and five-charge metals can be produced for stable porphyrins under specific conditions[80] summarized in Table 2.1.

Of all the metalloporphyrins listed in Table 2.1 the highest stability (in accordance with the classification given in reference 80, they belong to stability class I) is exhibited by complexes of metals Pd^{2+}, Al^{3+}, Cr^{3+}, Sn^{4+}, Si(IV), Ge(IV), V(IV), Ti(IV), W(V), Re(V), and Ta(V), which can be exposed to concentrated H_2SO_4 without the metal ion being split off. The use in synthesis reactions of porphyrins made spectroscopically pure by chromatographic separation[8] or extraction[23] does not require any additional purification of the complex.

If the starting porphyrin contains etherified carboxyl groups, for example H_2MP DME or H_2PP DME, etherification and chromatographic separation on a column are required after complex formation under stringent conditions, in some cases with subsequent recrystallization. Such procedures are described by Erdman *et al.*[81] for vanadyl mesoporphyrin synthesized from H_2MP DME and VCl_4 in a sealed glass tube at 165 °C, by Taylor[82] for Mn^{3+} mesoprophyrin synthesized from mesoporphyrin hydrochloride and $MnAc_2$ in glacial HAc with admission of air, and by other investigators.[23,83]

Many metals form metalloporphyrins with different degrees of oxidation of the central atom. They include Fe^{2+}–Fe^{3+}, Co^+–Co^{2+}–Co^{3+}, Ag^+–Ag^{2+},

Table 2.1 Conditions for synthesis of metalloporphyrins of multicharge ions

Metal ion	Reagent salt	Solvent	Conditions	Metallo-porphyrin
Al^{3+}, Cr^{3+}, Mn^{3+}, Fe^{3+}, Ga^{3+}, In^{3+}	$M(Acac)_3$	Ethanol	Heating in a sealed ampoule	(Acac)MP
Zr^{4+}, Hf^{4+}	$M(Acac)_4$	Ethanol	Heating in a sealed ampoule	$(Acac)_2MP$
TiO^{2+}, VO^{2+}, MoO^{3+}	$MO(Acac)_2$	Ethanol	Heating in a sealed ampoule	O=MP
Ta(V), Sb(V), Re(V)	$TaCl_5$, Sb_2O_5, Re_2O_5	Ethanol	Heating in a sealed ampoule	$O[(O)MP]_2$
Co^{3+}, Sc^{3+}	$M(Acac)_3$	Imidazole	Heating in a sealed ampoule	(Acac)MP
In^{3+}, Cr^{3+}, Rh^{3+}	MCl_3	Benzonitrile	Heating	(Cl)MP
W(V), Nb(V)	MCl_5	Benzonitrile	Heating	$(Cl_3)MP$
Sn^{4+}, Si(IV), Ge(IV)	$SnCl_2$, $SiCl_4$, $GeCl_4$	Pyridine	Heating	$(Cl_2)MP$
Fe^{3+}, Mn^{3+}	$FeCl_3$, $MnAc_2$	Glacial HAc	Heating	(X)MP[a]
Re(IV), V(IV)	$ReCl_5$, $VOSO_4$	Glacial HAc	Heating	$(Cl_2)ReP$, OVP
Rh^{3+}	$[Rh(CO)_2Cl]$	Glacial HAc	Heating	(Cl)RhP
Ru^{3+}	$RuCl_3$	Tetramethyl urea	Heating	ClRuP
Ir^{3+}	$[Ir(CO)_2Cl_2]$	Methyl cellosolve	Heating	ClIrP
Tl^{3+}	$Tl(CF_3COO)_3$	CH_2Cl_2	Heating	(X)TlP[a]

[a]X, anion of the respective salt.

Mn^{2+}–Mn^{3+}–Mn^{4+}, Sn^{2+}–Sn^{4+}, Cr^{2+}–Cr^{3+}, Ru^{2+}–Ru^{3+}. The least stable of all are Fe^{2+}, Co^{+}, Mn^{2+}, Ag^{+}, Cr^{2+}, and Sn^{2+} porphyrins. Cobalt usually yields complexes in the form of Co^{2+}; special oxidation conditions are required for it to become Co^{3+}.[84] However, some porphyrins promote oxidation in air; for example, protoporphyrin with $MnAc_2$ and $CoAc_2$ in hot glacial HAc forms (HO)MnPP and (HO)CoPP in the presence of HCl and air.[85]

A complex of mesoporphyrin DME with Co^{2+} prepared in methanol turns to (HO)CoMP when kept in air. The metal oxidation process is catalysed with olefins and acetylenes.[86,87]

2.3 Synthesis of Spectroscopically Pure Phthalocyanine, Polymeric Phthalocyanine, and Their Complexes

The synthesis of phthalocyanine and its complexes has been described at length by Linstead *et al.*[89–95] A specific feature of synthesizing phthalocyanine complexes (tetrabenzotetrazaporphin, **V**) is that they are seldom obtained from an available phthalocyanine ligand. More often the complex is formed from molecular fragments of phthalocyanine, namely phthalonitrile (**VI**), phthalimide (**VII**), phthalic anhydride (**VIII**), diiminoisoindoline (**IX**), and some other derivatives of *o*-phthalic acid in the presence of a metal ion source—chlorides, acetates, oxides of metals,

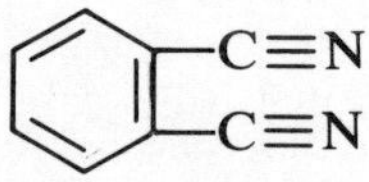

VI

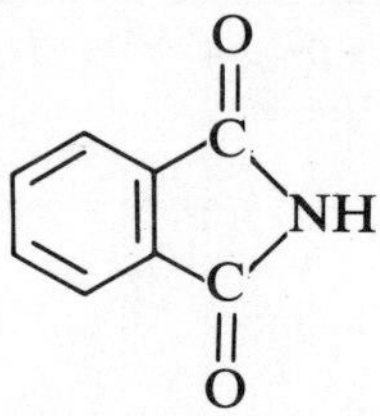

VII

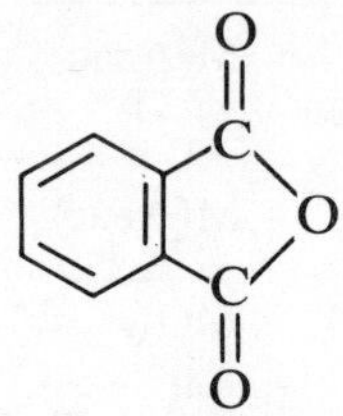

VIII

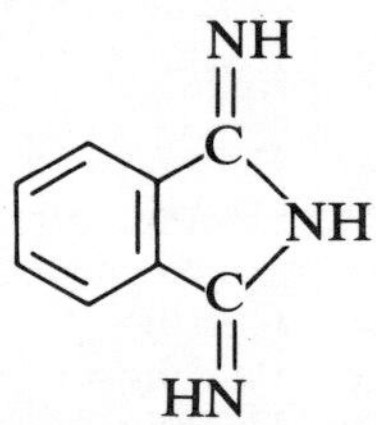

IX

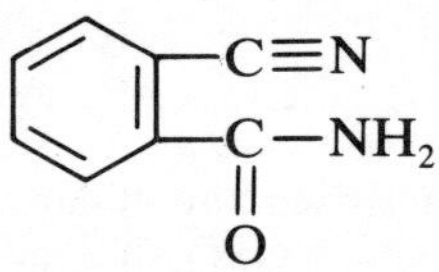

X

or free metals. Synthesis on the basis of **VII** and **VIII** necessitates the presence, in the reaction, of urea as an amminating agent and catalysts such as ammonium molybdate or boric acid. In any event, the reaction is conducted either in a melt of a solid mixture of dry reagents (the melt process) at 135–294 °C or in an inert high-boiling solvent such as nitrobenzene, *o*-dichlorobenzene, trichlorobenzene, or ethylene glycol.

The phthalonitrile molecules forming in the reaction polymerize, coordinating at the same time around the central metal ion, then close into a macrocycle:

$$4C_8H_4N_2 + M \rightarrow (C_8H_4N_2)_4M. \tag{2.2}$$

The hypothetical mechanisms of synthesis of phthalocyanine complexes from simple structural units are extensively covered in the literature.[96–99] The reaction that proceeds most smoothly and has the best yield is complexing with phthalonitrile.[94,97,100] In this way phthalocyanines of most metals except for Ag and Hg can be obtained. The reaction involves either metals (Zn, Cd, Co, Sn, Cu, Fe) or their salts ($NiCl_2$, $CoCl_2$, $ZnCl_2$, $AlCl_3$, $PtCl_2$, $RhCl_3$, $V(HCOO)_3$), oxides (CaO, PbO, OsO_4, V_2O_5, MnO_2), and sulphates ($PbSO_4$, $BaSO_4$). Mg, Sb, Fe, Co, and Ni react exothermally with *o*-cyanobenzamide (**X**) at 250 °C with a good yield. Syntheses involving phthalimide are more erratic.

Phthalocyanines of alkali and alkali earth metals can be prepared from phthalonitrile and alcoholates of metals in boiling anhydrous amyl alcohol.

Dimetallic salts Li_2Pc, Na_2Pc, and K_2Pc[94,95] are easily obtained in this fashion. Lithium with excess phthalonitrile may yield LiHPc which readily crystallizes, just as Li_2Pc, from acetone or absolute ethanol. Trace amounts of moisture result in removal of lithium and formation of H_2Pc insoluble in the overwhelming majority of solvents (except for quinoline and boiling α-bromonaphthalene). Complex Li_2Pc is interesting in that its solution with alcohol solutions of anhydrous salts incapable of intensive solvolysis, such as $CdCl_2$, $ZnCl_2$, $MnCl_2$, $PbCl_2$, $CoCl_2$, $HgCl_2$, and $AgNO_3$, immediately yields a residue of a pure complex of phthalocyanine with the other metal. With $AlCl_3$, $SnCl_2$, $ThCl_4$, $BiCl_3$, and other salts that lend themselves to solvolysis only H_2Pc is formed.

As one goes from alcohols to aprotic solvents, for example DMFA, even such phthalocyanines can be obtained on the basis of Li_2Pc as (Cl)AsPc ($AsCl_3$ is taken as the source of arsenic[100]) and (Cl)TiPc (from $TiCl_3$ in boiling quinoline[101]).

Some peculiarities can also be found in the synthesis of complexes GaPc[102], (Cl)GaPc, (Cl)InPc,[103] (Cl)MnPc,[104] (Cl_2)MnPc,[105] (Cl_2)GePc,[106] CrPc,[107] (Cl)RuPc, (Cl)IrPc,[108] (Cl)RhPc,[109] (Cl_2)OsPc,[110] $(HO)_2$TiPc,[111] $(HO)_2$ZrPc, $(HO)_2$HfPc,[112] UO_2Pc,[113] MoO_2Pc,[114] (Cl_2)SiPc,[115,116] RuPc,[117] AuPc,[118] and XEPc,[119,120] where E is a lanthanide and X is an anion, as well as complexes of phthalocyanine with metals in lower degrees of oxidation, such as Li_2CuPc,[121] LiMnPc, Li_2MnPc,[122] LiFePc, LiCoPc, and LiNiPc.[123]

Extensive and extremely valuable studies into the synthesis of phthalocyanine–lanthanide complexes have been conducted by Shklover and Plyushchev.[120] Later, phthalocyanine complexes of a radically new sandwich-type structure were synthesized (Fig. 2.1).[124–127]

Synthesis of a metal-free phthalocyanine, that is a phthalocyanine ligand, involves a step yielding its labile complexes with Na, Li, Mg, and Ca. The easiest way to produce phthalocyanines in a laboratory is by dissolving MgPc in a 100-fold excess of 96% H_2SO_4 with subsequent pouring of the solution on ice. However, even with all precautions taken the losses in H_2Pc may be as high as 50% for obvious reasons.[128] It is therefore

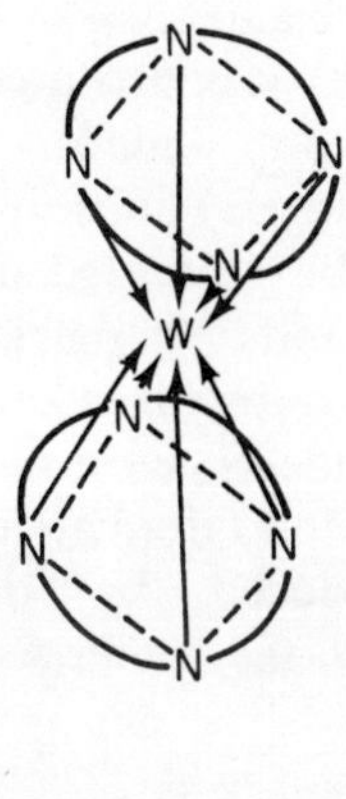

Figure 2.1 Structure of sandwich-type phthalocyanines

advisable to decompose MgPc or CaPc by boiling it with 2–6 M H_2SO_4 for 1–2 h.[129] Another possibility of obtaining H_2Pc is by boiling Na_2Pc with water.[92]

Phthalocyanines produced from phthalic acid derivatives both in a melt and in a solvent may contain a great amount of impurities such as phthalimide, phthalonitrile, their linear and cyclic polymers of a non-phthalocyanine nature, resinoid substances, salts of metals, and free phthalocyanine. To remove the impurities phthalocyanines are boiled alternately with acid and alkali solutions.[88–95] However, treatment with acids is possible only in the case of a stable group of phthalocyanines and impermissible in the case of labile complexes (with alkali and alkali earth metals, including BePc and MgPc, as well as with Sn^{2+}, Mn^{2+}, Pb^{2+}, Cd^{2+}, Hg^{2+}, Sb^{3+}, rare earth metals, Ag^+, Ga^{2+}, Cr^{2+}, As^{3+}, MoO_2^{2+}, and UO_2^{2+}, which easily lose the metal ion even when treated with diluted H_2SO_4 or HCl.

In the case of labile metal phthalocyanines satisfactory purification is achieved by multiple extraction of impurities with slightly heated acetone or benzene in a Soxhlet apparatus and by driving off the volatiles with phthalocyanine being heated to 300 °C in a vacuum. Further purification of stable phthalocyanines can be easily carried out by vacuum sublimation (10^{-3} mmHg, 300–500 °C) or, even better, by reprecipitation on ice from a 100-fold excess of 96% H_2SO_4. In this case phthalonitrile polymers and H_2Pc itself decompose completely and remain in the solution,[128,130] while vacuum sublimation fails to rid MPc fully of H_2Pc impurities. In recent years methods of fractional, multiple sublimation of MPc have been developed at the Ivanovo Chemical Technology Institute (ICTI) which yield excellent results in combination with reprecipitation from H_2SO_4. Metal phthalocyanines purified in this manner are characterized by spectroscopic purity and a constant conductivity value.[131–133]

Labile phthalocyanine complexes are practically impossible to obtain in a pure form. They contain a large amount of H_2Pc impurities which are not completely removed during sublimation of MPc, presumably because the range of the thermodynamic parameters of sublimation is very narrow.[134] In the case of CuPc the heat and entropy of sublimation are known and have the values $\Delta H_{\text{subl}} = 63.6$ kcal mol^{-1} and $\Delta S_{\text{subl}} = 67.2$ e.u.

A special group of phthalocyanines and their complexes is constituted by such derivatives as monochloro-, polychloro-, nitro-, amino-, and hydroxyphthalocyanines containing four and more (up to six) substituents in the benzene ring of phthalocyanine. They can be obtained from respective phthalic acid derivatives by conventional methods. Of particular interest is a new method of directed synthesis of such derivatives on the basis of polychloro derivatives of dimethylbenzene and its substituents, developed by Klyuev and Shiryaeva.[135–137] Chloro and sulpho derivatives of phthalocyanine and its complexes may be produced by direct chlorination[138] or sulphonation.[139] Sulpho derivatives of the general type

$CuPc(SO_3H)_n$, where $n = 2–4$, are of interest as water-soluble compounds.[140] They are industrially used as direct dyes or processed into sulphamides and their derivatives.[32]

Still more peculiar are phthalocyanine-like compounds whose molecule contains, instead of the benzene ring, a naphthalene,[141] pyridine, thiophene, pyrazine, or thionaphthene ring, derived for the first time by Linstead, Noble, and Wright.[142]

Note the close similarity in the synthesis methods and chemistry of the above analogues of phthalocyanine.[140–143] Their properties are very much like those of phthalocyanine, which is why they will not be mentioned in what follows.

A particular position among phthalocyanine derivatives is occupied by polymeric phthalocyanines and their complexes, studied for the first time by Marvell and Rassweiler.[144] These compounds have aroused interest because they are convenient models in investigations of the effect of the polymeric state of molecules on their physicochemical properties,[144–154] as well as in connection with the increased catalytic activity of some polymeric compounds[155–160] as compared to monomers.

These compounds are obtained from tetracarboxylic aromatic acids or their derivatives of the general structure **XI** or **XII**:

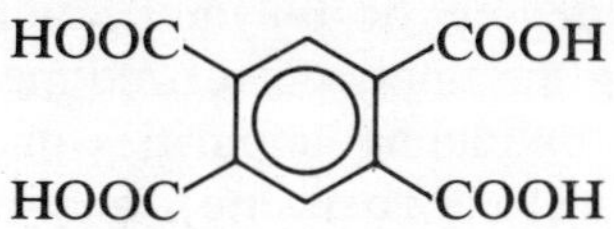

XI

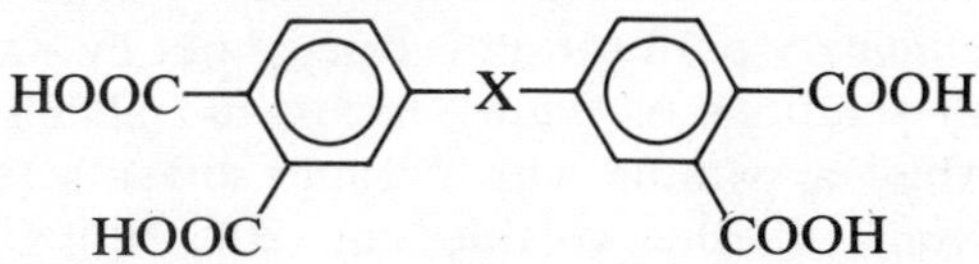

XII

The function of the aromatic nucleus can be performed not only by the benzene ring but also by any arene ring, and the bridging group X may be $—CH_2—$, $—(CH_2)_n—$, —CO—, —O—, or $—SO_2—$. The bridge X may be absent. A prerequisite for reactivity of the tetracarboxylic acid derivatives **XI** and **XII** as phthalocyanine-forming compounds is the ortho position of the carboxyls arranged in pairs. Instead of carboxylic acids their derivatives such as anhydrides, amides, imides, and nitriles are more often used. The principles of and conditions for synthesizing polymeric phthalocyanines are similar to those in the case of the above-mentioned monomers.

Fusion of pyromellitic acid (**XI**) or its derivatives with metals or metal salts in the presence, if necessary, of amination sources and catalysts leads to synthesis of complexes of cross-linked polymers (Fig. 2.2). The introduction into the system of a certain number of phthalic acid derivatives (**VI–X**) restricts the development of the polymer network of phthalocyanine units along one of the coordinate axes. In this case linear polymers are derived (see Fig. 2.2 between lines AA and BB). Similar cross-linked or linear complexes of polymeric phthalocyanines can be obtained from derivatives **XII**. The polymeric phthalocyanine ligands

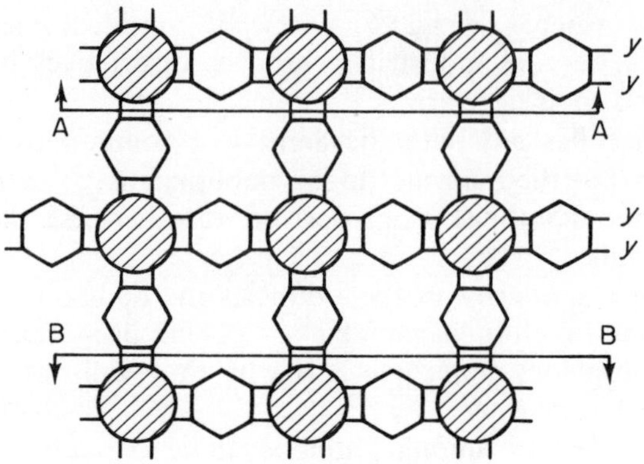

Figure 2.2 Molecule of polymeric phthalocyanine based on pyromellitic acid

themselves can be produced by decomposing their complexes with Ca^{2+}, Mg^{2+}, and Pb^{2+} using a diluted solution of $HClO_4$.

To isolate polymeric phthalocyanines from a melt and to purify them one may resort to reprecipitation of their solutions in concentrated H_2SO_4 on ice (stable complexes). This method is applicable to complexes of both linear and cross-linked structure with a low degree of polymerization. Preliminary purification is carried out by washing the finely pulverized melt with solutions of diluted acids and alkalis, by extracting impurities in a Soxhlet apparatus with organic solvents (e.g. acetone, benzene), and by driving off the volatiles at 200–300 °C at 10^{-3} mmHg for 2 or 3 days.[144,145,156,160] In view of the fact that some complexes of linear phthalocyanine polymers, as opposed to monomers, become soluble in DMFA, they can be purified by fractional precipitation with chloroform or water which is added to DMFA.

In addition to monometallic complexes of polymeric phthalocyanines, polymetallic chelates have also been prepared, containing different metals in the same macromolecule.[157,161,162] Such complexes are interesting as catalysts exhibiting synergy and selectivity in catalytic oxidation reactions and also as relatively simple model systems convenient for examining metal–metal exchange effects in conjugated systems. They can be produced by conventional techniques developed for polymeric phthalocyanines with the difference that the reaction involves salts of two or more metals at a time. Methods for their synthesis from polyphthalocyanine ligand in a solid phase and from aqueous solutions of salts have been developed.[161,162]

Polymeric complexes of monomeric M^{3+} and M^{4+} phthalocyanine in which polymerization involves the metal ion and counterion are of an entirely different nature. Such complexes are not considered here.

Chapter 3

Molecular structure of porphyrins and phthalocyanine

The various natural porphyrins and their synthetic analogues represented in Chapter 1 by the general formula **I** feature a unique molecular structure. This uniqueness stems from the conjugation over the entire inner 16-membered cycle known as the macroring, the planar configuration of the macroring (its complanarity), the possibility of various substitutions for the H atoms in the pyrrole rings (positions 1–8) and in the CH bridges (so-called meso positions α–δ), distribution of the occurring electronic substitution effects throughout the molecule, and, finally, the presence of proton-donor imino groups in the centre of the macroring.

The H atoms of the porphyrin imino groups being substituted by metals gives rise to additional donor–acceptor properties exhibited by the resulting metalloporphyrins, which determine the biological activity of many of them. The metal atom that has entered into coordination with porphyrin changes the properties of all atomic groups in the porphyrin molecule.

This effect is most pronounced on electronic energy levels of the molecules and manifests itself in varying electronic absorption spectra. The spectra also clearly show the functional substitutions in positions 1–8″ and α–δ of the molecule. Interestingly, the substitution of aza bridges (—N=) for four methine bridges (—CH=) in the porphyrin macroring, whereby a tetrazaporphyrin or porphyrazine macroring (phthalocyanine) is formed, results not only in entirely different electronic spectra[8,23] but also in a more pronounced aromaticity, sensitivity to oxidizers, more stable metal complexes, a substantial decrease in the proton-acceptor properties, the central imino groups losing their capacity for methylation, and other unusual properties. Many of these problems were discussed by the present author in earlier works.[163,164] Consider now the structural features of porphyrin and porphyrazine molecules in greater detail.

3.1 Aromatic Nature of Porphyrins

The aromaticity of molecules has been amply covered in the literature. In our discussion of this problem we shall restrict ourselves to references

163–166. By aromatic systems are meant cyclic compounds in which pronounced conjugation occurs in at least one cycle, that is π electrons are densely distributed over the entire closed contour of the cycle. From the standpoint of the molecular orbital theory these are systems featuring an extremely low-energy π-molecular orbital encompassing all atoms of the cycle. The symmetry or configuration of this orbital determines the conjugation contours in an aromatic molecule.

For complex aromatic systems the present author has introduced the concept of uniform-conjugation contour.[164] In such a contour all bonds carry the same π-electron charge. If the π-electron charges of bonds in a contour are not the same, such a system may be referred to as a variable-conjugation contour. In cases where insulator bonds with a π-electron charge approaching zero (the order of such a bond is close to unity) appear in the ring the conjugation contour transforms to a conjugation chain, and the aromaticity of the system disappears. This takes place in all cases where the conjugated cycle incorporates at least one atom having neither a π electron of its own, nor a π-electron pair, nor a low-lying vacant orbital. To establish the degree of aromaticity of porphyrins and their analogues the basic traits and features of the behaviour of aromatic systems will now be briefly outlined:

(1) The number of π electrons constituting the contour of an aromatic cycle must obey Hückel's rule.[167] Cyclic π systems inconsistent with this rule tend either to expel the extra π electron or to incorporate it.[168]

(2) All atoms included in the most stable π-molecular orbital must lie in the same plane. The non-coplanarity of a cyclic system reduces the overlap of π electrons and destabilizes the π orbital. As the system departs from coplanarity, its stability diminishes.

(3) To break down a stable aromatic orbital requires a sufficiently great amount of energy forming part of the energy of activation of a reaction in which π electrons convert to σ electrons. Therefore aromatic molecules do not have a tendency for addition reactions, that is their unsaturation is not pronounced although they are considered to be unsaturated compounds. They prefer reactions of substitution via C—H bonds.

(4) The redistribution of π electrons among all bonds forming part of the aromatic cycle results in fractional bond orders ($2 > n > 1$). In benzene the C—C bond order is $n = 1.5$. For this reason structural formulae with a bond order which is an integer are not suitable to describe the structure and properties of such molecules. More adequate for the purpose are special formulae in which the number of interacting π electrons is indicated, or molecular diagrams. Such formulae and diagrams will be given in what follows.

(5) Depending on the cycle size aromatic systems may have a large number of bonding and antibonding orbitals. Consequently the upper occupied orbital of π symmetry may be very high above the isolated

ethylene π orbital, while the lower empty π orbital may, on the contrary, have a very low energy. Hence a characteristic feature of extended aromatic π systems, namely their ability to give off an electron as easily as to acquire one, that is to behave as semiconductors.[169–173] This feature of π orbitals of extended aromatic systems is responsible for location of the bands of $\pi \rightarrow \pi^*$ transitions in the visible region. It is precisely for this reason that all aromatic systems are potential colour carriers, or chromophores.

In the case of aromatic molecules of the same symmetry the degree of their aromaticity can, in some instances, be determined with the aid of electronic absorption spectra. As can be inferred from the spectra, the molecule of phthalocyanine **(V)** is a more aromatic system than that of unsubstituted porphin **(I)** forming the basis of chlorophyll **(II)** and haem **(III)**. The energy of the electronic $\pi \rightarrow \pi^*$ transition in the first absorption band is 40 kcal mol^{-1} for metal-free phthalocyanine and 46 kcal mol^{-1} for free porphyrin. The high aromaticity of phthalocyanine provides for its outstanding thermal stability which is much lower in the case of porphin and its alkyl-substituted derivatives, its volatility at low pressure, and a weak tendency to break open the conjugated system of the macroring. The latter property is exhibited by phthalocyanine and its complexes only under the effect of such strong reducing agents as metallic Li,[174] hydrosulphite,[34,35] or when oxidized with hydrogen peroxide in an acid medium.[175]

Stable phthalocyanine complexes easily withstand exposure to Cl_2,[138] Br_2,[176] H_2SO_4, and oleum,[139] being involved in electrophilic substitution reactions in the benzene ring. It appears that the removal of a pair of phthalocyanine electrons from the upper bonding π orbital and their transformation to σ electrons in the presence of chemical reagents would have resulted in rearrangement of all bonding energy π levels and in their energy as a whole becoming higher due to losses in the conjugation energy or, to be more precise, due to the differences between the energy of the aromatic conjugated system and that of the open (chain) conjugated system. The amount of this difference is not known either for phthalocyanine or for porphyrin; however, the physical and chemical properties suggest that in the case of the latter this difference is small. Convincing proof can be found in Woodward's work.[165]

Reduction of porphyrins under mild conditions with Raney nickel in ethanol or ascorbic acid on nickel transforms the aromatic system of porphin to an open conjugated phlorin system **(XIII)**. A CH_2 group emerges as a conjugation insulator. Reduction under different conditions[177,178] yields chlorin **(XIV)** or bacteriochlorin **(XV)** systems in which the aromaticity of the starting porphin remains the same.

It should be borne in mind that aromaticity is an arbitrary concept. Changes in the environmental conditions may significantly affect the

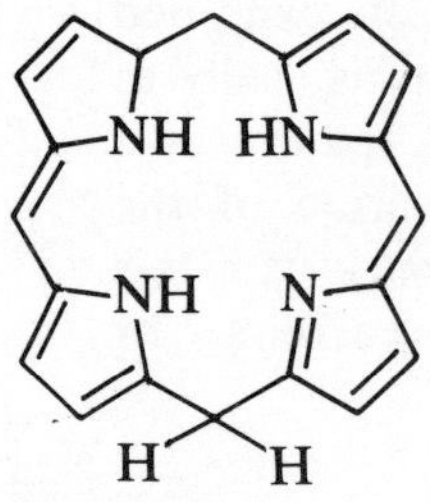

XIII

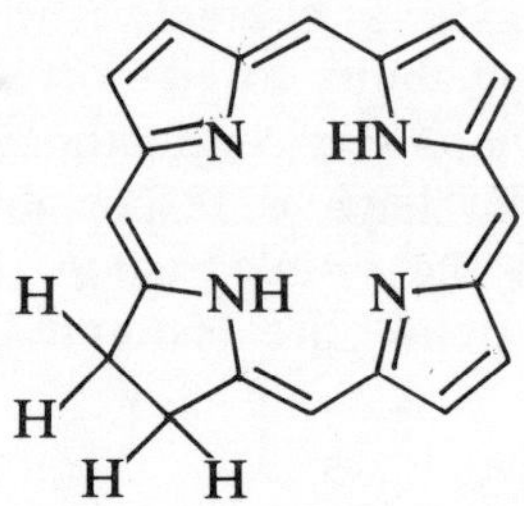

XIV

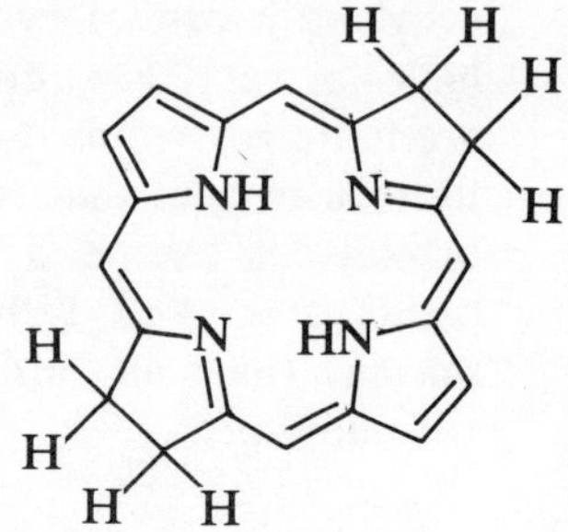

XV

stability of aromatic systems. One of the decisive factors is protonation. For example, porphyrin dication **XVI** rapidly and reversibly transforms to its respective phlorin (**XVII**) under the effect of the thioacetate ion at room temperature. If the conditions are right, porphyrin may be in

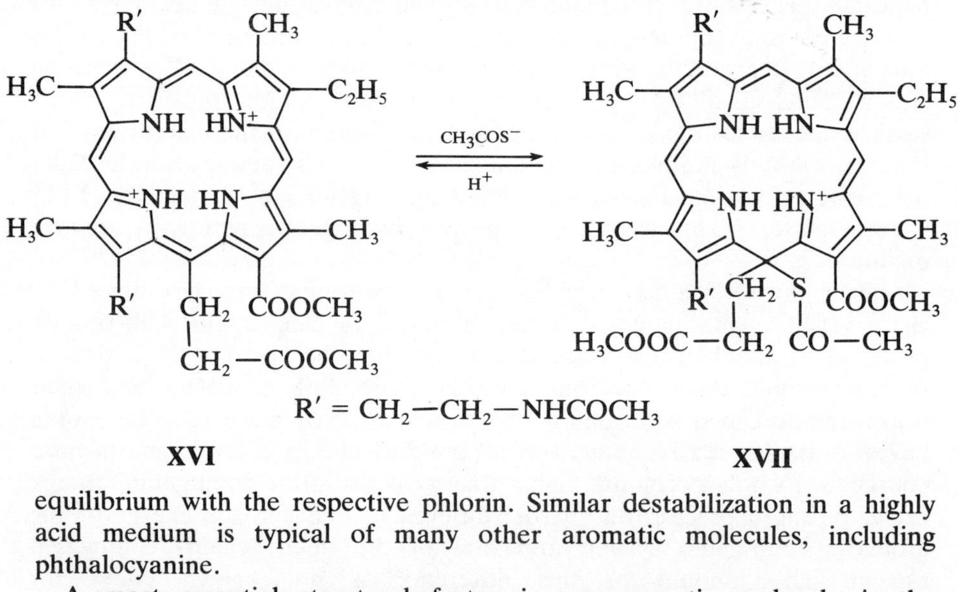

$R' = CH_2-CH_2-NHCOCH_3$

XVI **XVII**

equilibrium with the respective phlorin. Similar destabilization in a highly acid medium is typical of many other aromatic molecules, including phthalocyanine.

A most essential structural factor in an aromatic molecule is the distribution of the π-electron charge over the conjugation contours. It determines many physicochemical properties of such molecules,[163,164] and its knowledge is necessary to select an appropriate model for quantum-chemical calculations of conjugated systems.[179] Let us first examine this problem with reference to a simpler phthalocyanine molecule model.

3.2 Structural Features of Phthalocyanine and Porphyrin Molecules

The structural formula of phthalocyanine (V) as a model that has found extensive use in studies of the properties and structure of porphyrins was proposed for the first time by Linstead.[90] Robertson's X-ray analysis data[180–182] have corroborated its correctness as a whole. Phthalocyanine and its complexes turned out to be planar molecules. The coplanarity of the molecules follows from the classical formula (**V**) with due account for the resonance concept. At the same time Linstead pointed out that his formula offers no explanation for the equal lengths of the C—N bonds in the macroring and C—C bonds linking it with the benzene nuclei, the homogeneity of the oxidation products, the absence of isomeric forms of the metal derivatives, and many other things.

A more adequate presentation of the phthalocyanine structure may be formula **XVIII** proposed earlier[183] by the present author, in which dots

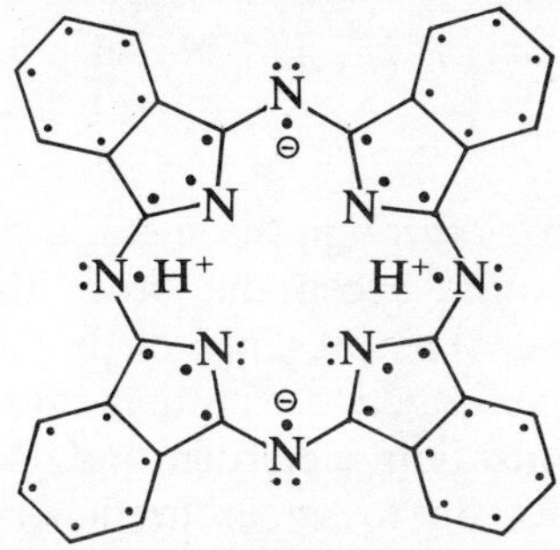

XVIII

stand for the 16 inherent π electrons of the macroring and the 24 π electrons of the benzene rings, and two circles represent the ionization electrons resulting from internal ionization of the two imino-hydrogen atoms of H_2Pc. As a consequence the protons are in the electrostatic field of three nitrogen atoms at once. The sextets of π electrons of the benzene rings form intrinsic (stable) π-electron shells participating in a very weak interaction with the π electrons of the macroring. The last 18 π electrons constitute an aromatic system independent of the benzene rings and obeying Hückel's rule ($4n + 2 = 18$).

The inherent π electrons of the macroring together with the two ionization electrons occupy nine bonding molecular orbitals (ψ_π) of the macroring. The lowest-energy orbital $1_{A_{1g}}$ corresponds to even distribution of the electron pairs among all C—N bonds. Corresponding to an ideal aromatic system of the H_2Pc macroring are nine antibonding π orbitals (ψ_{π^*}) which do not stay vacant due to the specific nature of conjugated aromatic π systems.

The relatively low-energy position of the orbitals is the reason why they are occupied by n electrons from the antibonding φ_N orbitals of the heteroatoms. This occupation is only partial due to the $\pi \rightleftarrows \sigma$ conversion of the n electrons. The formula proposed by the present author is consistent with the concept of delocalization of π electrons and, as will be shown, explains practically all physicochemical properties of phthalocyanine and its complexes. It is difficult though to account for the N—H bond vibration frequencies in infrared spectra with the aid of this formula. The nature of bonding of the imino-hydrogen atoms has always been one of the most interesting and difficult problems in the chemistry of porphyrins. Adherents of the proton delocalization concept,[26,184–186] irrespective of how the delocalization is achieved, arrive at chemical structures dissimilar to Linstead's formula and approaching formula **XVIII**. Careful analysis of spectroscopic and other physical data has convinced Solovyev[179] that phthalocyanine and its complexes are most likely to have formulae similar to **XVIII**, in which the equivalence of the four intracyclic nitrogen atoms results from the three-centre bonding of hydrogen atoms,

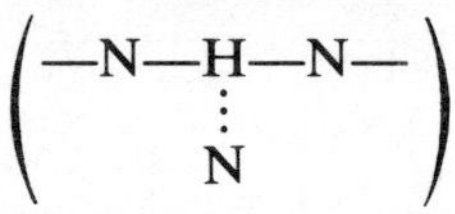

or redistribution of electrons through the metal orbitals.

Formula **XVIII** looks quite acceptable from the viewpoint of modern quantum-chemical concepts. The presence of protons in the field of several nuclei is energetically productive,[187] their formation resulting from liberation of the two ionization electrons and the conjugated system of phthalocyanine being built up to an aromatic one. Characteristically the dissociation electrons are not paired on the σ orbitals of nitrogen atoms but are distributed among the molecular orbitals just as in the case of simpler CO_3^{2-} and SO_4^{2-} ions. The absence of internal ionization of protons in H_2SO_4 is justified by the insufficient gain in energy because of the short length of the molecular orbitals of SO_4^{2-} and, as is apparent from electronic spectra, the comparatively high energy of the vacant π orbitals. In the case of the large unique conjugated system of phthalocyanine the process of coupling of ionization electrons is highly beneficial. This is corroborated by the closer affinity of phthalocyanines for electrons[188] and by their chemical properties.

The strongest arguments in favour of structure **XVIII** are as follows. It is known that H_2Pc and its metal complexes are protonated at a peripheral nitrogen atom rather than in the centre of the molecule.[189–192] As can be seen from formula **XVIII**, protonation in the centre of the H_2Pc molecule would not be energetically productive. Despite the presence of eight nitrogen heteroatoms phthalocyanine is a weak base and can attach only a single proton in solutions of concentrated H_2SO_4, HSO_3Cl, or oleum.[193,194] The only reason for such an interesting and even unique phenomenon may be the coupling of n-electron pairs of nitrogen atoms on vacant ψ_{π^*} orbitals of the macroring, that is the same phenomenon responsible for internal ionization and serving as an experimental proof of formation of an 18-membered stable aromatic system.

The insolubility of H_2Pc in most organic solvents, typical of ionic phthalocyanine complexes (e.g. SnPc, CaPc, Na_2Pc, HgPc, CdPc), is also indicative of the presence of ion-charge sources in the phthalocyanine molecule. The well known distortion of the H_2Pc molecule,[180–182] which is not observed in its complexes, can be easily explained from the standpoint of internal ionization of H atoms. The weak hydrogen bond forces would be insufficient for such a significant distortion of an aromatic system. Hydrogen ions stabilize the skeleton of the phthalocyanine macroring, therefore their removal, that is external ionization, involves serious difficulties. The phthalocyanine anion Pc^{2-} appears to be unstable. Its formation calls for stabilization of the whole molecule by way of intensive solvation, which is impossible because of the specific structure of phthalocyanine. The same applies to porphyrin molecules.[8] This problem will be considered at greater length in one of the following chapters.

Thus one of the arguments that may be adduced against formula **XVIII**, as to why H_2Pc is not a strong acid if it has already undergone internal ionization, is dismissed. The above considerations may also provide an answer to the relatively old baffling question: why does there not exist an 'ideally aromatic' (from the classical standpoint) dehydrophthalocyanine (**XIX**).[97] Its ideally symmetrical conjugated system (symmetry D_{4h}) is free of all drawbacks of H_2Pc save one. It does not contain the two hydrogen atoms which complement a 16 π-electron system to an 18 π-electron one, and protons stabilize the macroring skeleton. The deficiency of electrons in the aromatic system of dehydrophthalocyanine could have been compensated by *n* electrons of the heteroatoms. This, however, is not the case, and the molecule is not stabilized.

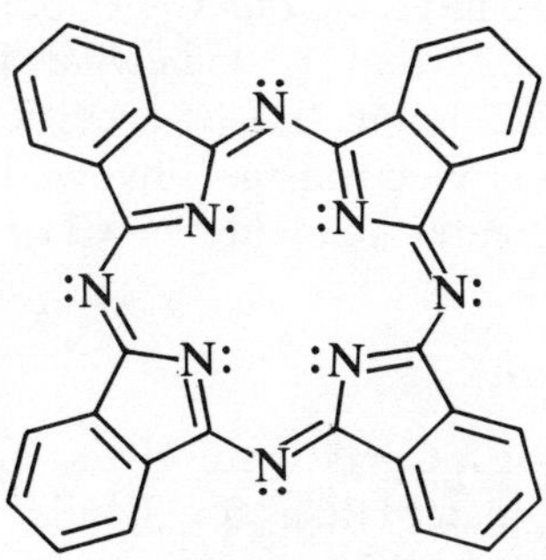

XIX

As has already been mentioned, an important aspect of aromatic molecules is their conjugation contours. Without finding a correct solution to this problem one cannot develop the theory of chromaticity, select appropriate models in quantum-chemical calculations, and elaborate the theory of reactivity of aromatic systems. Until recently many investigators involved in conjugation in phthalocyanine proposed conjugation contour **XX** by analogy with porphyrins, including two benzene rings in the chromophore system.[195] Calculation of π-electron charges of bonds, using

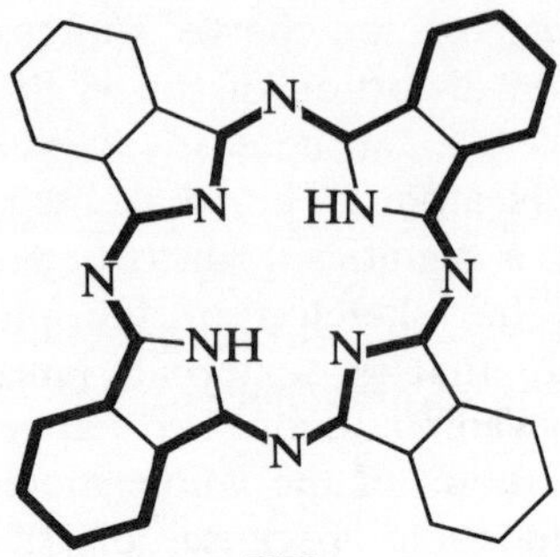

XX

semi-empirical equations[196] such as

$$l_{X-Y} = l^{\circ}_{X-Y} + \Delta l^{\pi}_{X-Y} \cdot A^{\pi}_{X-Y} \tag{3.1}$$

has proved this approach to be incorrect. Here l_{X-Y} is the length of bond X–Y (in ångstroms); A^{π}_{X-Y} is its π-electron charge, that is the number of π electrons participating in bonding; l°_{X-Y} is the standard length of ordinary bond X–Y with a π-electron charge equal to zero; and Δl^{π}_{X-Y} is the specific bond shortening with a π-electron charge equal to unity. Equation (3.1) implies that bond shortening depends on the nature of the bond and is proportional to its π-electron charge.

For a C—C bond the equation takes the following form:

$$l_{C-C} = 1.542 - 0.149A^{\pi}_{C-C}. \tag{3.2}$$

The averaged π-electron charge of the C—C bond in a benzene ring of phthalocyanine equals 1.02, that is somewhat higher than in benzene (0.93). In the case of C—C′ bonds linking a benzene ring to the macroring it is as low as 0.3, that is π electrons are involved in the formation of this bond to an extent not greater than in the CH_3—C_6H_5 bond of toluene. From the equation

$$l_{C-N} = 1.470 = 0.094A^{\pi}_{C-N} \tag{3.3}$$

we derive $A^{\pi}_{C-N} = 1.38$ for a C—N bond of the phthalocyanine macroring, that is a value much greater than for bonds in benzene. Hence the π-electron density at each of the macroring bonds is as high as in ethylene and much higher than in the benzene rings of phthalocyanine. The order of a C—N bond, equal to the sum of σ- and π-electron charges divided by two, is $3.38/2 \approx 1.7$. Thus the order of the C—N bonds in the phthalocyanine macroring exceeds 1.5, whereas in the benzene rings of phthalocyanine it is 1.5 for all bonds. The order of the bonds linking the benzene rings with the macroring is 1.16, that is they approach ordinary bonds.

Calculation data clearly suggest that the macroring and benzene rings of phthalocyanine are essentially independent aromatic systems, almost unrelated chromophores which are in a weak π-electron interaction with one another. The spectral data obtained for monofunctional substituted H_2Pc in benzene rings and spectra of protonated phthalocyanines strongly support the above supposition and indicate that the conjugation contour cannot take the form **XX**. This proposition made by Terenin in the 1940s[197] has been ignored in many works. For phthalocyanine the final form of the conjugation contour must be **XXI**. A similar conjugation contour is found in all phthalocyanine complexes with ions of metals of different valences.

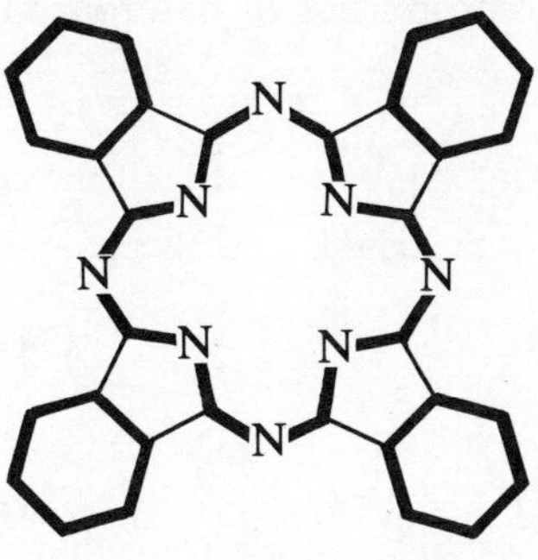

XXI

In his work,[179] Solovyev analysed the available physical data, including X-ray and spectroscopic data, and arrived at conjugation contours similar to **XXI** for conjugated systems of phthalocyanine and its complexes.

The phthalocyanine molecule is a multicontour conjugated system. Depending on the chemical reaction or physical process, the main contour may be either the macroring or the benzene rings. From the point of view of chromophoric properties, the main contour comprises the macroring. The benzene rings have a disturbing effect on the latter.

Determination of the distribution pattern of π electrons over the contours of phthalocyanine gives an insight into the extent of participation of heteroatomic *n* electrons in the formation of the π-electron cloud of the molecule. As was shown earlier[164] on the basis of addition of charges along individual bonds in H_2Pc, a total of 50–52 macroring electrons take part in the initiation of π-electron charges of the C—N, C—C, and C—H bonds. Of these electrons, 27 are distributed among the benzene rings, three among the C—C bonds, and 22 are in the macroring. In spite of the approximate character of this estimation it can be clearly seen that, apart from $18 + 24 = 42$ inherent π electrons of H_2Pc, another eight to 10 heteroatomic *n* electrons out of 16 participate in the conjugation. In other words about 50% of *n* electrons participate in the conjugation, that is the extent of their involvement is sufficiently great materially to affect the protonating capacity of H_2Pc. Indeed H_2Pc is an extremely weak base because *n* electrons have lost their σ nature as a result of the strong $\varphi_N \rightarrow \psi_{\pi^*}$ shift.

The apparent similarity of their formulae notwithstanding, porphyrins are substantially different from phthalocyanines as far as their properties are concerned.

Difficulties in defining many properties of porphyrins remain despite the numerous structural studies of these compounds.[198–208] This is particularly true in the case of rather complex optical properties of porphyrins.[8,179]

In dealing with general problems of the physical chemistry of porphyrins an effective approach is based on estimation of the π-electron charges of bonds and analysis of conjugation contours in porphyrins. It enables one to establish differences in the properties of heteroaromatic systems.

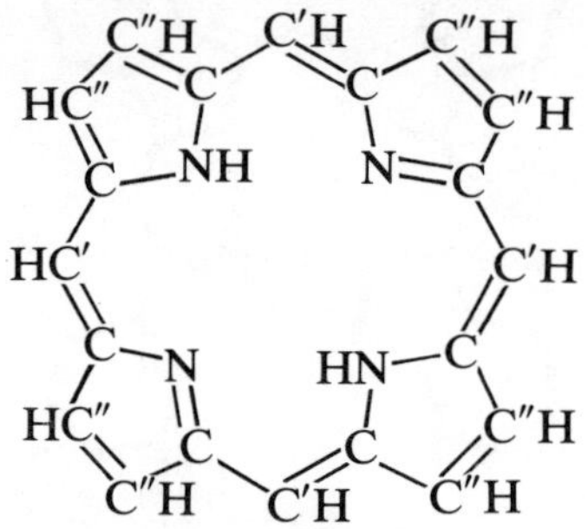

XXII

A calculation similar to that described above in connection with H_2Pc was made for porphin using X-ray data[202] on interatomic distances.

Bond length (Å)				
C″—C″	C″—C	C′—C	C—N	Reference
1.37	1.43	1.41	1.38	202
1.35	1.44	1.40	1.38	207

The latest X-ray data for porphin[207] give practically the same general picture of distribution of π-electron charges of bonds as in an earlier source[163] ($A^{\pi}_{C''-C''} = 1.15$, $A^{\pi}_{C''-C} = 0.74$, $A^{\pi}_{C'-C} = 0.87$, and $A^{\pi}_{C-N} = 0.96$ Å). The π-electron charge of the C″—C″ bond was found to be halfway between those in benzene (0.93) and ethylene (1.40). This is indicative of a stronger π-electron interaction between the pyrrole C″H=C″H groups of porphin and the 16-membered macroring as compared to a similar interaction between the benzene nuclei in phthalocyanine. Consequently, charge $A^{\pi}_{C''-C}$ has increased from 0.33 (phthalocyanine) to 0.74 (porphin). It would hardly be appropriate to assume, for this reason, that the CH=CH groups are isolated from the macroring. The strong effect of 'pyrrole' substituents (positions 1–8) on the electronic absorption spectra indicates that the entire porphin molecule is integrated into a single chromophore system. Hence the wide variety of absorption spectra in the case of porphyrins.

The much lower π-electron charges of the porphin macroring bonds (about 0.93 on the average instead of 1.38 in H_2Pc) are responsible for a less pronounced aromatic nature of porphyrins, their poorer thermal stability, and a stronger tendency of their macroring to open under the action of reagents.[165] The lower aromaticity, in turn, provides for a higher energy of the vacant ψ^*_π orbitals and a less pronounced electron affinity of porphyrins, which manifests itself in their decreased capacity to form dative π bonds M → N with metals which are t_{2g}-electron carriers and in a lower stability of Cu^{2+}, Ni^{2+}, Co^{2+}, Pd^{2+}, and other complexes as compared to similar phthalocyanines. The lower π-electron charges in the porphyrin macroring result in its perimeter being extended by a total of 0.8 Å, which leads to a somewhat larger central open portion and may be one of the main reasons for the extremely reduced stability of σ complexes of porphyrins, such as Zn^{2+}, Al^{3+}, Ga^{3+}, and Sn^{4+} complexes, as compared to analogous phthalocyanines. While the attachment of these ions to phthalocyanine in the course of coordination does not cause any deformation of its skeleton, a different picture is observed in porphyrins.

One of the most important features of porphyrin molecules is the minor participation of non-covalent electron pairs of the nitrogen atoms in conjugation. The addition of π-electron charges of all bonds in a porphyrin molecule after subtraction of the number of inherent π electrons gives the number of conjugated *n* electrons. Rough estimates[163] indicate that only one (or about two, according to more accurate estimates) out of eight *n* electrons takes part in a conjugation with 22 inherent π electrons of porphin. Hence anywhere between 15 and 25% of *n* electrons are involved in conjugation. In the case of phthalocyanine their involvement is as great as 50%. These figures give every reason to expect that the basic properties of porphyrins would be more pronounced. Indeed H_2Ps are easily diprotonated ($H_2P + 2H^+_{solv} \rightleftarrows H_4P^{2+}_{solv}$) in diluted strong acids or aqueous carboxylic acids, that is in media where phthalocyanines exhibit only a weak hydrogen bond.

The higher σ-electron density in φ_N porphyrins is responsible for their extremely weak acid properties (the total ionization constant of H_2P is *c.* 10^{-32}). The acid properties of phthalocyanines are much more pronounced (their acid ionization constant is greater by 10 orders of magnitude). Because of the weaker aromatic properties of the porphyrin macroring (to be more precise, the aromatic part of the porphyrin molecule comprises the entire 22 π-electron system rather than the macroring!), no energetically favourable conditions are created for internal ionization of the imino-hydrogen atoms and coupling of two ionization electrons. The porphyrin molecule is therefore characterized by a multicontour conjugation system and localization of H atoms in the form of an N—H bond. Their localization is best demonstrated in a methylation reaction[209,210] as a result of which the imino-hydrogen atoms are replaced by CH_3 groups, and *N*-methyl- then *N*-dimethylporphyrins are formed.

If etioporphyrin is heated together with CH_3I in a sealed tube for several days, *N*-CH_3NEP[211] is formed capable of complexing and exhibiting, due to the disturbed coplanarity of the starting porphyrin molecule, high solubility in organic solvents as well as high basicity. Methylation of octaethylporphin with CH_3I in hot dichlorobenzene yields *N*,*N*-dimethyloctaethylporphin characterized by a still greater departure from the porphyrin structure. The low π-electron charges of bonds in the aromatic system of porphyrins provide for their greater stability to oxidizing agents as compared to phthalocyanines.[149,212,213] The gradual filling of the ψ_π^* orbitals with *n* electrons of the nitrogen atoms, which takes place as one, two, three, or four methine bridges of tetrabenzoporphin (**XXIII**) are replaced by nitrogen bridges, eventually resulting in phthalocyanine, leads to an experimentally established phenomenon—gradual decrease in the molecule's resistance to oxidizing agents.[214] It is precisely in porphyrins that the tendency to oxidation and hydrolytic destruction of the macroring is minimum. In an acid medium phthalocyanine is very easily oxidized and its macroring is hydrolytically split open.[130,215]

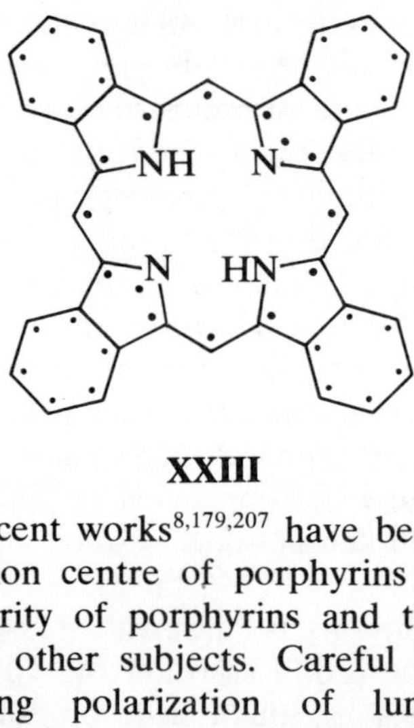

XXIII

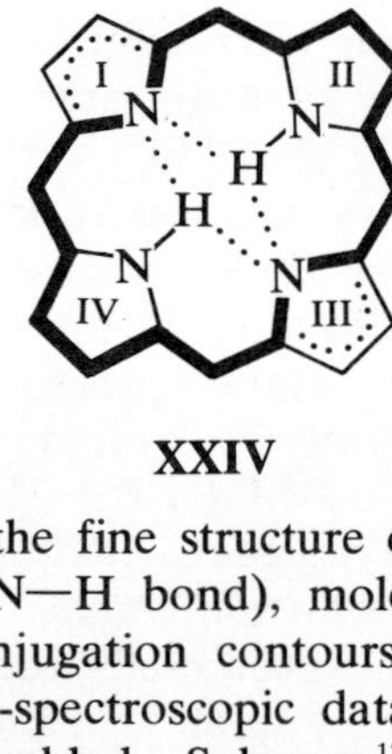

XXIV

Recent works[8,179,207] have been concerned with the fine structure of the reaction centre of porphyrins (the state of the N—H bond), molecular planarity of porphyrins and their complexes, conjugation contours, and some other subjects. Careful analysis of infrared-spectroscopic data and limiting polarization of luminescence has enabled Solovyev[179] to substantiate the concept of intramolecular hydrogen bond >N—H···N< in porphyrins and to hypothesize a conjugation contour (**XXIV**) with symmetry D_{2h} for free porphin. In solution the molecule of porphin and its C derivatives is probably planar, whereas X-ray analysis data sometimes[216] attest to a deviation from coplanarity in the crystalline state. As was established later, the non-coplanarity of ordinary C-substituted porphyrins is in fact a crystalline effect as a consequence of the dense packing of molecules in the presence of C substituents.[179] In the case of porphin[217,218] and extra complex of Zn-tetraphenylporphin with pyridine,[219] the porphin macroring has been found to be completely coplanar. The distribution of π-electron charges of bonds in porphin gives no reason to conclude that freely solvated molecules are not coplanar.

Interestingly, Tulinsky[207] has concluded from analysis of a wealth of X-ray analysis data concerning porphyrins that two pyrrole rings are not equivalent to the other two because of the contribution of the H atom, that is because of the NH group being excluded from the conjugation. Tulinsky proposes conjugation contours similar to **XXIV** but completely excludes two ethylene bonds from the conjugation, with which the present author certainly cannot agree. These 'half-isolated' double bonds are involved in a sufficiently strong π-electron interaction with the rest of the conjugated system of prophin. A legitimate question arises here: is there any reason to exclude NH groups from the conjugation? It must be remembered that in the pyrrole molecule, as can be inferred from its chemical and physicochemical properties, the *n* electrons of an NH group are in a strong conjugation with four inherent π electrons of the molecule and form an aromatic sextet, that is a sufficiently stable closed π-electron shell. The $n \rightleftarrows \pi$ conjugation results in the H atom of pyrrole being capable of protonation under the effect of strong bases.

Why, then, cannot a similar phenomenon occur in the porphin molecule? Infrared spectra[8] indicate that the conditions for an $n \rightleftarrows \pi$ conjugation in porphin are more favourable than in pyrrole. It can be assumed that a sufficiently strong π-electron interaction brings the C—NH bonds of porphin into a conjugation. Therefore formula **XXIV** must be modified so that all porphin bonds are included in the conjugation contours. It should be borne in mind, however, that the π and *n* electrons of the nitrogen atoms in the pyrrole rings of porphin are not completely equivalent. This is what makes porphin different from phthalocyanine.

However, as is correctly pointed out by Solovyev *et al.*,[8,179] in porphin complexes with M^{2+}, M^{3+}, and M^{4+} all four pyrrole rings become equivalent owing to the redistribution of *n* and π electrons, that is owing to conjugation. In these complexes the conjugation contours incorporate all C—C and C—N bonds, that is the aromatic system comprises all 22 π electrons. All of the eight *n* electrons are excluded from the conjugation, that is the system becomes less aromatic, which is indicated by an increase in the energy of the first electronic transition. This does not mean at all that a metalloporphyrin is always less thermally stable than the starting porphyrin. Porphyrin complexes have an ideally symmetrical π-electron system D_{4h}. Naturally its symmetry is disturbed in the case of asymmetrical alkyl C substitution as well as hydrogenation of one of the two ethylene double bonds, resulting in chlorins and bacteriochlorins. This more specific problem is treated at length elsewhere.[8]

One of the most interesting problems of the structural chemistry of porphyrins is the structure of the π-electron system and reaction centre of tetrabenzoporphin (**XXIII**) and tetrazaporphin in which aza bridges replace methine ones. As has been concluded[179] from infrared- and electron-spectroscopic data, the structures of tetrabenzoporphin and tetrazaporphin are similar to that of porhin. Such a conclusion seems to be premature since it may only be drawn after studying the physicochemical properties of these compounds, their complexes, and their functional substituents, which, unfortunately, have not been elucidated so far and, what is more, no X-ray analysis data are yet available for these molecules.

It is doubtful that the conjugation contours of porphin and its tetrabenzo derivative are similar. It is more likely that H_2TBP features two different aromatic systems—a stable benzene sextet and a 14 π-electron macroring. The benzene and macrocyclic chromophores are equivalent as far as the force of the π-electron interaction is concerned, which follows from calculations of the π-electron charges of bonds in phthalocyanine and porphin. Therefore the breakdown of one at the expense of the other is hardly possible. At the same time H_2TBP has a porphin macroring substantially different from the tetrazaporphin ring of phthalocyanine. As was shown above, conditions are provided in phthalocyanine for a strong $n \rightleftarrows \pi$ conjugation and internal ionization of protons. In the porphin macroring no such conditions exist, which is why the H atoms are arranged as in formula **XXIV**. One can predict, to a first approximation, conjugation contours of type **XXI** in tetrabenzoporphin with the presence of NH groups, that is an aromatic system occupying an intermediate position between phthalocyanine and porphyrin. This conclusion is in agreement with reference 8. Making predictions about tetrazaporphin is more difficult. I shall come back to the discussion of this problem in the chapter dealing with electronic spectra.

3.3 Quantum-chemical Data on the Structure of Porphyrins and Phthalocyanine

The results of quantum-chemical calculations of molecules provide a sound theoretical foundation for predictions based on the relationship between the experimentally studied properties of molecules and their structure. In the case of porphyrins and phthalocyanines these calculations are complicated because of the complexity of the molecules themselves and the large number of non-equivalent atomic nuclei, bonds, and electrons. The fit of the calculation results to the experiment is ensured by selection of an appropriate calculation method, semi-empirical parameters, and, particularly, π-electron model. All these problems have been profoundly, critically, and lucidly treated in one of the chapters of a previous monograph.[8] Therefore, in the present volume, the dependence of the calculation results on the model selection will only be considered briefly.

As was pointed out earlier,[163,164,179] the results of calculating π-electron systems of the phthalocyanine and porphyrin types must be strongly dependent on proper selection of a π-electron model, that is the structure of multicontour conjugated systems, localization or delocalization of hydrogen atoms, and allowance for the mutual disturbing action of chromophores. As a rule such models vary from one work to another, which leads to ambiguities in the calculation results.[8,220,221]

In one of his works[220] Chen critically assessed the results of calculating free phthalocyanine on the basis of various models. The results of calculations by the LCAO–MO method using a model of localized H atoms were found to be completely unsatisfactory. The ratio of the position of the longest-wave band in the electronic spectra of $H_2Pc(\upsilon_1')$ to the difference between this band and its doublet $(\Delta\upsilon')$ turns out to be $\upsilon_1'/\Delta\upsilon' = 4.50$ instead of the experimental value of 19.54. The use of a delocalized H-model with a bonding scheme proposed earlier[164] by the present author has shown good agreement with the experiment (ratio $\upsilon_1'/\Delta\upsilon'$ lies within the range 19.35–20.41). Chen[220] has shown the inconsistency of the four-orbital model of Simpson, Platt, and Gouterman as applied to phthalocyanine. This model, however, adequately describes the porphyrin spectra.

Thus quantum-chemical calculations by the molecular-orbital method corroborate the ionization of the imino-hydrogen atoms in phthalocyanine and the lack of such ionization in the case of porphyrins, that is they show a fundamental difference between these two types of molecules. The use of real conjugation contours and π-electron charges of bonds for phthalocyanine[164] in the potential-well model of the molecule yields satisfactory results in calculations of the position of the first band in the spectra of phthalocyanine complexes.[222] The results of quantum-chemical calculations of other properties of porphyrins will be considered later in conjunction with the properties of these molecules and their complexes.

Chapter 4

Ionization and coordination properties of porphyrins and phthalocyanine

This chapter deals with the ionization and coordination properties of porphyrins and phthalocyanine, which are the most important properties of these molecules and determine their ability to enter into reactions with metal salts and yield coordination compounds.

IONIZATION PROPERTIES

The notion of ionization properties of porphyrin molecules implies generation of ionic forms, that is processes involving charge changes. The ionization of porphyrin molecules may proceed along different pathways. As compounds containing two most protonated N—H bonds they may form, in highly proton-acceptor media, a monoanion (HP^-) and a dianion (P^{2-}) in equilibrium processes

$$H_2P + X \rightleftarrows XH^+ + HP^-, \tag{4.1}$$

$$HP^- + X \rightleftarrows XH^+ + P^{2-}. \tag{4.2}$$

In this case ionization may involve acid substituents of porphyrin side chains containing a carboxy, hydroxy, sulpho, or other groups whose ionization is similar to that of the corresponding functional derivatives of organic compounds and will not be considered here.

Porphyrin molecules are characterized by the presence of four to eight heteroatoms that are *n*-electron carriers. Therefore their protonation may be expected at the N atoms, yielding cationic forms of porphyrins, namely a monoanion (H_3P^+), a dication (H_4HP^{2+}), and higher-order cations, in reactions

$$H_2P + H^+ \rightleftarrows H_3P^+, \tag{4.3}$$

$$H_3P^+ + H^+ \rightleftarrows H_4P^{2+}. \tag{4.4}$$

The ionization, as in reactions (4.1)–(4.4), is in fact well known and is generally referred to as 'anionic–cationic' ionization.

In recent years extensive studies have been under way into the generation of ionic forms of porphyrins of an entirely different nature. Anionic forms are obtained[223–228] by causing a porphyrin to react, in a solution of an inert solvent (e.g. tetrahydrofuran), with an alkali metal (or anions of arenes and heterocycles) in the absence of oxygen:

$$H_2P + n\text{Na} \rightarrow n\text{Na}^+ + H_2P^{n-}. \qquad (4.5)$$

In reaction (4.5) metalloporphyrin (or free porphyrin) may yield mono-, di-, tri-, and tetra-anions. However, as opposed to anions, as in (4.2), these are anion radicals. They are formed in a reaction in which an electron is transferred from a reducing agent (hydrazine, hydroquinone, ascorbic acid, etc.) to the prophyrin molecule.

The photoreduction reactions have been discovered and carefully studied by Krasnovsky.[9,15]

Under the effect of strong electron acceptors radical-cationic forms may be obtained. The photochemical processes of their formation have been extensively investigated by Yevstigneev *et al.*[228–231] and Dilung *et al.*[232–236] Cation-radicals in porphyrin can be isolated electrochemically with subsequent examination of their absorption spectra.[237–239]

$$\text{MP} + \text{oxidizing agent} \rightleftarrows \text{MP}^+ + \text{reducing agent}. \qquad (4.6)$$

The ionic forms produced in reactions (4.5) and (4.6) result from anion-radical and cation-radical ionization. The role and importance of these forms in various processes involving porphyrins seem to be quite significant. However, they belong to a highly specific field of porphyrin chemistry, related but remotely to the subject matter of this monograph.

In view of this only processes of acid ionization (4.1), (4.2) and protonation (4.3), (4.4) of prophyrins will be discussed in what follows since they are directly associated with the structure and complexing of porphyrin ligands, the stability of metalloporphyrins, and the mechanism of their formation and dissociation. These processes are the ultimate result of the acid–base interaction[240] on free porphyrins and, as a result of H^+ and metal cations exhibiting many common properties, belong to processes competing with reactions of coordination of porphyrins to metal salts.

4.1 Acid Dissociation of Porphyrins and Phthalocyanine

The existence and properties of anionic forms of porphyrins belong to many other unsolved problems in the chemistry of porphyrins. The possibility of existence of porphyrin monoanion HP^- has not yet been definitively proved. It is not certain whether the spectroscopically detected particle P^{2-} is a dianion or a complex salt of an alkali metal, for example, $Na_2^+P^{2-}$.[8]

Another thing that puzzles investigators is the fact that even in extremely concentrated solutions of alkalis (KOH, NaOH) in thoroughly dehydrated alcohols there persists the four-band visible spectrum of free porphyrin H_2P with the band intensities diminishing as follows: IV > III > II > I

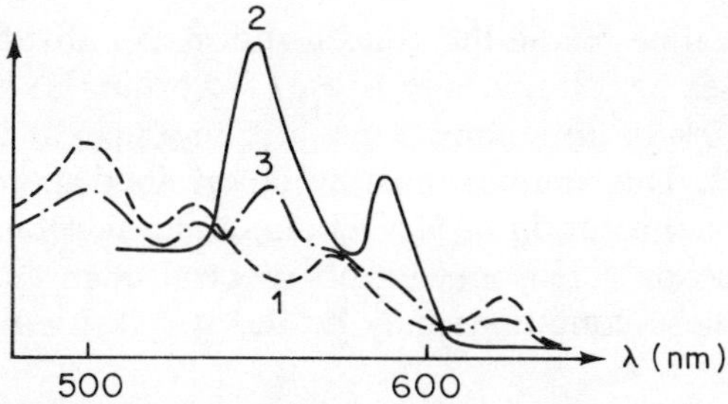

Figure 4.1 Electronic spectrum of etioporphyrin in a sodium alcoholate solution. Water addition: curve 1, 10%; curve 2, 0%; curve 3, 1%

(Fig. 4.1). A two-band spectrum appears only in a solution of sodium alcoholate and similar highly basic aprotic media, which may be ascribed to dianion P^{2-} or its disodium salt $Na_2^+P^{2-}$. Addition of a small amount of water to the alcoholate solution of porphyrin immediately produces an intermediate spectrum of an H_2P and P^{2-} mixture. Further addition of water gives the spectrum of pure porphyrin. From the point of view of chemistry and thermodynamics there is nothing surprising in the above phenomena which suggest that porphyrin dianion P^{2-} is an extremely strong base, probably as strong as alkyl sodium carbanions and similar compounds.

This is why the introduction of a proton-donor molecule (e.g. water, ethanol) first yields ion-molecular associates such as $P^{2-} + H_2O \rightleftarrows P^{2-}\cdots H{-}O{-}H$ with subsequent instantaneous and almost complete transfer of a proton from the water molecule to the dianion: $P^{2-}\cdots H{-}O{-}H \rightleftarrows PH^-\cdots PH^- \rightleftarrows PH^- + OH^-$. When water or an alcohol is added, as the basicity decreases the acid–base interaction proceeds further:

$$PH^- + H_2O \rightleftarrows PH^-\cdots H{-}O{-}H \rightleftarrows PH_2\cdots OH^- \rightleftarrows PH_2 + OH^-.$$

Simple calculations show that a 20 M solution of NaOH will yield, in the presence of a moderate amount of water, not more than 1% of porphyrin dianion which escapes spectroscopic detection in the case of the usually used 10^{-5}–10^{-4} M solutions of prophyrin. However, the equilibrium theory tells us that in an alkaline solution of even the weakest acid there is always present a certain (no matter how negligible) amount of all of its anionic forms.

Even in alkaline solutions of alcohols and other organic solvents porphyrins enter into a number of equilibrium reactions:

$$(HN)_2P + OH^- \underset{}{\overset{K_{OH}}{\rightleftarrows}} \underset{\mathbf{a}}{(HN)P{-}N{-}H \cdots H{-}O^-} \overset{K_{S_1}}{\rightleftarrows}$$

$$\underset{\mathbf{b}}{(HN)P{-}N^-\cdots H{-}O{-}H} \rightleftarrows \underset{\mathbf{c}}{(HN)PN^- + H_2O \cdots \text{solv.}} \quad (4.7)$$

In moderately alkaline media the concentration of individual ionic forms of prophyrin increases as follows: **a** $\gg$ **b** $>$ **c**. According to rough quantitative estimates up to 10% of an anionic form of **b** or **c** appear in a 10 M aqueous solution of NaOH. This amount must be determined spectroscopically. In a concentrated alcohol solution of NaOH almost all porphyrin must be in the form of PH^-. The only reason why no spectral changes in porphyrin are observed under these conditions may be the spectral indistinguishability of H_2P and PH^-.

An attempt has been made[8] to provide a reasonable explantion for the spectral similarity of different ionic forms of porphyrins. Proceeding from the authors' assumption that the removal of a single proton does not alter either the number of active π electrons in the conjugated system or the D_{2h} symmetry of the starting H_2P molecule, one can understand the spectral indistinguishability of H_2P and PH^-. For this reason, which is beyond any doubt, the problem of existence of porphyrin monoanions remains open to dispute. As far as the concentrations of individual forms of the first stage of porphyrin ionization (**a**, **b**, and **c**) are concerned, the ratio between the solvated (**c**) and hydrated (**b**) forms of the monoanion will depend on the nature of the organic solvent and the degree of its dehydration. At present these forms are not distinguishable in the thermodynamics of solutions, and, for this reason, all acid-ionization constants derived so far are arbitrary. The ratio between the H_2P and **a** concentrations will depend on the alkalinity of the medium. As a result of the high ion-molecular association constant K_{OH} practically all H_2P molecules will take the form **a** at high alkali concentrations.

The redistribution of the PH^- charge among all atoms of the macroring will render the ionization of the second proton as difficult as that of the first one:

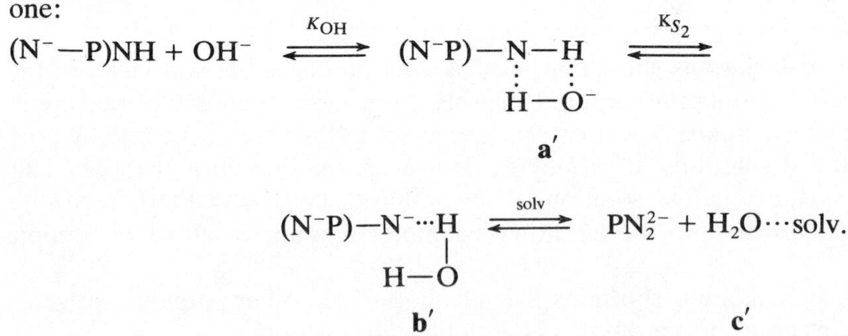

(4.8)

The same applies to the ratio between the concentrations of individual forms as to reactions (4.7). The experimentally determined second-stage porphyrin ionization constant K_{S_2} will also be arbitrary or relative, that is related to strictly preset experimental conditions. In modern literature there are practically no data on the thermodynamics of porphyrin ionization processes, which may be explained by the enormous difficulties encountered in experimental determination of the acid-ionization constants of porphyrins at various temperatures. Of the available reliable constants we can mention only one derived by McEwen[211] 40 years ago. He determined the total ionization constant of etioporphyrin by the method of photometric titration of its solution in methanol with a benzene solution of triphenylmethyl sodium. Assuming that $pK_{CH_3OH} = 16$, he obtained $pK_{S_1} + pK_{S_2} \approx 32$.

The lack of ionization constants of porphyrin seriously hinders the understanding of these highly important processes and holds back the chemistry of porphyrin complexes because it does not permit calculation of their instability constants. In this connection attempts have been made over the past decade to define ionization constants at least approximately. Even this, however, is extremely difficult. The extremely high alkali concentrations required to initiate ionization and the very slight changes in pH in this process do not permit the use of electrometric methods.

The spectral indistinguishability of H_2P and PH^- makes it impossible to resort to spectrophotometry which is widely employed in most applications. Therefore investigators had to use the distribution[240] and solubility[241–244] methods. The distribution method was used in studying the acid ionization of chlorophyllic acid ($a + b$).

To this end, 15 ml of a buffer with a pH of 9.0–11.28 was added to 15 ml of a 1×10^{-4} M solution of chlorophyllic acid in butyl alcohol. The mixture was shaken for several minutes in a separating funnel at 20–22 °C and allowed to stand for 15 min for stratification. Then the optical density of the butanol solution was measured with the aid of a photometer at effective wavelengths of 465, 533, and 618 nm, and the H_2Chl concentrations in butanol and the total concentration of HChl and Chl^{2-} in the aqueous layer were determined.

The results obtained by this method[240] are summarized in Table 4.1. The author of the last-mentioned reference presents the following picture of ionization in a two-layer system:

$$H_nChl_{(w)} \rightleftarrows H_nChl_{(but)} \rightleftarrows nH^+ + Chl^{n-}. \tag{4.9}$$

Table 4.1 Distribution of chlorophyllic acid between butanol and aqueous buffer

	$[H_2Chl]$ (%)				
pH	Buffer	Butanol	$[Chl]_w^{n-}/[H_nChl]_{but}$	$K'_{S_1} \times 10^{11}$	$K'_{S_2} \times 10^{20}$
9.0	1.8	98.2	0.018	1.8	1.8
9.2	4.1	95.9	0.043	2.7	1.7
9.4	8.6	91.4	0.094	3.7	1.5
9.6	11.4	88.6	0.129	3.2	0.8
9.8	16.7	83.3	0.200	3.2	0.5
10.0	22.6	77.4	0.292	2.9	0.3
10.2	29.4	70.6	0.416	2.6	0.2
10.6	38.3	61.7	0.621	1.6	0.04
11.0	50.3	49.7	1.010	1.0	0.01
11.3	58.1	41.9	1.390	0.7	0.004
			Averages	$2.3 \pm 0.9 \times 10^{-11}$	$0.97 \pm 0.60 \times 10^{-20}$

The constants of Table 4.1 are written by Kukhtevich[240] as follows:

$$K'_{\mathrm{H_2Chl(o/w)}} = [\mathrm{H^+}]_w[\mathrm{HChl^-}]_w/[\mathrm{H_2Chl}]_0 \tag{4.10}$$

He arrives at this form by combining the ionization constant of H_2Chl in water with the constant of distribution of H_2Chl between water and the organic phase (butanol):

$$K = [\mathrm{H_2Chl}]_o/[\mathrm{H_2Chl}]_w. \tag{4.11}$$

A relation similar to (4.10) is written by Kukhtevich for the total ionization constant of H_2Chl, which he terms the extraction constant. As was pointed out earlier,[244] Kukhtevich had not taken into account corrections for the distribution of H^+ and ionic forms of chlorophyllic acids in a two-phase system, and had made some errors in the experiment. Therefore quantity K'_{S1} ($K'_{\mathrm{H_2Chl(o/w)}}$) makes some sense, bearing in mind that under given experimental conditions monoanion $HChl^-$ primarily occurs. However, without correction of K'_S for ionization in an aqueous medium even this constant is of no practical value. Moreover, it relates to a mixture of two different compounds, H_2Chl*a* and H_2Chl*b*, exhibiting different ionization and extraction properties.

The calculation of the value of K'_{S_2}, made by Kukhtevich,[240] is not permissible at all because the sum of concentrations of $[HChl^-]$ and $[Chl^{2-}]$ is substituted into the expression of the total ionization constant, the concentration of $[Chl^{2-}]$ being negligibly small in the entire range of experimental pH values. This is why the value of K'_{S_2} in this range differs by 2.5 orders of magnitude. It should be admitted that the above-described extraction method may be useful, provided a correct quantitative approach is tried, for rigorous determination of the ionization constants of various porphyrins.

A more rigorous determination of the first-stage ionization constant of various porphyrins was attempted in the present author's laboratory.[241–243] For this purpose a microversion of the solubility method had been developed.

To provide for a required amount of the solid phase the porphyrin of interest was deposited from an ether or chloroform solution of polyethylene beads, then the solvent was evaporated. After repeating this step a number of times the beads were placed in a glass ampoule, flooded with anhydrous methanol or a methanol solution of KOH, taken in an amount of 2–8 ml depending on the solubility of the porphyrin, and the ampoule was sealed. Then the ampoule was shaken in a special thermostat until equilibrium was established between the solid phase and the solution, which was checked by the optical density of the latter.

Blank tests have shown that at 25 °C the equilibrium was attained after 6 h for H_2Chl*a* and H_2Chl*b* and after 36 h for H_2TPP. The solubility of porphyrins was determined from the optical density of the solution separated from the solid phase on a Schott funnel No. 4, with the aid of calibration graphs representing porphyrin concentration versus optical density, plotted in advance. As the alkali concentration grew, the solubility of porphyrins increased linearly, which is indicative of the predominant role of the first-stage ionization of porphyrins (4.7).

Figure 4.2 shows solubility $\bar{S} = \bar{c}_{H2P} + \bar{c}_{HP^-} = S_0 + \bar{c}_{HP^-}$ versus KOH concentration in the case of mesoporphyrin. Similar relationships were obtained for two other porphyrins. The first acid ionization constant was calculated from the equation

$$K_{S_1} = [H^+](\bar{S} - S_0)/S_0. \quad (4.12)$$

The value of $[H^+]$ was calculated assuming that under the experimental conditions practically all of the KOH turns to $CH_3O^-K^+$. The concentration of $CH_3OH_2^+$, which was identified with that of H^+, was calculated from the methanol autoprotolysis constant ($pK_{CH_3OH} = 16.7$). The same constant values of K_{S_1} for the examined porphyrins—chlorophyllic acids *a* and *b*, mesoprophyrin DME, protoporphyrin, etioporphyrin, tetraphenylporphin, and tetramethoxyphenylporphin (Tables 4.2–4.8)—attest to the correctness of the assumptions made. Table 4.9 lists mean values of the ionization constants of the above porphyrins.

As can be inferred from the tabulated data, chlorophyllic acid *a* is stronger than chlorophyllic acid *b*. This extremely interesting fact may be ascribed to the increased electron density at the nitrogen atoms in H_2Chl*b*, which results in a weaker protonation of the N—H bond, that is a somewhat greater covalence of this bond. As will be shown below, the higher kinetic and thermodynamic stabilities of the H_2Chl*b* complexes have the same explanation. The increase in the electron density at the N atoms of H_2Chl*b* may be the consequence of either a different arrangement of protons in it as compared to H_2Chl*a*, which is scarcely probable in view of what has been found out[8] about the ionic forms of chlorins, or the +C effect of the formyl group, which compensates for its −I effect. The assumption as to the +C

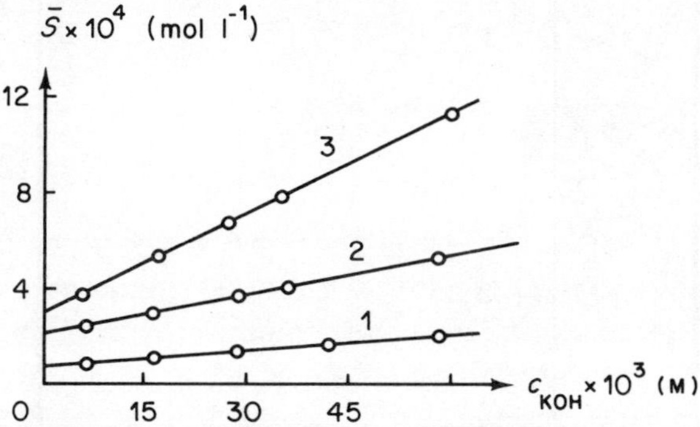

Figure 4.2 Solubility of mesoporphyrin in a methanol solution of KOH. Temperature: curve 1, 15 °C; curve 2, 25 °C; curve 3, 35 °C

Table 4.2 First-stage acid ionization of chlorophyllic acid *a* in methanol[a]

c^0_{KOH} (M)	$c_{H^+} \times 10^{15}$ (g-ion l^{-1})	15 °C		25 °C		35 °C	
		$\bar{S} \times 10^4$ (M)	$K_{S_1} \times 10^{15}$	$\bar{S} \times 10^3$ (M)	$K_{S_1} \times 10^{15}$	$\bar{S} \times 10^3$ (M)	$K_{S_1} \times 10^{15}$
0	—	0.30	—	1.08	—	1.42	—
0.0016	12.50	—	—	1.28	2.30	1.85	3.79
0.0018	11.11	7.14	1.48	—	—	—	—
0.0026	7.69	—	—	1.41	2.35	2.14	3.90
0.0027	7.41	7.60	1.53	—	—	—	—
0.0034	5.88	—	—	—	—	2.37	3.94
0.0042	4.76	8.38	1.57	—	—	—	—
0.0048	4.17	—	—	1.67	2.28	2.77	3.96
0.0054	3.70	8.82	1.48	—	—	—	—
0.0056	3.57	—	—	1.76	2.25	—	—
0.0072	2.78	9.64	1.47	—	—	—	—
0.0077	2.60	—	—	2.04	2.31	3.50	3.81
0.0114	1.75	11.77	1.52	—	—	4.54	3.85
0.0133	1.50	—	—	2.71	2.26	5.08	3.87
0.0152	1.32	13.70	1.55	2.97	2.31	—	—

[a]The data given in reference 241 contain calculation errors.

Table 4.3 First-stage acid ionization of chlorophyllic acid *b* in methanol[a]

c^0_{KOH} (M)	$c_{H^+} \times 10^{15}$ (g-ion l^{-1})	15 °C		25 °C		35 °C	
		$\bar{S} \times 10^4$ (M)	$K_{S_1} \times 10^{15}$	$\bar{S} \times 10^4$ (M)	$K_{S_1} \times 10^{15}$	$\bar{S} \times 10^3$ (M)	$K_{S_1} \times 10^{15}$
0	—	5.23	—	7.35	—	1.30	—
0.0011	18.20	5.46	8.01	—	—	—	—
0.0014	14.30	—	—	7.97	1.20	1.48	1.99
0.0023	8.70	5.71	8.00	—	—	—	—
0.0028	7.14	—	—	8.61	1.22	1.67	2.04
0.0037	5.40	—	—	9.14	1.32	1.80	2.08
0.0046	4.35	6.20	8.05	9.44	1.24	—	—
0.0062	3.23	—	—	—	—	2.10	1.99
0.0067	2.99	6.65	8.13	—	—	—	—
0.0073	2.74	—	—	10.73	1.26	—	—
0.0079	2.53	—	—	—	—	2.36	2.06
0.0092	2.17	7.18	8.09	—	—	—	—
0.0095	2.11	—	—	—	—	2.54	2.01
0.0104	1.92	—	—	12.06	1.23	—	—
0.0108	1.85	7.51	8.07	—	—	—	—
0.0116	1.72	—	—	—	—	2.79	1.97
0.0121	1.65	7.80	8.10	—	—	—	—
0.0124	1.61	—	—	13.05	1.25	—	—

[a]The data given in reference 241 contain calculation errors.

Table 4.4 First-stage acid ionization of mesoporphyrin DME in methanol

c^0_{KOH} (M)	$c_{H^+} \times 10^{15}$ (g-ion l^{-1})	15 °C		25 °C		35 °C	
		$\bar{S} \times 10^5$ (M)	$K_{S_1} \times 10^{16}$	$\bar{S} \times 10^4$ (M)	$K_{S_1} \times 10^{16}$	$\bar{S} \times 10^4$ (M)	$K_{S_1} \times 10^{15}$
0	—	7.21	—	1.60	—	2.61	—
0.0047	42.55	8.06	5.02	—	—	—	—
0.0056	35.70	—	—	1.92	7.14	3.39	1.07
0.0149	13.42	9.83	4.87	—	—	—	—
0.0155	12.90	—	—	2.47	7.02	4.68	1.02
0.0253	7.91	11.68	4.90	—	—	—	—
0.0264	7.58	—	—	3.12	7.20	5.97	0.98
0.0310	6.45	—	—	3.35	7.06	7.07	1.10
0.0342	5.85	13.43	5.05	—	—	—	—
0.0530	3.77	16.80	5.01	—	—	—	—
0.0550	3.64	—	—	4.75	7.17	9.55	0.97
0.0628	3.18	—	—	5.18	7.12	11.12	1.04
0.0641	3.12	18.64	4.95	—	—	—	—

Table 4.5 First-stage acid ionization of protoporphyrin DME in methanol

c^0_{KOH} (M)	$c_{H^+} \times 10^{16}$ (g-ion l^{-1})	25 °C $\bar{S} \times 10^4$ (M)	25 °C $K_{S_1} \times 10^{16}$	35 °C $\bar{S} \times 10^4$ (M)	35 °C $K_{S_1} \times 10^{16}$
0	—	1.48	—	2.07	—
0.010	20.00	—	—	2.44	3.56
0.017	11.70	1.85	2.95	—	—
0.030	6.67	—	—	3.17	3.54
0.041	4.88	2.41	3.07	—	—
0.051	3.92	—	—	3.93	3.53
0.073	2.74	—	—	4.51	3.23
0.081	2.47	3.27	2.99	—	—
0.995	9.95	5.67	2.79	—	—

Table 4.6 First-stage acid ionization of etioporphyrin in methanol

c^0_{KOH} (M)	$c_{H^+} \times 10^{16}$ (g-ion l^{-1})	25 °C $\bar{S} \times 10^6$ (M)	25 °C $K_{S_1} \times 10^{16}$	35 °C $\bar{S} \times 10^6$ (M)	35 °C $K_{S_1} \times 10^{16}$
0	—	1.83	—	2.74	—
0.0160	12.50	2.057	1.54	3.16	1.91
0.0203	9.85	2.13	1.62	3.28	1.94
0.0261	7.66	2.21	1.59	3.44	1.95
0.0340	5.88	2.32	1.58	3.66	1.98
0.0400	5.00	2.40	1.55	3.81	1.95
0.0461	4.34	2.48	1.54	3.98	1.97

Table 4.7 First-stage acid ionization of tetraphenylporphin in methanol

c^0_{KOH} (M)	$c_{H^+} \times 10^{16}$ (g-ion l^{-1})	25 °C $\bar{S} \times 10^6$ (M)	25 °C $K_{S_1} \times 10^{17}$	35 °C $\bar{S} \times 10^6$ (M)	35 °C $K_{S_1} \times 10^{17}$
0	—	4.29	—	5.51	—
0.0180	11.11	4.49	4.44	5.77	5.22
0.0379	5.28	4.66	4.54	6.06	5.28
0.0541	3.70	4.82	4.55	6.31	5.36
0.0674	2.97	4.93	4.44	6.47	5.17
0.0858	2.33	5.14	4.61	6.74	5.20
0.1012	1.98	5.24	4.38	6.99	5.33

Table 4.8 First-stage acid ionization of tetramethoxyphenylporphin in methanol

		25 °C		35 °C	
c^0_{KOH} (M)	$c_{H^+} \times 10^{16}$ (g-ion l^{-1})	$\bar{S} \times 10^6$ (M)	$K_{S_1} \times 10^{17}$	$\bar{S} \times 10^6$ (M)	$K_{S_1} \times 10^{17}$
0	—	3.66	—	5.11	—
0.0162	12.34	3.77	3.70	5.28	4.07
0.0264	7.58	3.83	3.49	5.39	4.17
0.0476	4.20	3.97	3.57	5.62	4.20
0.0742	2.70	4.15	3.62	5.87	4.02
0.0916	2.18	4.25	3.51	6.09	4.19
0.1012	1.98	4.33	3.62	6.17	4.10

Table 4.9 Constants of first-stage acid ionization of porphyrins in methanol (K_{S_1})[a]

Porphyrin	15 °C	25 °C	35 °C
H_2Chl*a*	$(1.5 \pm 0.03) \times 10^{-15}$	$(2.29 \pm 0.03) \times 10^{-15}$	$(3.87 \pm 0.05) \times 10^{-15}$
H_2Chl*b*	$(8.06 \pm 0.04) \times 10^{-16}$	$(1.25 \pm 0.03) \times 10^{-15}$	$(2.02 \pm 0.04) \times 10^{-15}$
H_2MP DME	$(4.97 \pm 0.06) \times 10^{-16}$	$(7.12 \pm 0.06) \times 10^{-16}$	$(1.30 \pm 0.04) \times 10^{-15}$
H_2PP DME	—	$(2.95 \pm 0.08) \times 10^{-16}$	$(3.47 \pm 0.01) \times 10^{-16}$
H_2EP	—	$(1.57 \pm 0.03) \times 10^{-16}$	$(1.95 \pm 0.03) \times 10^{-16}$
H_2TPP	—	$(4.49 \pm 0.07) \times 10^{-17}$	$(5.26 \pm 0.06) \times 10^{-17}$
H_2TMOPP	—	$(3.59 \pm 0.06) \times 10^{-17}$	$(4.13 \pm 0.06) \times 10^{-17}$

[a]The figures are averaged from Tables 4.2–4.8.

effect of the formyl group in porphyrins, made earlier by the present author,[245–247] is therefore not surprising.

The sign of the electronic effect of a substituent is determined by the nature of its partner, therefore every aromatic molecule will feature its own series of substituents exhibiting an electronic C effect.

Benzene and similar arenes are characterized by a −C effect of the formyl group, probably due to the lower energy of the antibonding ψ^* C=O molecular orbitals as compared to benzene ones.[248]

Porphyrins are characterized, as can be seen from the electronic absorption spectra,[8] by lower-energy ψ^*-molecular orbitals as compared to the orbitals of the CHO and CH=CH_2 groups, therefore the transfer of π electrons from these substituents into the macroring of porphyrins is energetically justified. An increase in the π-electron density invariably leads to a higher σ-electron density at the N atoms and to stabilization of the N—H bond.

The difference in N—H bond stabilities does not directly affect the enthalpies of ionization of porphyrins (Table 4.10) because a major role in these processes is played by the solvent molecules that stabilize the emerging anionic form by way of solvation. Because of the less pronounced solvation of HChl*b* in methanol the enthalpy of the first-stage ionization of HChl*b* is lower (by 1 kcal mol^{-1}) than in the case of HChl*a*. In all examined

Table 4.10 Thermodynamic parameters of ionization of porphyrins at 25 °C

Porphyrin	ΔH (kcal mol^{-1})	ΔZ (kcal mol^{-1})	ΔS (e.u.)	K_{S1}^{25}
H_2Chl*a*	9.6	19.9	−35	2.29×10^{-15}
H_2Chl*b*	8.7	20.3	−39	1.25×10^{-15}
H_2MP DME	6.7	20.6	−47	7.12×10^{-16}
H_2PP DME	3.0	21.2	−61	2.95×10^{-16}
H_2EP (I)	4.0	21.5	−59	1.57×10^{-16}
H_2TPP	2.9	22.3	−65	4.49×10^{-17}
H_2TMOPP	2.5	22.4	−67	3.59×10^{-17}

porphyrins the ionization is accompanied by a sharp decrease in the entropy (by 35–67 e.u.) and a perceptible increase in the enthalpy (by 9.6–2.5 kcal mol^{-1}) of the system, depending on the porphyrin species.

Thus both the enthalpy and entropy components of free energy hinder the ionization of porphyrins, which is accompanied by its increase (by 20–22 kcal mol^{-1}). The role of individual free energy components of various chemical processes and the importance of these components for analysis of the phenomena occurring in solutions have been covered at length in the Soviet literature.[249,250]

The next pair of structurally similar porphyrins, namely mesoporphyrin DME (**XXV**) and protoporphyrin DME (**XXVI**), features a similar system of side substituents. At the same time both porphyrins differ widely from chlorophyllic acids **XXVII** and **XXVIII** is having a chlorin π system, a cyclopentanone ring, and an extremely large phytol residue ($C_{20}H_{39}$) in the ester group. This is why one cannot compare the state of the N—H bonds in systems **XXV**, **XXVI** and **XXVII**, **XXVIII** in terms of ionization constants. However, comparing the state of bonds inside these systems is possible and useful.

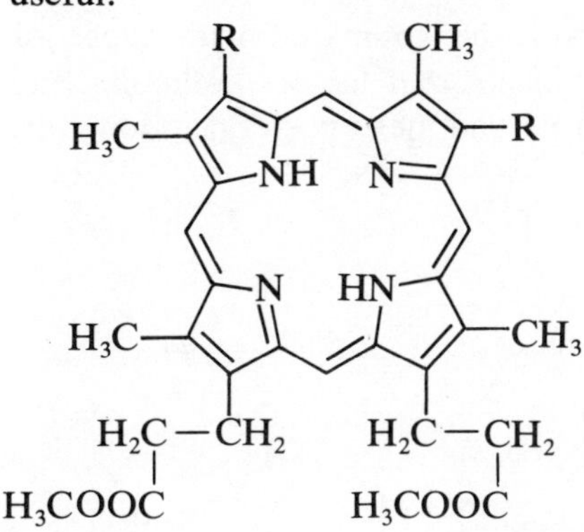

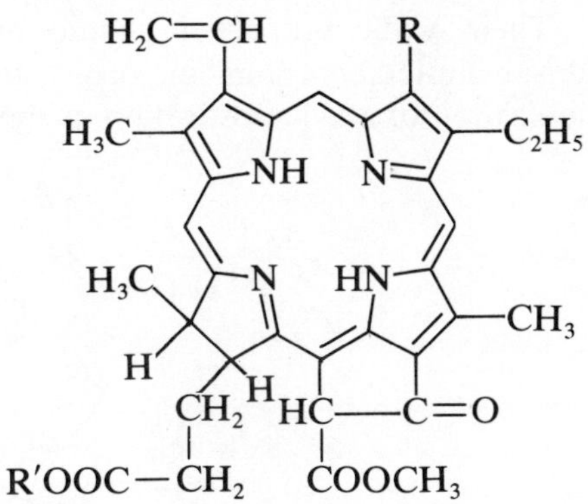

XXV: R = C_2H_5 **XXVII**: R = CH_3; R′ = $C_{20}H_{39}$ (*a* series)

XXVI: R = CH=CH_2 **XXVIII**: R = H—C=O; R′ = $C_{20}H_{39}$ (*b* series)

H_2MP DME contains only alkyl and so-called pseudo-alkyl substituents (methyalted residues of propionic acid). The pseudo-alkyl substituents may be ranked with alkyl ones but with a proviso if we are dealing with electronic substitution effects ($\pm$I and $\pm$C effects). They are not equivalent in the case of solvation, steric, and field effects. The oxygen atoms of the ester group of propionic acid are readily solvatable in alcohol solutions, and the group itself produces a certain screening effect on the two adjacent nitrogen atoms. Mesoporphyrin has a lower first-stage ionization constant (7.12×10^{-16}) than even chlorophyllic acid *b*.

In view of the non-isoentropic nature of ionization of porphyrins and variations in ΔH of this process due to solvation it is difficult to make any predictions as to changes in the bond state in mesoporphyrin. It may be assumed that the decrease in entropy in the course of ionization of porphyrins in the series from H_2Chl*a* to tetramethoxyphenylporphin is due to the decreasing stability of the emerging monoanionic forms HP^- and their stabilization in solution due to solvation since the solvation conditions and the state of H^+_{solv}, as well as the structural changes in the solvent during ionization, are the same for all porphyrins examined.

Interestingly, as one goes from mesoporphyrin to protoporphyrin with asymmetric incorporation of two vinyl groups instead of ethyl ones, the first-stage ionization constant becomes twice as low, ΔS decreases by 14 e.u., and ΔH by 3.4 kcal mol^{-1}. There is every reason to believe that, just as in the case of chlorophyllic acids *a* and *b*, the introduction of vinyl groups brings about a +C effect with the corresponding transfer of π electrons into the macroring and with an increase in the π-electron density at the nitrogen atoms, hence a lower degree of protonation of the N—H bond of protoporphyrin.

The lowest values of K_{S_1} are observed in such symmetrical alkylporphyrins as etioporphyrin (**XXIX**), $\alpha,\beta,\gamma,\delta$-tetraphenylporphin (**XXX**), and $\alpha,\beta,\gamma,\delta$-tetramethoxyphenylporphin (**XXXI**).

Their acidic properties should preferably be compared with those of unsubstituted porphin in order to establish the factors affecting the ionization of alkyl substitution in the pyrrole and meso positions. However,

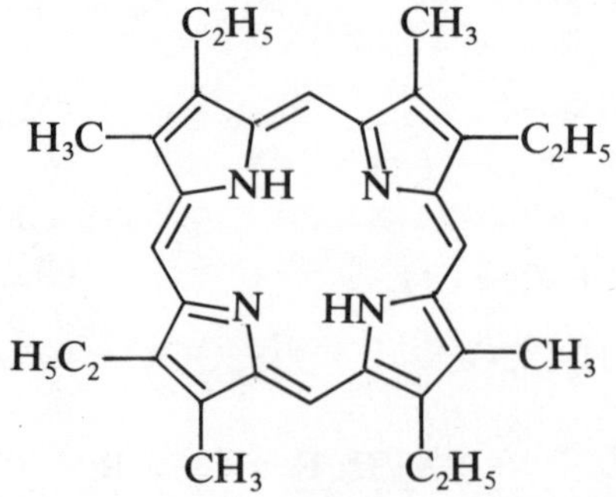

XXIX

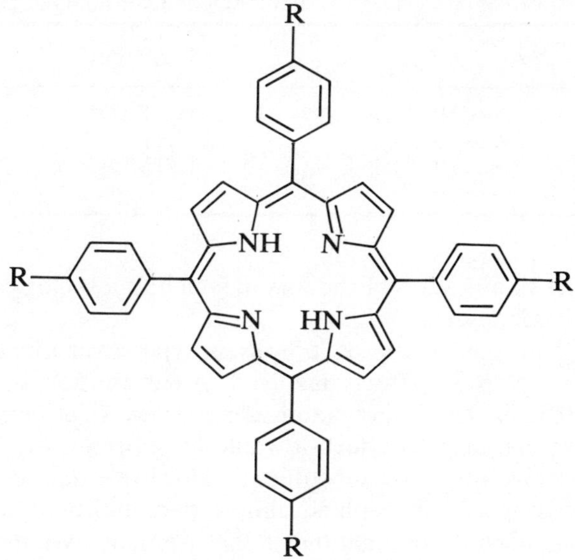

XXX: R = H

XXXI: R = OCH_3

the solubility method, even in its microversion, requires sufficiently large amounts of porphyrins, which is difficult to achieve in the case of porphin.

It is interesting to note that K_{S_1} of etioporphyrin is 4.5 times lower than in the case of mesoporphyrin and twice as low as in the case of protoporphyrin. The physical factors responsible for alkyl substituents influencing the ionization of N—H bonds are not yet known. It is important to remember that the value of $K_{S_1} \times K_{S_2}$ obtained for etioporphyrin by an entirely different method[211] is equal to 10^{-32} (at $K_{S_1} \approx K_{S_2}$). If so, the square of K_{S_1} for etioporphyrin, derived by the solubility method[242,243] gives $K_{S_{1,2}} \approx K_{S_1}^2 = 2.5 \times 10^{-32}$.

Such an agreement between the constants obtained by different methods gives conclusive evidence of the applicability of both methods to porphyrins. Besides, it suggests that in the case of all porphyrins the stage of detachment of the second proton structurally and energetically duplicates the first stage, that is $K_{S_1} \approx K_{S_2}$ for all porphyrins. For approximate values of the total acid ionization constants of porphyrins, obtained by the solubility method, see Table 4.11.

The weakest acid of all examined porphyrins was found to be tetraphenylporphin and its methoxy derivative. This fact is not surprising and has the following two possible explanations:

(1) The introduction of phenyl groups into the meso positions of porphin probably disturbs the planar structure of the macroring (which disappears in the case of covalent complex formation[219]). Any deviation of the aromatic system from coplanarity will counteract the N—H ionization as a result of

Table 4.11 Approximate values of total ionization constants of porphyrins

Porphyrin	$K_{S_{1,2}} = K_{S_1}^2$	Porphyrin	$K_{S_{1,2}} = K_{S_1}^2$
H_2Chl*a*	5×10^{-30}	H_2EP (I)	2.5×10^{-32}
H_2Chl*b*	1.6×10^{-30}	H_2TPP	2×10^{-33}
H_2MP DME	5×10^{-31}	H_2TMOPP	1.2×10^{-33}
H_2PP DME	9×10^{-32}		

weaker electronic stabilization of the anion, which follows logically from the data presented in Chapter 3.

(2) Probably, in the series of porphyrin-type aromatic systems, the direction of the electronic effects involved in substitution is substantially different from that in the aliphatic and arene series. If tetraphenylporphin had exhibited any conjugation effect, it would have been a +C effect, just as in the case of formyl and vinyl substituents. However, due to the fact that the *meso*-phenyl rings lie in a plane almost perpendicular to that of the macroring, no significant conjugation of the aromatic systems is possible. That leaves the induction effect, the direction of which is yet to be determined. Possibly, the σ interaction between the benzene ring and the macroring is compensated. Another possibility is that the phenyl group displays a weak +I effect with respect to the porphyrin system. This assumption requires experimental proof. At any rate the minimum values of acid ionization constants for H_2TPP, and particularly for H_2TMOPP, are consistent with the above assumption. The necessary proof will be provided in what follows. One thing that is beyond doubt at this juncture is the higher covalence of the N—H bond in H_2TPP as compared to other porphyrins (see Table 4.11). In this connection, the earlier conclusion[251] as to the strength of the N—H bond in H_2TPP seems to be premature.

Concluding this discussion of the quantitative data on acid ionization of porphyrins (see Table 4.10), it should be pointed out that the values of ΔH calculated from the temperature dependence of ionization constants are approximate in character. It would be more appropriate to obtain data directly from calorimetric measurements, which is what the present author is now contemplating doing in his laboratory. Nevertheless, the general picture presented by the approximate values of Table 4.10 is correct and correlates well with the structural and reactivity data. Note the extremely important role of solvation, that is participation of the solvent in all processes occurring in porphyrin solutions. Ionization is no exception. This process is characterized by a strictly solvation-thermodynamic compensation effect (Fig. 4.3).

The concept of solvation-thermodynamic compensation effect was introduced in earlier works[252–254] and implies, in the case of processes in a homogeneous system, compensation for the energy expended in breaking of chemical bonds or in disturbances of other interactions due to solvation of

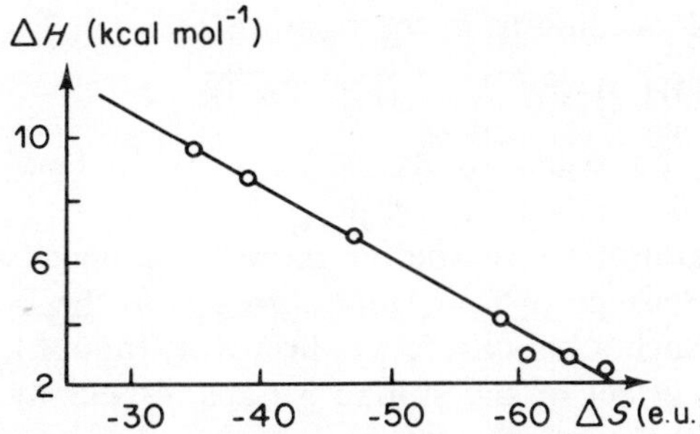

Figure 4.3 Solvation-thermodynamic compensation effect of porphyrin ionization

the forming particles. The occurrence of an overall solvation-thermodynamic compensation effect is mathematically confirmed by the linear relation

$$\Delta H = a + b\Delta S. \tag{4.13}$$

Such effects are typical of a system of similar chemical reactions taking place in one or several solvents, in which ionization is compensated at the porphyrin monoanions and associated with their stabilization.

Analysis of the above thermodynamic parameters of acid ionization of porphyrins clearly suggests that they are highly important in elucidating many aspects of porphyrin chemistry: structural, solvation, reactivity, and others. Therefore accumulation of data on the ionization constants of structurally different porphyrins at various temperatures and in various solvents is a task of paramount importance in the chemistry of porphyrins. Unfortunately no other literature concerning the quantitative aspects of acid ionization of porphyrins is available apart from the references cited above.

In one[256] of the few interesting works[140,255–257] on the properties of tetrasulphophthalocyanines the authors tried to determine the ionization constant of the N—H bonds of a tetrasulphophthalocyanine completely ionized via the sulpho groups ($H_2Pc(SO_3^-)_4$). It was found that in an alkaline aqueous solution of this compound ($pH \geqslant 10$) a significant change in its spectrum is observed, which is attributed to detachment of one of the protons:

$$H_2Pc(SO_3^-)_4 \rightleftarrows [HPc(SO_3^-)_4]^- + H^+.$$

Since $H_2Pc(SO_3^-)_4$ in an aqueous solution is associated and present in the form of stable dimers, the authors[256] determined, with the aid of the associate stability constant ($K_D^{20} = 9.3 \times 10^7$) found by them and the molar extinction coefficients ($\varepsilon = 2.4 \times 10^5$ for HPcTS and $\varepsilon = 9.6 \times 10^4$ for $(H_2PcTS)_2$ at a wavelength of 680 nm in the range of pH values at which the concentration of HPcTS is proportional to that of the OH^- ions, that is the second N—H proton is not detached), the acid ionization constant of tetrasulphophthalocyanine from the relation

$$K_{S_1} = (K_w/[OH^-])\sqrt{2K_D/(c - [HPcTS^-])}.$$

It was found to be equal to $K_{S_1} = 2.5 \times 10^{-10}$. This value exceeds the respective ionization constants of porphyrins by five to seven orders of magnitude. The authors[256] ascribe the more pronounced acidic properties of H_2PcTS to the presence of four sulpho groups in the benzene rings of the phthalocyanine. Such a big difference, however, cannot be explained merely by the induction effect of the sulpho groups, especially since the benzene rings of the phthalocyanine are in a very weak conjugation with the macroring. Of course, sulpho groups do make a certain contribution to ionization, but it is hoped that the ionization constant of unsubstituted phthalocyanine is not much higher than that of H_2PcTS.

The much more pronounced acidity of phthalocyanine as opposed to porphyrins can be easily understood by taking into account the specific features of electronic π systems of these molecules, described in Chapter 3. On the other hand, the wide differences in the values of K_S of these aromatic systems serve as a vivid proof of internal ionization of the imino-hydrogen atoms of H_2Pc and a strong conjugation of the n electrons of the nitrogen atoms with the inherent π electrons of the macroring. To the high values of K_S of phthalocyanine there must correspond the lowest values of electron density at the N atoms of phthalocyanine as compared to all other porphyrins. The experimental data on protonation of phthalocyanine and its complexes, to be presented in this chapter, will show that this is true and that phthalocyanines are extremely weak bases. Indirect data on the state of the N—H bonds in porphyrins and the effect of substituents on them can be obtained from infrared spectra.[258–262] The results presented in these references have been closely and critically examined.[8] However, these data have never been discussed in connection with acid ionization of N—H bonds. In this case, the data listed in Table 4.12 and taken from reference 8 are discussed precisely in this context.

Despite the presence of a nitrogen atom the pyrrole molecule **(XXXII)** does not exhibit the typical properties of a base, whereby it differs significantly not only from aliphatic secondary amines but also from pyridine. At the same time pyrrole is a very weak acid whose acid ionization constant in water is $K_S \approx 3 \times 10^{-16}$.[263] Analysis of the structure and properties of six-membered nitrous heterocycles (azines, diazines, and triazines) and five-membered heterocycles with O, N, and S atoms gives an insight into the distribution pattern of electrons in pyrrole, which is fully consistent with its physicochemical properties.

A nitrogen atom forms two sp^2-hybrid $\sigma_{C\text{–}N}$ bonds highly polarized toward the heteroatom. The decrease in the energy of the antibonding orbitals of the conjugated system, due to a σ shift, facilitates filling of one of them with n electrons of the nitrogen atom. As a result the n-electron pair loses its σ character completely, as well as the ability to be protonated with acids. This

Table 4.12 Frequencies of symmetrical valence vibrations of the N—H bond in pyrrole and its macrocyclic derivatives[8]

Compound	ν_{N-H} (cm^{-1})
Pyrrole	3495
Hexamethylpyrromethene	3455
Bilirubin	3410
Octamethylporphinogen	3445
Porphin	3305
Octaethylporphin	3310
Octamethylporphin	3311
Octamethyltetrazaporphin	3299
Synthetic chlorin	3348
Tetrahydroporphin	3355
Tetraphenylporphin	3315
Tetrabenzoporphin	3305
Tetrazaporphin	3300
Phthalocyanine	3290
Chlorophyllic acid *a*	3400
Etioporphyrin	3320
Protoporphyrin	3315
Mesoporphyrin	3315

yields a six-electron aromatic system obeying Hückel's rule, which loses its aromaticity and stability in highly acidic media because of the $\pi \rightleftarrows \sigma$ conversion of the electron pair of the nitrogen atom and its being bound by a proton.

In neutral and alkaline media the molecule is a stable aromatic system. The σ shift of the n-electron pair towards the ring is considerably greater than the σ shift due to the $N \leftarrow C$ bonds, therefore the pyrrole molecule features a dipole moment equal to 1.80 D with the positive pole at the N atom, which results in a marked protonation of the N—H bond.

The incorporation of pyrrole in a conjugated system, for example, with formation of pyrromethene or its alkyl derivatives (e.g. hexamethylpyrromethene (**XXXIII**)), stabilizes to a considerable extent the integral conjugated system by lowering the energy of the ψ and ψ^* molecular orbitals. As a consequence, the protonation of the N—H bond may become more pronounced, the ν_{N-H} vibrations may diminish, and K_S may increase if the ionization of pyrrole and pyrromethene is isoentropic. The frequency of ν_{N-H} vibrations of hexamethylpyrromethene (see Table 4.12) is in fact lower by 40 cm^{-1}. The more developed conjugated system of bilirubin accounts for a reduction by another 45 cm^{-1}. In the case of the conjugated system of porphin the difference becomes as great as 190 cm^{-1} with respect to pyrrole. The cycle closure is of course a sterically hindering factor[8] which, even in the case of the mobile cyclic system of *meso*-octamethylporphinogen (**XXXIV**), accounts for a reduction in frequency by 50 cm^{-1} (in all cases pyrrole is used as the reference point). In

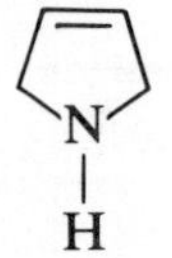

XXXII

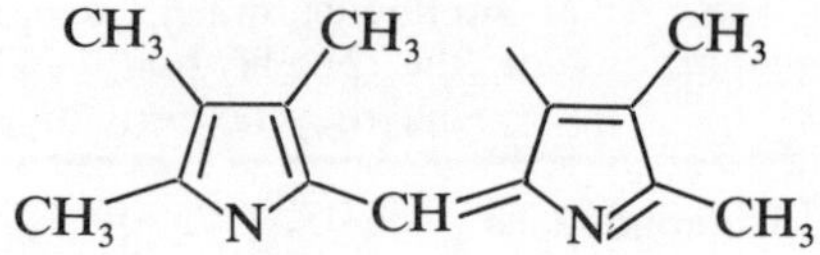

XXXIII

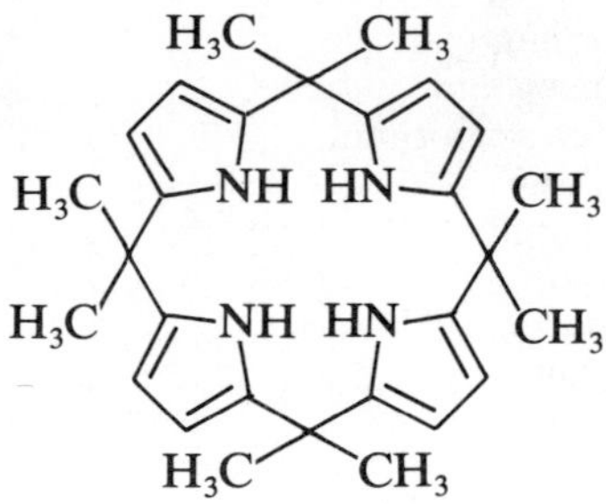

XXXIV

the case of porphinogen such a decrease is not quite clear. It is difficult to tell what are the steric hindrances associated with the vibration of N—H bonds but it is absolutely clear that the emergence of a unique conjugated aromatic molecule with a system of low-lying ψ^* orbitals must favour the transfer of the *n* electrons of nitrogen atoms into the macroring, protonation of the N—H bond, and a decrease in the $\nu_{N\text{-}H}$ frequency.

The data presented in Chapter 3, the numerical values of Table 4.12, and other physicochemical properties of porphyrins are in full agreement with this conclusion. Corresponding to the phthalocyanine molecule, which is unique among porphyrin compounds, is the maximum decrease in frequency by 205 cm^{-1}. It can be clearly seen that alkylation of porphin with a great number of alkyl groups in any positions enhances the electron density at the nitrogen atoms and perceptibly increases the value of $\nu_{N\text{-}H}$, that is leads to a greater covalence of the N—H bond. A similar effect is observed during meso substitution by phenyl groups. Hydrogenation of pyrrole rings from porphin to chlorin and even to bacteriochlorin increases the value of ν_{N-H} by 43–50 cm^{-1}.

If the process of acid ionization had been isoentropic one could have ascribed more acidic properties to porphin as compared to all of its alkyl and pseudo-alkyl derivatives and hydrogenated forms. Tetrabenzoporphin would have been of similar acidity, tetrazaporphin would have been more acidic, and phthalocyanine would have been the strongest acid. Unfortunately, the available values of ν_{N-H} for etio-, meso-, and protoporphyrins as well as tetraphenylporphins make it impossible as yet to distinguish between the states of their N—H bonds. This is probably due to the insufficiently accurate determination of some values of ν_{N-H} and to the different conditions under which they are determined.

It is difficult at present to see why the vibration frequencies of the N—H bonds in meso- and protoporphyrin are equal in value (3315 cm^{-1}). This may be due to the overwhelming steric effect of the CH_2—CH_2—$COOCH_3$ groups on these bonds. The very high frequency (3400 cm^{-1}) in the case of chlorophyllic acid *a* seems to be associated with the presence of a phytol residue rather than the weak protonation of the N—H bonds. We are not discussing here the possible complications of the states of the N—H bonds in different pyrrole molecules because of intramolecular hydrogen bonding for this problem has been exhaustively covered elsewhere,[8,179] In addition, the effect of such bonds on the vibration frequency of the N—H bonds must be practically the same not only in porphyrins but also in other N—H-containing molecules, including pyrrole itself, if we are speaking of infrared spectra in the condensed phase.

4.2 The State of Porphyrin and Phthalocyanine Molecules in Proton-donor Media

Neutral porphyrin molecules, that is unaltered porphyrin ligands, exist in the solutions of most organic solvents, including such proton-donor media as monocarboxylic acids and their mixtures with aprotic solvents.

Protonation with subsequent ionization yielding cationic forms in accordance with equations (4.3) and (4.4) usually takes place when strong mineral or organic acids $HClO_4$, H_2SO_4, HCl, H_3PO_4, CCl_3COOH, or RSO_3H in an organic solvent are added to the H_2P solution; this process is promoted by aqueous solutions, for example 90% and more diluted CH_3COOH or 95% HCOOH.[264] In cases where a porphyrin molecule contains carboxyl groups it is assumed[265] that protonation and salt formation take place in mixtures of organic solvents and acetic acid.

The depth of the protonation reaction is strongly dependent on the structure of the ligand molecule: porphyrins are protonated relatively more easily,[266–268] while phthalocyanines are protonated only in strong H_2SO_4 solutions. This difference in the basic properties of porphyrins and phthalocyanine stems from the dissimilar electronic structure of their molecules (see Chapter 3) and is now quite clearly understood from the viewpoint of the molecular orbital method.

The normally rigorous investigation of acid–base equilibria, whereby the equilibrium constants and other thermodynamic parameters are determined first, the various forms in which particles exist in solutions are established in this way, and, finally, their physical properties are studied, was abandoned in the case of porphyrins. Studies into the protonation of porphyrins and occurrence of their ionic forms have so far been conducted predominantly by spectroscopic methods: examination of absorption spectra in solutions of different degrees of acidity which in most cases was difficult to determine quantitatively. Direct thermodynamic measurements have been very few.

Up until the present many investigators have been and still are trying to guess, using absorption spectra, what is actually taking place in solutions. The serious difficulties encountered when a similar approach is used are shown by Gurinovich *et al.*[8] in their work, which gives comprehensive analysis of the spectroscopy of ionic forms of porphyrin in the context of their molecular structure.

Of greatest interest for present purposes are thermodynamic measurement results. The first work of this kind was accomplished more than 40 years ago.[269] It involved potentiometric titration of some porphyrins and chlorins in glacial acetic acid with perchloric acid and calculation of the ionization constants of mono- and diprotonated forms of porphyrins:

$$pK_{PH_3^+} = pH - \log(PH_2/PH_3^+), \tag{4.14}$$

$$pK_{PH_4^{2+}} = pH - \log(PH_3^+/PH_4^{2+}). \tag{4.15}$$

The derived constants are listed in Table 4.13, from which a number of interesting conclusions can be drawn. Methylphaeophorbide *b* differs from chlorophyllic acid *b* by the absence of the phytol residue (substituted by CH_3). It is a weaker base than methylphaeophorbide *a*. They are different in that a CHO group replaces CH_3 (**XXVIII**).

Rhodine g_7 (**XXXV**) is also a weaker base than chlorin e_6 (**XXXVII**). Their constants $K_{PH_4^{2+}}$ are different almost by a factor of two. Thus it can be seen that the substitution of a CHO group for CH_3 results in poorer basicity. This conclusion cannot be considered final since the values of $K_{PH_3^+}$ are not known for these molecules, although in another work[268] it is stated categorically.

It should be noted that a similar conclusion is at variance with the data on acid ionization of chlorin and rhodine systems and their coordination properties. It is difficult to imagine that in certain porphyrin molecules the N—H groups exhibit more acidic properties while at the same time the nitrogen atoms become more basic. On the contrary, the integrity of the entire π- and σ-electron system of the macroring is maintained if the

Table 4.13 Ionization constants of acidic forms of porphyrins

Porphyrin	Acetic acid[269]		Water[270]	
	$pK_{PH_3^+}$	$pK_{PH_4^{2+}}$	$pK_{PH_3^+}$	$pK_{PH_4^{2+}}$
Methylphaeophorbide *a*	—	−1.4	—	+0.2
Methylphaeophorbide *b*	+0.3	−1.7	+1.9	−0.1
Rhodine g_7	—	0.0	—	+1.6
Rhodochlorin	—	−0.7	—	+0.9
Chlorin e_6	—	+0.3	—	+1.9
Chlorin p_6 (**XXXIX**)	—	−0.2	—	+1.4
Phyllochlorin	—	−0.1	—	+1.5
Phaeopurpurine	+0.5	−1.8	+2.1	−0.2

following regularity is observed: the more pronounced the acidic properties, the lower the basicity of the molecule, ion, or radical.

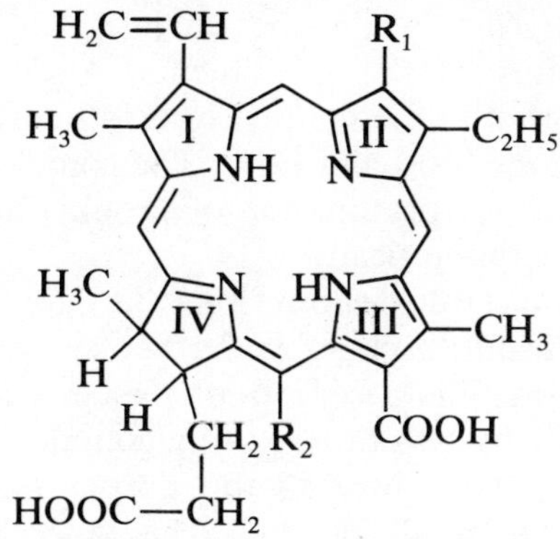

XXXV: R_1 = CHO; $R_2 = CH_2COOH$
XXXVI: R_1 = CHO; $R_2 = CH_3$
XXXVII: $R_1 = CH_3$; $R_2 = CH_2COOH$
XXXVIII: $R_1 = CH_3$; $R_2 = CH_3$
(chlorin e_4)
XXXIX: $R_1 = CH_3$; R_2 = COOH
(chlorin p_6)
XL: $R_1 = CH_3$; R_2 = H
XLI: $R_1 = CH_3$; $R_2 = CH_3$; COOH (III) = H

As can be inferred from the above, one of the primary tasks in porphyrin chemistry is systematic accumulation of thermodynamic data, including the first- and second-stage protonation constants of porphyrins, obtained in a broad temperature range under strictly constant experimental conditions. The data available at present are far from being accurate. Table 4.13 suggests that a stabilization of PH_3^+ does not necessarily entail a stabilization of PH_4^{2+}. Transition from methylphaeophorbide *b* to phaeopurpurine methyl ether, accompanied by incorporation of an additional carboxyl group in the first pyrrole nucleus instead of a vinyl group,[61–62] results in $K_{PH_3^+}$ becoming almost twice as high whereas $K_{PH_4^{2+}}$ perceptibly decreases. Therefore, judging from the first stage of protonation, phaeopurpurine is a stronger base than methylphaeophorbide *b*. When the values of $K_{PH_4^{2+}}$ are compared, the reverse is true.

Thus no conclusions can be drawn from the data of Table 4.13 as to the changes in the basic properties of a porphyrin molecule as a whole, caused by changes in its structure. What we can speak of is the first and second protonation stages separately.

Transition from chlorin e_6 and rhodine g_5 (**XXXVI**) to rhodochlorin (**XL**) involves replacement of the CH_3 and CH_2COOH groups by an H atom in the γ position. In this case the value of constant $K_{PH_4^{2+}}$ increases 10-fold that is the protonating capacity decreases when the pseudo-alkyl group in the meso position is removed. This phenomenon finds no explanation in terms of the electron theory. Solvation probably plays a major role in the protonation of porphyrins, while the changes in entropy are strongly dependent on the nature of the substituents. A similar phenomenon has already been encountered in the context of acid ionization of porphyrins in accordance with reactions (4.3) and (4.4). The complete lack of data on the enthalpies and entropies of protonation of prophyrins makes it difficult to analyse these processes theoretically.

A few indisputable facts do follow from the data of Table 4.13 though, which should be taken into account in further experimental studies of the ionization of porphyrins. First, ionization constants $K_{PH_3^+}$ and $K_{PH_4^{2+}}$ differ but insignificantly, only by two orders of magnitude, that is the attachment of the first proton to the centre of the porphyrin molecule reduces its basicity to a relatively small degree. This may be attributed to the insufficient stability of the monoprotonated particle. Second, the dependence of ionization constant $K_{PH_4^{2+}}$ on the molecular structure is sufficiently great, that is almost as great as in the case of acid ionization of PH_2 and PH^-. Third, ionization constants $K_{PH_3^+}$ and $K_{PH_4^{2+}}$ in glacial acetic acid and in water differ only by a factor of 40, that is the ability of the medium in which protonation of H_2P and ionization of the protonated particles occur to accept donors exerts a minor influence.

Ionization constants $K_{PH_3^+}$ and $K_{PH_4^{2+}}$ for some porphyrins were found by way of spectrophotometric titration in nitrobenzene[271] as well as in aqueous solutions of a detergent, namely 2.5% sodium dodecylsulphate.[272,273] Some of the found values are presented in Table 4.14.

Comparison of Tables 4.14 and 4.13 indicates that in aqueous solutions of detergents ionization constants $K_{PH_3^+}$ of porphyrins have a much lower value

Table 4.14 Ionization constants of acidic forms of porphyrins in sodium dodecylsulphate at 20 °C

Porphyrin	Substituents in position C2	C4	$pK_{PH_3^+}$ (± 0.05)
Mesoporphyrin	CH_2CH_3	CH_2CH_3	5.94
Protoporphyrin	$CH{=}CH_2$	$CH{=}CH_2$	4.89
Coproporphyrin[a]	$(CH_2)_2COOCH_3$	$(CH_2)_2COOCH_3$	5.58
Deuteroporphyrin	H	H	5.63
4-Formyldeutero-porphyrin	H	CHO	3.90
2,4-Diformyldeutero-porphyrin	CHO	CHO	2.90
2,4-Diacetyldeutero-porphyrin	$COCH_3$	$COCH_3$	3.50

[a] 1,3,5,7-Tetramethylporphin-2,4,6,8-propionic acid.

(by one or two orders of magnitude), which is indicative of their high stability due, probably, to the higher degree of solvation of PH_3^+ ions in the detergent solutions. It is well known that most porphyrins are practically insoluble in water if they do not contain a large number of non-etherified COOH and SO_3H groups, owing to the weak solvation of the hydrophobic part of the molecule.

The regularities determining the strength of acids and bases in various solvents are well known[274,275] if we are dealing with acids of an inorganic nature or organic acids with a simple hydrocarbon residue. These regularities cannot be applied to porphyrins because of the complexity of the electronic and solvation phenomena typical of them.

Analysis of Table 4.14 suggests[23,268] that the basicity of porphyrins is reduced when alkyl groups are substituted by hydrogen atoms, vinyl groups, and, to a greater extent, acetyl and formyl groups. Pseudo-alkyl groups CH_2—CH_2COOH are equivalent to hydrogen atoms.

Of particular interest is comparison of these results with the data of Table 4.10. The effect of the substituents on the acid ionization of H_2P and HP^- is opposite to their effect on the ionization of PH_3^+. Inclusion of the —CHO and —CH=CH_2 groups hinders the ionization of H_2P and favours that of PH_3^+. In other words, substituents reducing the acidity of a porphyrin molecule also reduce its basicity. This unusual phenomenon can be explained by the opposite effect of the substituents on the electronic stabilization of PH^- and PH_3^+ or on the solvation of these particles.

The lack of data on the enthalpic and entropic changes in the process of protonation of porphyrins does not permit the problem to be resolved unambiguously. The results given in Table 4.10 suggest that the effect of substituents on the protonation of porphyrins (see Table 4.14) is purely entropic rather than energetic. The protonation is probably accompanied by changes in entropy ($+\Delta S$) which become more positive as the polarizing action of the substituent on the porphyrin molecule or, to be more precise, on its π-electron cloud increases in intensity (CHO > $COCH_3$ > CH=CH_2 > CH_2CH_3).

In any event the need to conduct fundamental systematic studies into the equilibrium in porphyrin–acid systems at various temperatures is emphasized once more. One fact is beyond any doubt—all porphyrins containing a macroring are protonated in the centre of the molecule, attaching one and then a second proton:

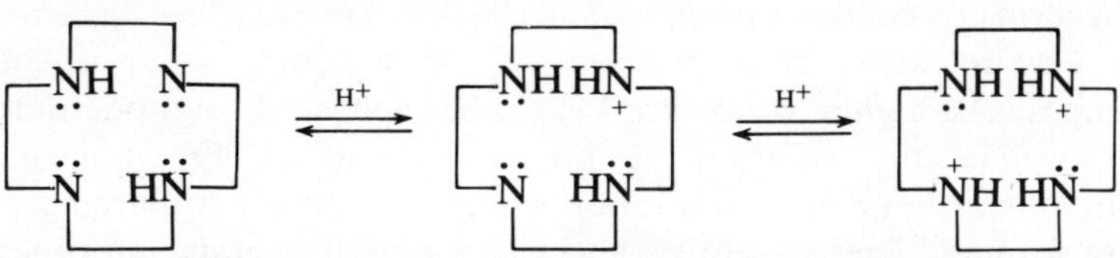

(4.16)

The positive charge of the dication in symmetrically substituted porphyrins is evenly distributed among all nitrogen atoms with the result that the π-electron cloud becomes symmetrical, which manifests itself in regular changes in the electronic absorption spectra.[8] The attachment of a third and a fourth proton with formation of PH_5^{3+} and PH_6^{4+} cations does not occur. No more protonation takes place despite the presence of two more unbound electron pairs at nitrogen atoms and the relatively weak acidity of PH_4^{2+} is not surprising. Both *n*-electron pairs of the PH_4^{2+} ion enter into a strong conjugation with the π electrons of the macroring, which is promoted by the accumulation of a positive potential at the macroring skeleton atoms. The conjugated system of PH_4^{2+} is as unique and stable, owing to the high degree of symmetry, as in covalent porphyrin complexes. This is indicated by the visible four-band spectrum of H_2P becoming a two-band spectrum of the PH_4^{2+} ion and MP, which is closer to the spectra of phthalocyanines featuring the most symmetrical and stable conjugated system.[3]

Before proceeding to the ionic forms of more complex molecules, such as tetrabenzoporphin, tetrazaporphin, and its derivatives, let us dwell on the protonation of phthalocyanine and its complexes, for which a wealth of thermodynamic data have been obtained.[37,192,193,239] These data combined with spectroscopy and knowledge of the electronic structure of molecules provide a clear picture of the process.

Phthalocyanine (**V**) and its numerous stable complexes are practically insoluble in water and organic solvents because they are more hydrophobic and feature a higher molecular lattice energy as compared to porphyrins. Even in quinoline and α-chloronaphthalene their solubility seldom exceeds 10^{-5} M and in supersolvents such as sulpholane, DMSO, and DMFA they are not soluble. Phthalocyanines do not react with aqueous solutions of most strong mineral acids and do not pass into solution.

The most effective and practically the only solvent for them is strong H_2SO_4. The solubility of all stable phthalocyanines is high in it and may vary widely with its concentration. Being a highly protogenic substance H_2SO_4 also serves as a solvent and a reagent, which permits study of the acid–base interaction, dissociation kinetics, and electronic absorption spectra of phthalocyanine under rigorously identical conditions. Significantly, aqueous sulphuric acid exhibits, in the range of concentrations from 15 to 18.8 M (the so-called Brand region), the properties of an 'ideal' solvent, that is the ability to maintain the constancy of the activity coefficients of ions and polar molecules irrespective of the ionic strength of the solution.[276] This property permits reliable use of the equilibrium and rate constants measured in it and expressed in terms of concentration.

Sulphuric acid with concentration $c < 8$ M does not dissolve phthalocyanines. At higher concentrations, when physical methods detect the appearance in the solution of non-dissociated H_2SO_4 molecules, phthalocyanines start passing into solution whereby intensely coloured green solutions are formed. The absorption spectra of phthalocyanines in organic solvents and in H_2SO_4 differ to such an extent that one can speak of a strong chemical interaction.[189] The spectra of stable phthalocyanines in H_2SO_4

remaining unchanged in time suggests that no slow reactions like sulphonation or oxidation take place in the solution. The different spectra in organic solvents and in H_2SO_4 and the dissolution of phthalocyanines result from an acid–base interaction that can be expressed as follows for phthalocyanine complexes and the ligand itself:

$$MPc_{(s)} + H_2SO_4 \rightleftharpoons MPcH^+ \cdot HSO_4^-, \quad (4.17)$$

$$MPc_{(s)} + H_2SO_4 \rightleftharpoons MPcH^+ + HSO_4^-, \quad (4.18)$$

$$MPc_{(s)} + 2H_2SO_4 \rightleftharpoons MPcH_2^{2+} + 2HSO_4^-, \quad (4.19)$$

$$H_2Pc_{(s)} + H_2SO_4 \rightleftharpoons H_2PcH^+ + HSO_4^-, \quad (4.20)$$

$$H_2Pc_{(s)} + 2H_2SO_4 \rightleftharpoons H_2PcH_2^{2+} + 2HSO_4^-. \quad (4.21)$$

The absorption spectra of H_2Pc and MPc remain the same as one goes from α-chloronaphthalene to H_2SO_4 (Fig. 4.4). A strong bathochromic shift (by 80–120 nm) of all bands occurs, except for one band at a wavelength of 225–230 nm.[190,215] The spectral changes during protonation are absolutely the same for all phthalocyanine complexes as well as the H_2Pc ligand. This is indicative of the same type of proton attachment and the same sites of its localization. The absence of electron-donor centres in the molecules of CuPc and other stable MPcs at the intracyclic nitrogen atoms (because of the covalent binding of their electron pairs by the metal ion) and the strong bathochromic spectral shift in H_2SO_4 leave no doubt that H^+ is attached to the external, so-called extracyclic nitrogen atoms of MPc.

Thus the same manner in which spectral changes occur in H_2Pc and MPc is sufficient proof that the H_2Pc ligand attaches protons not to the centre of the molecule (otherwise a hypsochromic shift would have been observed just as in the case of porphyrins) but to its periphery. The attachment of H^+ to the periphery of the macroring **XLII** strongly polarizes the phthalocyanine

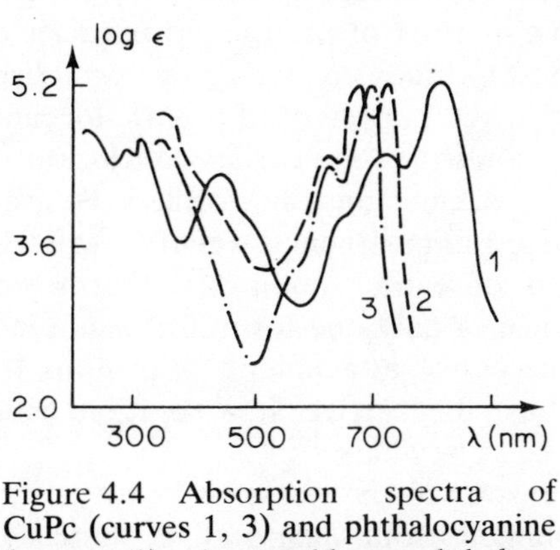

Figure 4.4 Absorption spectra of CuPc (curves 1, 3) and phthalocyanine (curve 2) in α-chloronaphthalene (curves 2, 3) and 18 M H_2SO_4 (curve 1)

molecule. The resulting strong dipole decreases the energy of all electronic transitions associated with the macroring.

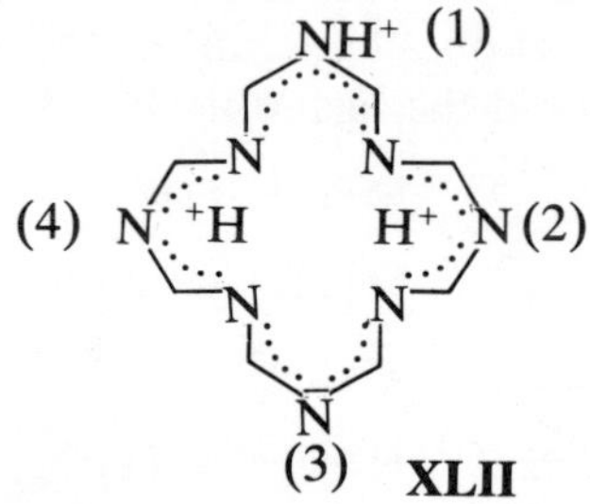

XLII

As was mentioned in Chapter 3, the protonation of H_2Pc at the periphery rather than in the centre of the molecule is one of the many arguments in favour of the internal ionization of the two imino-hydrogen atoms of H_2Pc. The position of bands in the electronic absorption spectra of all stable phthalocyanine complexes with doubly charged metals does not depend on the H_2SO_4 concentration (within the range 8–18.8 M) and remains practically the same even in oleum.[193] This means that phthalocyanines yield only a single protonated form in aqueous H_2SO_4 and in oleum.

4.3 Thermodynamics of Acid–Base Interaction of Phthalocyanine and Porphyrin Complexes

As has already been pointed out, one cannot unambiguously determine the degree of protonation of porphyrins and phthalocyanines from spectroscopic studies alone.[287] Gurinovich *et al.*[8] give a detailed analysis of all contemporary spectroscopic approaches to solving the problem of the type of protonated phthalocyanines. The most fruitful approach involves polarized luminescence. However, even this method fails to provide a single solution if the possible number of protonated particles exceeds two.

By using this method Gurinovich *et al.*[8] have established that protonation in H_2SO_4 takes place at the periphery of a phthalocyanine molecule, that is they have definitively confirmed the earlier conclusions.[189,192] They also contend that the $MPcH_4^{4+}$ cation cannot be formed. Hence, as can be assumed on the basis of polarized luminescence, $MPcH^+$, $MPcH_2^{2+}$, and $MPcH_3^{3+}$ are formed. On the basis of earlier findings[277] Gurinovich *et al.*[8] reject the possibility of occurrence of only one form, for example $MPcH^+$ or $MPcH_2^{2+}$, and maintain that successive attachment of protons to phthalocyanines is what takes place. Thus, they arrive at a solution which is a compromise between those proposed elsewhere.[192,277]

An unambiguous answer to the question as to the nature of the forming particles can be obtained from quantitative studies of the equilibria of solubility of phthalocyanine in protonating media, which have been done by many investigators.[37,192,193,239] The solubilities of H_2Pc and about 20 of its stable complexes have been examined in the Brand region in the temperature range 20–70 °C. Phthalocyanine is, as was shown in Chapter 3, an extremely weak base in comparison with porphyrins. Neither phthalocyanine nor its complexes pass into solutions of strong acids containing H_3O^+ ions and having no stronger protogenic particles. This gives every reason not to take into account the following reaction in considering equilibria (4.17)–(4.20):

$$MPc_{(s)} + H_3O^+ \rightleftarrows MPcH^+ + H_2O. \tag{4.22}$$

The lack of a rectilinear relationship between the solubility (S, M) and the concentration of the molecular form of H_2SO_4 (Fig. 4.5) permits reaction (4.17) to be excluded. In solutions of concentrated aqueous H_2SO_4 the salt-like compounds are completely ionized,[276] which is due to the extremely high dielectric constants of these solutions. Thus the possibility of formation of $MPcH^+$, $MPcH_2^{2+}$, $MPcH_3^{3+}$, and similar protonation products of H_2Pc in sulphuric acid solutions has been studied.

The calculation equations were derived as follows. Application of the law of mass action to equilibrium (4.18) gives

$$K'_{MPc} = ([MPcH^+]f_3[HSO_4^-]f_2)/([H_2SO_4]f_1). \tag{4.23}$$

The values of the activity coefficients f_1, f_2, and f_3 remaining constant in the Brand region permits expression (4.23) to be written as

$$K_{MPc} = ([MPcH^+][HSO_4^-])/[H_2SO_4]. \tag{4.24}$$

Substituting the ratio of $[HSO_4^-]$ to $[H_2SO_4]$, taken from Brand's equation,[277]

$$H_0 = pK_{H_2SO_4} + \log([HSO_4^-]/[H_2SO_4]), \tag{4.25}$$

into (4.24) and performing the necessary transformations, we derive an

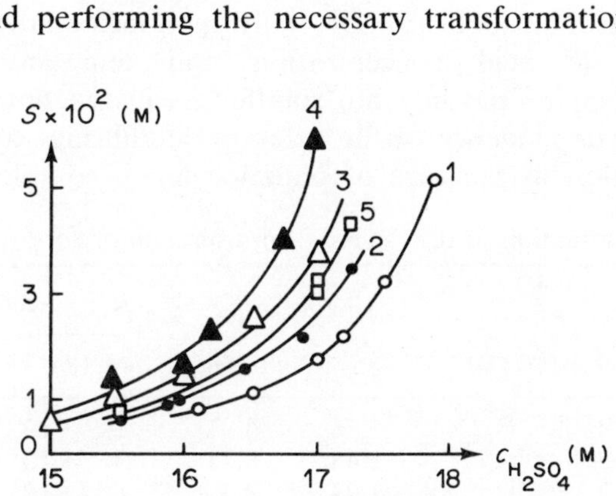

Figure 4.5 Solubility of CuPc in sulphuric acid at various temperatures. Temperature: curve 1, 20 °C; curve 2, 30 °C; curve 3, 40 °C; curve 4, 50 °C; curve 5, 70 °C

equation suitable for calculating the solution equilibrium constant (pK_{MPc}), interrelating the solubility exponent (pS) with the acidity constant of sulphuric acid ($pK_{H_2SO_4}$) and Hammett's acidity function (H_0):

$$pK_{MPc} = pS + pK_{H_2SO_4} - H_0. \quad (4.26\text{–}27)$$

The values of H_0 were taken from reference 278 and those of $pK_{H_2SO_4}$ from reference 279. These values necessary for calculations were generalized in reference 194.

For an n-stage protonation reaction

$$MPc_{(s)} + nH_2SO_4 \rightleftarrows MPcH_n^{n+} + nHSO_4^- \quad (4.28)$$

one can derive the equation[240]

$$pK_{MPc} = pS + npK_{H_2SO_4} - H_0 \quad (4.29)$$

suitable for calculating the solution equilibrium constants of phthalocyanines with due account for formation of any one of the $MPcH_n^{n+}$ particles. The constant value of K_{MPc} in the acidity range can be obtained only if the value of the number of attached protons is true. Table 4.15 lists solution equilibrium constants of copper phthalocyanine, calculated from equation (4.29) at $n = 1$ and 2.

As can be inferred from the tabulated data, only at $n = 1$ does the equilibrium constant remain strictly invariable, that is reaction (4.18) obeys the mass action law. At $n = 2$ the constants are not invariable, and reaction (4.19) does not obey the mass action law.

The situation is further aggravated when phthalocyanine dissolution processes are calculated at $n = 3$ and 4, as can be seen in the case of H_2Pc and its 14 complexes examined in a broad temperature range. The available data[108–110,113,183,192,280–284] clearly indicate that in aqueous H_2SO_4, beginning with an 8 M concentration, only monoprotonation is observed irrespective of the acid concentration and temperature with the phthalocyanine complex passing into solution as in reaction (4.18) (Table 4.16). The strong dependence of the solution equilibrium constants on the H_2SO_4 concentration in the case of phthalocyanine complexes with Sn^{4+},

Table 4.15 Determination of the degree of protonation of copper phthalocyanine

				K_{CuPc} according to reaction	
$c^0_{H_2SO_4}$ (M)	$S_{CuPc} \times 10^5$ (M)	pS	$-H_0$	(4.18)	(4.19)
17.88	5500	1.26	8.91	$2.29 \cdot 10^{-2}$	$9.55 \cdot 10^{-3}$
17.44	3300	1.48	8.71	$2.19 \cdot 10^{-2}$	$1.45 \cdot 10^{-2}$
16.98	1950	1.71	8.49	$2.14 \cdot 10^{-2}$	$2.35 \cdot 10^{-2}$
16.50	1140	1.94	8.30	$1.95 \cdot 10^{-2}$	$3.30 \cdot 10^{-2}$
16.09	977	2.01	8.12	$2.51 \cdot 10^{-2}$	$6.45 \cdot 10^{-2}$
15.50	552	2.26	7.88	$2.45 \cdot 10^{-2}$	$1.10 \cdot 10^{-1}$
15.00	245	2.61	7.62	$2.00 \cdot 10^{-2}$	$1.62 \cdot 10^{-1}$

Table 4.16 Solution equilibrium constants (pK_{MPc}) of phthalocyanine and its complexes at 25 °C in accordance with reaction (4.18)

$c^0_{H_2SO_4}$ (M)	$S \times 10^5$ (M)	pK_{MPc}
Copper phthalocyanine, CuPc		
17.88	5500	1.64
17.44	3300	1.66
16.98	1950	1.67
16.50	1140	1.71
16.09	977	1.70
15.00	245	1.70
15.50	552	1.61
11.18	3.50	—
9.05	0.17	
Zinc phthalocyanine, ZnPc		
17.25	560	2.34
16.60	302	2.29
16.09	191	2.31
14.52	32.5	2.32
13.17	5.50	2.28
13.10	2.90	—
12.33	1.70	—
Nickel phthalocyanine, NiPc		
14.52	252	1.43
13.17	39.4	1.42
13.17	38.9	1.43
12.33	11.8	1.50
12.33	11.9	1.49
Copper pentachlorophthalocyanine, $CuPcCl_{15}$		
17.88	350	2.84
17.44	214	2.85
16.98	124	2.87
16.09	51.1	2.88
16.00	45.5	2.87
15.27	26.4	2.82
15.23	30.1	2.74
16.50	85.2	2.84
Platinum phthalocyanine, PtPc		
16.00	932	1.53
15.00	537	1.59
15.50	567	1.57
14.50	202	1.49
Phthalocyanine, H_2Pc		
17.91	5520	1.65
17.92	6000	1.61
17.65	4470	1.62
17.65	4070	1.66
17.18	2400	1.67
16.00	55.8	—
Iridium phthalocyanine, $(HSO_4)IrPc$		
17.00	4120	1.38
16.50	2890	1.31
16.00	1490	1.35
15.50	1040	1.33
15.00	1060	1.07
14.50	640	1.01
Uranyl phthalocyanine, UO_2Pc		
17.60	340	2.66
17.00	141	2.73
16.00	83.2	2.63
Rhodium phthalocyanine, $(HSO_4)RhPc$		
16.50	5480	1.03
16.00	3150	1.03
16.00	3980	0.93
15.50	1915	1.06
15.00	1310	0.97
14.50	670	0.99
14.00	370	0.96
15.50	2390	0.97
Tin phthalocyanine, $(HSO_4)_2SnPc$		
17.65	605	2.49
17.43	454	2.50
17.00	331	2.43
16.90	208	2.57
16.75	224	2.49
16.50	63.1	2.97
16.30	26.0	3.25
16.25	20.6	3.33
16.10	11.5	3.52
16.00	8.10	3.59

Table 4.16 *continued*

$c^0_{H_2SO_4}$ (M)	$S \times 10^5$ (M)	pK_{MPc}
Nickel chlorophthalocyanine, NiPcCl		
17.88	10000	1.38
17.44	5750	1.42
16.93	2950	1.49
16.09	1320	1.47
15.75	850	1.51
15.00	467	1.42
Palladium phthalocyanine, PdPc		
17.70	5300	1.58
16.50	1660	1.52
16.00	780	1.61
15.50	719	1.46
15.50	680	1.49
14.50	161	1.58
14.00	90.1	—
13.50	57.5	—
Gallium phthalocyanine, $(HSO_4)GaPc$		
18.00	5450	1.81
17.50	2455	1.86
17.00	1600	1.79
16.50	870	1.83
15.50	340	1.81
15.00	191	1.81
14.50	161	—
14.00	123	—
Ruthenium phthalocyanine, $(HSO_4)RuPc$		
17.50	17400	1.01
17.00	11100	0.94
17.00	11300	0.94
16.50	7800	0.88
15.50	2620	0.93
15.00	1410	0.94
Cobalt phthalocyanine, CoPc		
17.88	1910	2.10
14.44	1290	2.07
16.98	721	2.10
16.09	322	2.08
15.75	251	2.04
15.00	120	2.01
Dichlorosilicon phthalocyanine, Cl_2SiPc		
17.75	542	2.63
17.50	266	2.79
17.25	224	2.75
16.06	67.5	2.75
Aluminium phthalocyanine, $(HSO_4)AlPc$		
17.88	2240	2.03
17.44	1060	2.13
16.98	1060	1.92
16.50	745	2.39
16.09	85.5	2.60
15.27	35.8	2.66
15.23	28.0	2.74
14.80	17.9	2.71
14.50	14.8	2.62
14.03	7.5	2.65
13.98	7.1	2.65
10.70	1.6	—
Osmium phthalocyanine, $(HSO_4)_2OsPc$		
17.00	5020	1.29
16.50	6270	0.97
16.00	6450	0.72
16.00	6530	0.78
15.50	5200	0.58
15.00	2470	0.65
14.50	5070	0.07
14.00	2790	0.04
Vanadium phthalocyanine, $(HSO_4)_2VPc$		
17.00	1050	1.97
16.50	665	1.92
16.45	648	1.91
16.09	478	1.89
16.00	447	1.85
15.50	254	1.92
15.45	198	2.09
15.00	142	1.91
14.50	80.9	1.88
14.00	38.4	1.92
12.00	2.88	1.94
10.00	1.00	—

Al^{3+}, and Os^{4+} is due to changes in the extra coordination at the central atom.

The thermodynamic data of Table 4.16 and the independence of the absorption spectra of phthalocyanines of the H_2SO_4 concentration suggest that phthalocyanine and all of its complexes are extremely weak monoacid bases in an aqueous solution of H_2SO_4 of any concentration. The very low values of pK_{MPc} indicate that phthalocyanines are bases as weak as, for example, nitrobenzene. The attachment of the first proton to them, yielding monoprotonated phthalocyanines H_2PcH^+ or $MPcH^+$ (**XLII**) which are already extremely strong acids—much stronger than H_2SO_4, due to delocalization of the positive charge over the macroring and polarization of its electron cloud toward the attached proton—rules out the possibility of a secondary protonation yielding $H_2PcH_2^{2+}$ and $MPcH_2^{2+}$.

The attachment of a second proton, if it ever takes place, is most probable at the N3 atom but never at N2 or N4 (see **XLII**). Otherwise counterpolarization of the phthalocyanine chromophore would take place, which, in accordance with the theory of chromaticity of organic molecules,[22] would have resulted in an extremely strong hypsochromic shift of the first long-wave band with respect to $MPcH^+$. In experiments an insignificant bathochromic shift of this band is observed as the H_2SO_4 concentration increases from 14 to 18 M and higher.[193]

In the case of MPc complexes, where M is a doubly charged metal ion, this shift is as small as 2–3 nm. As was corroborated earlier,[239] this bathochromic shift is due to a greater depth of protonation of the molecule, that is a greater depth of penetration of the proton into the electronic shell of the base (in this case an N atom), as a result of its diminishing contact with the environment. Indeed, as the H_2SO_4 concentration grows, the solvent gradually loses its solvating capacity, hence its association with the attached proton. Quantum-chemical calculations[285] show that in the case of protonation of heterocycles such as pyridine, quinoline, and acridine the association of the proton with the solvent materially affects the electronic spectra and depends on the nature of the heterocycle involved. The protons of pyridinium and quinolinium remain in a strong interaction with the anion of the solvent, while in the acridinium ion the proton is practically out of contact with the anion because of the great protonation depth. The same calculations[286] indicate that the bond between the proton and nitrogen in aromatic heterocyclic compounds is predominantly ionic.

Hence the proton in the $MPcH^+$ complex will respond to changes in the solvating capacity of the medium. As one passes from sulphuric acid to oleum (18 M $H_2SO_4 \rightarrow$ 2% oleum), there occurs a small hypsochromic shift of the first absorption band of ligand H_2PcH^+ and its complex $MPcH^+$,[193] equal to 4–5 nm. This shift may be interpreted as a result of hydrogen bonding of acid $H_2S_2O_7$ with atoms N2, N3, and N4 of complex $MPcH^+$ (**XLII**).

The controversy concerning the degree of protonation of phthalocyanines[8] stem, as was shown earlier,[239] from different interpretations of the term 'protonation' by different investigators, from the fact that this term lacks a clear definition when used in many works dealing with porphyrins and phthalocyanines, and from its application to widely dissimilar conditions of acid–base interaction in aqueous sulphuric acid solutions, in organic solvents having a low dielectric constant,[287] and in a gaseous phase.[8,288] Ignoring the medium, the state of the particles, and the solvation when acid–base interaction is compared in different, often incompatible conditions is precisely what leads to misinterpretation of spectroscopic data.[287]

In view of the above it is appropriate to introduce the concepts of complete and partial protonation. By analogy with oxidation–reduction processes it has been proposed[239] to use the term 'complete protonation' to mean a process of practically complete transfer of a proton from an acid to a base

$$B + HA \rightleftarrows BH^+ + A^-. \tag{4.30}$$

Complete protonation must be accompanied by ionization of the salt. If interaction (4.30) is restricted to the stage of formation of molecular-ionic associates of the $B\cdots H{-}A$, $B{-}H^+\cdots A^-$ type, it is called partial protonation. The formation of a hydrogen bond marks the beginning of protonation, that is the process of proton transfer.[274,275] It is clear that, depending on the medium in which protonation occurs, the reacting system may be in any phase of proton transfer from ions BH^+ and A^- to neutral particles B and HA. In other words, the depth of protonation of base B will vary from 1 to 0.

The direction of the process, that is the protonation depth, is determined by the difference in the proton affinity of particles B and A, which can be written[24] as

$$\Delta K = K_B - K_{A^-} = (\Delta H^\circ_{HA} - \Delta H_{HB^+}) + (\Delta H^\circ_B - \Delta H^\circ_{A^-}). \tag{4.31}$$

Here ΔH° is the enthalpy of formation of protonated and non-protonated particles in a gaseous state.

Depending on their protonating capacity in the gaseous phase, acids form the following series: $HClO_4 > H_2SO_4 > HI > HBr > HNO_3 > HCl > CH_3COOH > HF$. In solution this series may undergo considerable changes because of the participation of the solvent in the acid–base interaction between acid HA and protonated base BH^+ as well as in solvation.

By the above-mentioned 'protonation depth' is here meant the depth of penetration of the proton into the electronic shell of the atom exhibiting basic properties, measured experimentally by the degree of equilibrium (4.30) shifting to the right. Therefore one can determine the depth of protonation of a base, for example phthalocyanine, primarily from the equilibrium constants of reactions (4.30). As will be shown later, one can distinguish between particles according to their protonation depth, that is distinguish between B···H—A and $BH^+ \cdots A^-$ according to their reactivity in coordination to metal salts. The particles behave differently not only in chemical reactions but also in physical processes, particularly in an interaction with electromagnetic radiation, that is they have different spectra.

It is now appropriate to analyse in greater detail the spectral changes of water-soluble copper tetrasulphophthalocyanine occurring in a medium with increasing acidity. These data are presented in a major and interesting work by Ahrens and Kuhn.[287] In an aqueous solution of H_2SO_4 perceptible protonation of water-soluble phthalocyanines starts only at 7 M H_2SO_4 when non-dissociated molecules of sulphuric acid appear. Hence CuPcTS molecules are protonated but partially and poorly in acids containing only H_3O^+ ions and weaker acid particles because of the extremely weak hydrogen bonds between all four peripheral nitrogen atoms N1, N2, N3, N4 (**XLII**) and H_3O^+ ions. Although protonation does take place, it is spectrally indistinguishable. Sulphuric acid (7 M H_2SO_4) molecules protonate phthalocyanine complexes more readily. Under such conditions CuPc produces a long-wave peak at $\lambda = 792$ nm[190] which shifts, at 18 M H_2SO_4, to 794 nm. CuPcTS in 7 M H_2SO_4 produces a long-wave peak at $\lambda = 726$ nm which persists when the concentration is increased up to 10 M H_2SO_4.

In an aqueous solution with an infinitesimally low concentration of 1×10^{-7} M monomeric unprotonated particle $CuPc(SO_2O^-)_4$ features a peak at a wavelength of 674 nm in the first absorption band. Consequently, in the range of concentrations from 7 to 10 M H_2SO_4, the shift ($\Delta\lambda$) of the first protonated spectrally active particle is as small as *c.* 50 nm rather than *c.* 112 nm, as in the case of $CuPcH^+$. Hence the spectral feature at 726 nm does not fully correspond to a protonated particle. Because atoms N1, N2, N3, and N4 are bound by the protons of the neighbouring SO_3H groups and the macroring is deactivated by inclusion of four electron-acceptor groups in the benzene rings, associates of the $CuPcTS \cdots H_2SO_4$ type appear within the above H_2SO_4 concentration range with a strong hydrogen bond $N \cdots HO_3S$ reducing the basicity of the nitrogen atoms. Such associates are, of course, also formed on the solid phase of CuPc which is in contact with H_2SO_4 solutions under the described conditions.[192] However, the insufficiently strong interaction in particles $CuPc \cdots H_2SO_4$ does not permit them to pass into solution from the solid phase. This is why such particles have not been found in sulphuric acid solutions in the case of water-insoluble phthalocyanines.

The peak of the first ionic form of the monoprotonated copper tetrasulphophthalocyanine $[CuPcTS]H^+$ appears only at a concentration of 11 M H_2SO_4. Its intensity increases up to a concentration of 0.2% oleum, and the absorption spectrum becomes similar to the typical spectra of $MPcH^+$.[190] The peak of the first band occurs at a wavelength of 762 nm, that is it does not reach the peak of $CuPcH^+$ ($\lambda = 792$–794 nm). The hypsochromic shift of $[CuPcTS]H^+$ with respect to $CuPcH^+$ is due to the fact that the basicity of the macroring diminishes as a result of its deactivation by the sulpho groups and the competition between the adjacent N atoms and sulpho groups of associate $N{-}H^+{\cdots}OSO_2H$ for a proton, which leads to a smaller depth of protonation of the microring and a weaker effect of the proton on the molecular orbital energies. In what follows (Chapter 7) absorption spectra of metal phthalocyanines will be used as an example to illustrate the effect of the electronic nature of the coordinated metal on the basicity of phthalocyanine, protonation depth, and position of the first band in H_2SO_4 solutions.[37,289]

In a solution of 0.5–15% oleum the peak at $\lambda = 762$ nm in the spectra of tetrasulphophthalocyanine gradually shifts to 772 nm, which corresponds to the protonation depth increasing with the protonating capacity of the medium. Ahrens and Kuhn[287] ascribed the bands at wavelengths of 726, 762, and 769 mm to di-, tri-, and quadriprotonated forms of tetrasulphophthalocyanine, respectively. The monoprotonated form, which is the only one to occur, could not be found in the absorption spectra of aqueous sulphuric acid solutions. This form was considered to be responsible for the shoulder at 709 nm in the absorption spectra of CuPcTS taken in H_2SO_4–H_2O–dioxane and H_2SO_4–H_2O–HAc mixtures. However, in this case the picture is complicated by the aggregation of tetrasulphophthalocyanine and association of ions to such a degree that quantitative interpretation of the spectra taken in these mixtures is not yet possible. Direct thermodyanic measurements have shown that unsubstituted phthalocyanines form only $MPcH^+$. Naturally, a weaker base, as for example CuPcTS, cannot yield any other forms but a monoprotonated one (I am only speaking of ionic forms).

Protonation is much more complicated in mixtures of acids with organic solvents due to ionic association. For example, when hydrochloric acid is added to CuPc solutions in dimethylphthalate,[8] the band at 680 nm shifts first to 715 nm then to 740 nm. The peak of the first band ($\lambda = 792$ nm), corresponding to $CuPc^+$, appears only after dimethylphthalate is removed from the system. The low dielectric constant of dimethylphthalate is responsible for the lower protonating capacity of acids and virtually precludes ionization of the acid–base interaction product. This is why it is most likely that the peak at 715 nm corresponds to CuPC···HCl and the peak at 740 nm corresponds to $CuPcH^+{\cdots}Cl$. The protonation depth gradually increases with the amount of HCl in the organic solvent but never reaches its limit characteristic for $CuPcH^+$. Besides, the strength of HCl is not sufficient under the above conditions. Gurinovich *et al.*[8] have not taken into consideration the possibility of non-ionized forms of CuPc appearing during acid–base interaction, which is quite natural for a base as weak as phthalocyanine.

It should be emphasized that the polarized luminescence data fully correlate with the thermodynamic data obtained for phthalocyanines in an aqueous solution of H_2SO_4. The complexity of spectra in the case of copper sulphophthalocyanine in H_2SO_4 suggests that in considering the protonation of nitrous heterocyclic compounds one should ignore their functional substituents with groups having atoms capable of being protonated. Studies into the protonation of porphyrins, and especially azaporphyrins, in organic proton-donor media are still in their infancy. Systematic quantitative investigations in this field will elucidate the spectral and chemical differences between the non-ionized and ionized protonated forms of porphyrins and give a better insight into the phenomena occurring at heteroatoms in acidic media.

The protonation of porphyrin molecules in the gaseous phase is of particular interest. First data for phthalocyanine complexes were communicated by Sidorov and Terenin.[288] They have shown that a solid phthalocyanine complex–gaseous acid (HBr, HCl, HAc) system yields products of sufficiently deep protonation. The gaseous acid forms a protonation product across the entire phthalocyanine film at once. Infrared spectra show that no unprotonated molecules are left. Judging from the position of the first absorption band (λ = 790 nm) the protonation in the $MgPc_{(s)}1–HBr_{(g)}$ system is complete, that is the $MgPcH^+$ ion is formed. In this case Br^- occupies an intermediate position in the molecular crystal lattice, probably being localized near the Mg^{2+} atom rather than near the N–H bond. Weaker protonation is observed in acetic acid (the first absorption band is at λ_1 = 770 nm) and is still weaker in HCl (bands at λ_1 = 745 nm).

In this process the determining role is played by the basic properties of the acid anion and its ability to coordinate to the central atom. Owing to the greater tendency of the magnesium atom in MgPc to coordinate to the anion in the presence of gaseous acids, MgPc is more readily protonated than CuPc or H_2Pc. Their protonation depth is less because λ_1 is more hypsochromic. According to Sidorov and Terenin,[288] phthalocyanines in a solid phase attach not more than one proton. The spectral manifestations of the incomplete stepwise protonation of phthalocyanines in a solid phase are not yet known. One can only assume that in weakly ionizing media several acid molecules may be attached, yielding associates $MPc\cdots nHA$ ($n < 4$) via the hydrogen bond, which will be spectrally equivalent to formation of the cation $MPcH^+$. However, this hypothesis needs new experimental data to verify it.

Another interesting problem is the depth, or degree, of protonation and the site of proton localization in tetrapyrrole molecules occupying an intermediate position between porphyrin and phthalocyanine in so far as their molecular structure is concerned. Initially, tetrabenzoporphin and alkyl derivatives of tetrazaporphin will be discussed (unsubstituted H_2TAP is unstable in an acidic medium). No quantitative studies into the acid–base properties of these compounds have been reported so far. The literature only contains data[290] on the spectroscopic behaviour of these compounds in acidic media.

The hypsochromic shift of the first absorption band of tetracyclohexenetetrazaporphin (**XLIII**) and the fewer bands in an acidic medium indicate that the ligand itself behaves like porphyrins, attaching two protons to the centre of the molecule:

$$H_2TCHTAP + 2H^+ \rightarrow H_4TCHTAP^{2+}.$$

Its stable metal complexes feature a bathochromic shift of the first band during protonation, which suggests that the peripheral nitrogen atoms are protonated. The spectral picture is the same as in the case of phthalocyanines;[8] however, the slight shift of the first absorption band (*c*. 30 nm in DMFA) is indicative of partial protonation without ionization, yielding associate CuTCHTAP···HX. A similar behaviour is observed in the case of the copper complex of tetrazaporphin. Thus the aza substitution in porphin alone is not sufficient for the ligand to be protonated at the periphery, similarly to H_2Pc.

The condensation of porphin together with benzene residues, that is transformation of porphin to tetrabenzoporhin (**XLIV**), does not alter the overall picture of protonation[8] typical of porphyrins. The protonation takes place in the centre of the molecule. The absorption spectrum features a single band but, instead of the expected hypsochromic shift during protonation, a slight bathochromic shift occurs (by 6 nm in DMFA saturated with HCl[290]). No conclusion can be drawn from the available data as to the number of attached protons (one or two). For this species of intermediate porphyrins the absolute lack of thermodynamic data is particularly acute.

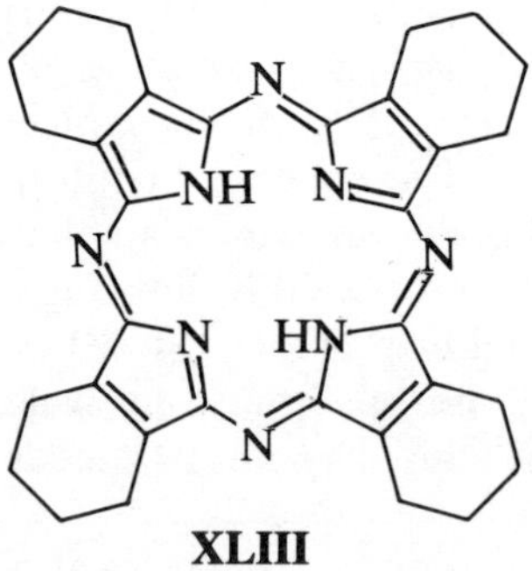

XLIII

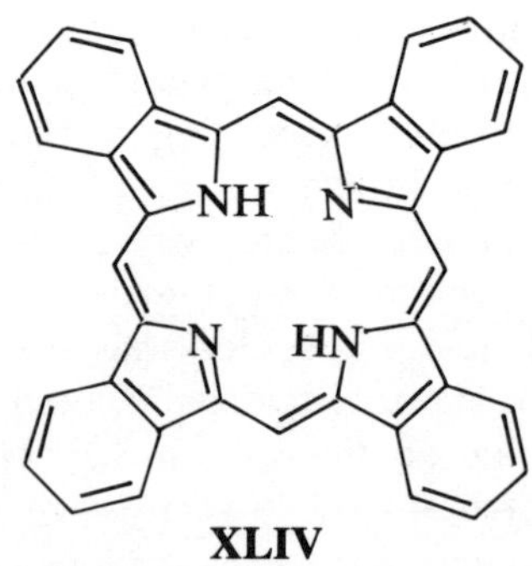

XLIV

4.4 Enthalpy–Entropy Characteristics of Processes of Protolytic Dissolution and Protonation of Phthalocyanine Complexes

The equilibria of dissolution of phthalocyanines in an aqueous solution of H_2SO_4 depend strongly, as was shown in Table 4.16, on the nature of the central metal atom coordinated to phthalocyanine. Table 4.17 lists the mean values of constants derived from Table 4.16 and the values of free

Table 4.17 Equilibrium constants (pK^{25}_{MPc}) and free energies of dissolution (ΔG_S) of phthalocyanine and its complexes in H_2SO_4

Complex	pK^{25}_{MPc}	ΔG_S (kcal mol^{-1})	Complex	pK^{25}_{MPc}	ΔG_S (kcal mol^{-1})
H_2Pc	1.65 ± 0.02	2240 ± 30	$(HSO_4)AlPc$	2.03–2.65	2760–3640
ZnPc	2.31 ± 0.02	3150	$(HSO_4)GaPc$	1.82 ± 0.02	2480
CuPc	1.64 ± 0.03	2240	$(HSO_4)RuPc$	0.94 ± 0.03	1280
CoPc	2.07 ± 0.03	2820	$(HSO_4)RhPc$	0.99 ± 0.04	1350
NiPc	1.45 ± 0.04	1980	$(HSO_4)IrPc$	1.34 ± 0.02	1820
PdPc	1.52 ± 0.05	2070	$(HSO_4)_2SnPc$	2.49–3.59	3390–4880
PtPc	1.50 ± 0.06	2040	$(HSO_4)_2VPc$	1.92 ± 0.02	2620
$CuPcCl_{15}$	2.84 ± 0.03	3880	$(HSO_4)_2OsPc$	0.04–1.29	55–1760
UO_2Pc	2.67 ± 0.04	3640	$(Cl_2)SiPc$	2.73 ± 0.05	—

energies of protolytic dissolution of phthalocyanine complexes, calculated from these constants.

Depending on the electronic structure and charge of the metal ion the values of pK_{MPc} vary, at 25 °C, from 0.94 to 2.73. This attests to a tangible effect of the metal atom on the entire phthalocyanine molecule, which will be considered later in the book. The value of the equilibrium constant is considerably affected by all hydrogen atoms in the benzene rings being substituted by chlorine ($pK_{CuPc} = 1.64$ and $pK_{Cl_{15}CuPc} = 2.84$). The low values of the protolytic dissolution equilibrium constants of phthalocyanines are due to their negligibly small basicity. The free energy of dissolution increases by 2–5 kcal mol^{-1}.

At first the rise in temperature promotes dissolution, hence protonation of phthalocyanine complex molecules, then inhibits it (Table 4.18, Fig. 4.6). From the data of Table 4.18, representing the temperature dependence of the dissolution equilibrium constants derived earlier[37,284] for some phthalocyanine complexes, the enthalpies and entropies of dissolution at temperatures ranging from 20 to 50 °C were calculated (Table 4.19). The rectilinear relationship between log K_{MPc} and inverse temperature (see

Table 4.18 Equilibrium constants of dissolution of phthalocyanine complexes (pK_{MPc}) in H_2SO_4 at various temperatures

t (°C)	CuPc	$CuPcCl_{15}$	CoPc	ZnPc	$(HSO_4)AlPc$	
20	1.75 ± 0.03	—	—	2.31 ± 0.01	—	—
25	1.64 ± 0.03	2.84 ± 0.03	2.07 ± 0.03	2.31 ± 0.02	2.67[a]	2.03[b]
30	1.52 ± 0.02	2.81 ± 0.04	1.93 ± 0.05	—	2.61	2.14
40	1.43 ± 0.03	2.70 ± 0.02	1.78 ± 0.01	—	2.52	2.38
50	1.29 ± 0.06	2.65 ± 0.02	1.66 ± 0.06	1.91 ± 0.01	2.36	2.52
70	1.54 ± 0.02	2.70 ± 0.01	1.72 ± 0.01	1.94 ± 0.08	2.52	2.86

[a]Dissolution in the entire temperature range in 16 M H_2SO_4.
[b]Same, in 17 M H_2SO_4.

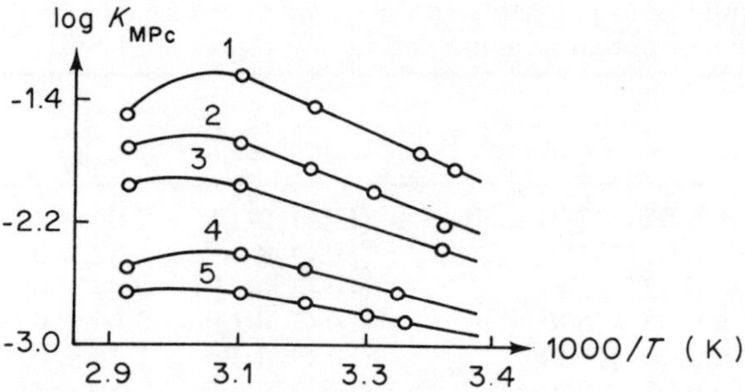

Figure 4.6 pK of phthalocyanines versus temperature. Phthalocyanines: curve 1, CuPc; curve 2, CoPc; curve 3, ZnPc; curve 4, ClAlPc; curve 5, $Cl_{15}CuPc$

Fig. 4.6) in this range is indicative of the enthalpy of dissolution being constant and permits calculation of this quantity from the van't Hoff isobar equation. Application of the second law of thermodynamics has made it possible to calculate the entropy variations occurring in the course of protolytic dissolution of phthalocyanine complexes.

Since constants K_{MPc} are independent of the ionic strength of the solution in the Brand region, they may be considered as themodynamic rather than concentration constants, and the values of parameters ΔG, ΔH, and ΔS, calculated from these constants, can be ascribed to 80–99% H_2SO_4 and 25 °C as standard conditions.[37] Regardless of the nature of the doubly charged metal and the H_2SO_4 concentration phthalocyanine complexes pass into solution endothermally with a moderate increase in entropy.

The variations in enthalpy during protolytic dissolution may be grouped according to individual energy components of elementary processes if the following thermodynamic cycle is written:

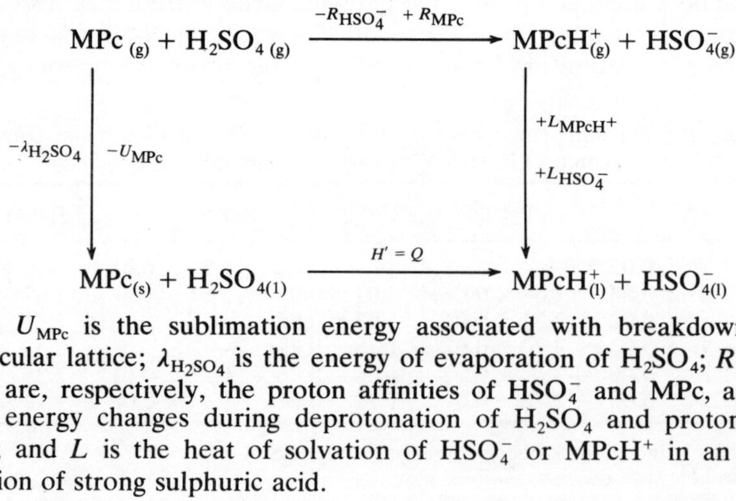

Here U_{MPc} is the sublimation energy associated with breakdown of the molecular lattice; $\lambda_{H_2SO_4}$ is the energy of evaporation of H_2SO_4; $R_{HSO_4^-}$ and R_{MPc} are, respectively, the proton affinities of HSO_4^- and MPc, associated with energy changes during deprotonation of H_2SO_4 and protonation of MPc; and L is the heat of solvation of HSO_4^- or $MPcH^+$ in an aqueous solution of strong sulphuric acid.

Table 4.19 Enthalpy (ΔH), free energy (ΔG_S), and entropy (ΔS) of dissolution of phthalocyanine complexes in H_2SO_4

Parameter	CuPc	$CuPcCl_{15}$	CoPc	ZnPc	$(HSO_4)AlPc$	
ΔH (kcal mol^{-1})	5.70 ± 0.6	3.66	6.81	5.23	5.68[a]	−8.85[b]
ΔG (kcal mol^{-1})	2.24 ± 0.04	3.88	2.82	3.15	3.64	2.76
ΔS (e.u.)	11.6 ± 2	−0.7	13.4	7.0	7.0	−39

[a] Dissolution in 16 M H_2SO_4.
[b] Dissolution in 17 M H_2SO_4.

The following relation is derived from the above thermodynamic cycle:†

$$\Delta H' = R_{MPc} + L_{MPcH^+} + L_{HSO_4^-} - \lambda_{H_2SO_4} - U_{MPc} - R_{HSO_4^-}. \tag{4.32}$$

Since some terms in expression (4.32) are independent of the phthalocyanine species, we have

$$\Delta H' = R_{MPc} + L_{MPcH^+} + \text{const.}, \tag{4.33}$$

$$\text{const.} = L_{HSO_4^-} - \lambda_{H_2SO_4} - U_{MPc} - R_{HSO_4^-}. \tag{4.34}$$

The enthalpy of dissolution of phthalocyanine (and porphyrin) complexes in acids is completely dependent on the relationship between the protonation depth (R_{MPc}), solvation efficiency (L_{MPcH^+}), and constant.

In the overwhelming majority of cases involving dissolution of phthalocyanine complexes the sign of parameter ΔH is determined by the values of equation (4.34). The value of ΔH decreases perceptibly (see Table 4.19), and ΔS becomes less positive under conditions conducive to a greater degree of solvation of the protonated particle due to the tendency of the central atom to extra coordination of ligands (ZnPc) or due to polarization of the molecule as a result of substitution in a benzene ring ($CuPcCl_{15}$). An interesting phenomenon is observed during dissolution of aluminium phthalocyanine $(HSO_4)AlPc$ in H_2SO_4. Thermodynamic and spectroscopic data[190,280,284] indicate that extra ligand HSO_4^- forms a complex with $AlPc^+$ primarily through an electrostatic interaction. Sharp changes in the solvating capacity of sulphuric acid at a concentration approaching 16 M[276] bring about variations in the molecular state of aluminium phthalocyanine in 17–18.8 M H_2SO_4.

In sulphuric acid of lower concentrations the predominant form of the complex is ionized $AlPc^+ + HSO_4^-$ corresponding to which in Table 4.16 are weak protonation and $pK \approx 2.65$. In a 17 M and stronger solutions of H_2SO_4 aluminium phthalocyanine is predominantly in the form of extra complexes $HSO_4 \cdot AlPc$ or $[(HSO_4)_2AlPc]^-$ with $pK \approx 2.00$ in Table 4.16. The extra coordination may be the reason why the value of ΔH of aluminium phthalocyanine varies within these concentration limits from 5.68 to −8.85 kcal mol^{-1}, that is the endothermal dissolution process becomes

†Here thermodynamic notation is used ($\Delta H' = Q$).

exothermal, and the value of ΔS drops from 7 to -39 e.u. (see Table 4.19). This example suggests that extra coordination in porphyrin-type molecules may considerably complicate the thermodynamics of their protonation. The same applies to the protonation of solid phthalocyanine complexes with gaseous acids. The role of the solvent in the protonation of phthalocyanine molecules with sulphuric acid is highly important. The data of Table 4.19 are indicative of occurrence of the solvation-thermodyanic compensation effect (4.13) which is typical of ionization of porphyrins in media of any acidity and alkalinity. The changing entropy in the case of dissolution of phthalocyanines may be expressed as

$$\Delta S = \Delta S_b + \Delta S_s - \Delta S_{solv}. \tag{4.35}$$

Here ΔS_b is associated with breakdown of the crystal lattice, ΔS_s with the structural changes in H_2SO_4 due to incorporation of large phthalocyanine molecules, and ΔS_{solv} with the solvation of HSO_4^- and $MPcH^+$ in solution. The first two terms of equation (4.35) are practically the same for all phthalocyanines, while the changes in the value of ΔS are determined by those in the solvation of ion $MPcH^+$.

Of particular interest is the changing sign of the thermal effect of protolytic dissolution of phthalocyanines, occurring at 50–70 °C. This change may be attributed to the accumulation of water in a strong H_2SO_4 solution with the temperature rising as a result of reaction $HSO_4^- + H_3O^+ \rightleftarrows H_2O + H_2SO_4$ and with increasing solvation of HSO_4^- and $MPcH^+$, which compensate for the thermal effect introduced by the terms of equation (4.33). In the following section it will be shown how the thermodynamic parameters of protonation of phthalocyanine (and porphyrin) molecules correlate with the nature of the chemical bond between the metal atom and ligand.

COORDINATION PROPERTIES

4.5 Electronic Effects of Coordination

The concept of coordination properties of ligands of the porphyrin and tetrazaporphyrin types includes the following factors:

(1) the reactivity of the ligand during coordination of metal salts at the central nitrogen atoms (the rate of formation of metalloporphyrins is actually meant);

(2) the kinetic and thermodynamic stability of metalloporphyrins, that is the rates and equilibria of dissociation of these complexes;

(3) the capacity of metalloporphyrins for additional coordination of simple and complex ligands (so-called extra coordination at the central metal atom);

(4) the secondary coordination of salts at the functional substituents of natural and synthetic complex metalloporphyrins.

One of the most important properties of porphyrin and phthalocyanine molecules is their ability to coordinate to metal ions, yielding stable intercomplex salts with structure **XLV**. Stable complexes of porphyrins or

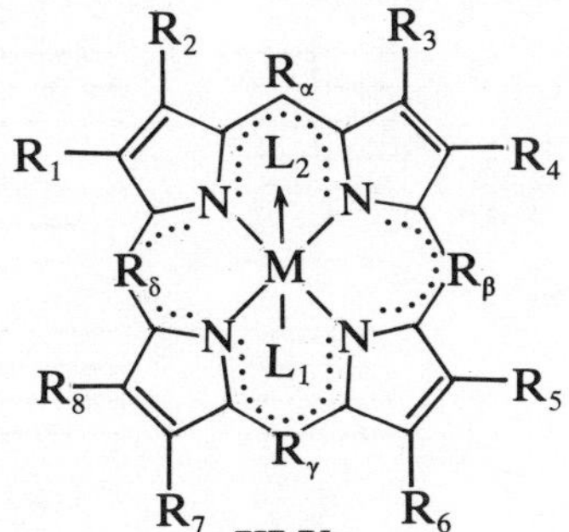

XLV

metalloporphyrins and complexes of phthalocyanines or metal phthalocyanines result from formation of four equivalent σ bonds N → M, that is filling of vacant s, p_x, p_y, and $(n-1)d_{x^2-y^2}$ or $nd_{x^2-y^2}$ orbitals of the cation with σ electrons of the central nitrogen atoms. We shall refer to the latter as coordinating nitrogen atoms. The resulting σ bonds are so strong in the case of metals Cu^{2+}, Ni^{2+}, Co^{2+}, and Zn^{2+} that in solid complexes practically no replacement of the metal ions by protons takes place at H_2SO_4 concentrations ranging from 2 to 7 M.[291] In the case of coordination of a triply or quadruply charged ion the formation of a covalent complex involves, in addition to d, s, and p^2 orbitals, p_z and d_{z^2} orbitals that participate in the attachment of the extra ligand along the z axis perpendicular to the plane of molecule **XLV**. The extra ligands may be anions (HSO_4^-, Cl^-, OH^-, Ac^-, O^{2-}, and others) or molecules (H_2O, NH_3, NH_2CONH_2, and others). All phenomena involved in extra coordination will be discussed later (see Chapter 6).

The formation of σ bonds M ← N in phthalocyanine and porphyrin complexes leads to four electron pairs of the coordinating nitrogen atoms being excluded from the conjugation. This is vividly illustrated in Fig. 4.7, which shows schematically the filling of antibonding π orbitals of the phthalocyanine macroring with the heteroatomic n electrons. The attendant decrease in the energy of the ψ_π^* orbitals favours conjugation of the heteroatomic n electrons which have not taken any part in the coordination to the metal. As a result the remaining n electrons acquire a π rather than a σ character and are not easily protonated with acids.

It is suggested that the above redistribution of electrons and the associated changes in the properties of porphyrin molecules should be termed the *σ-electron effect of coordination* or, simply, the σ effect. It is similar in nature to the positive induction effect (+I) of substitution in conjugated organic molecules. The electronic σ effect reduces the capacity of phthalocyanine and porphyrin complexes for protonation and dissolution

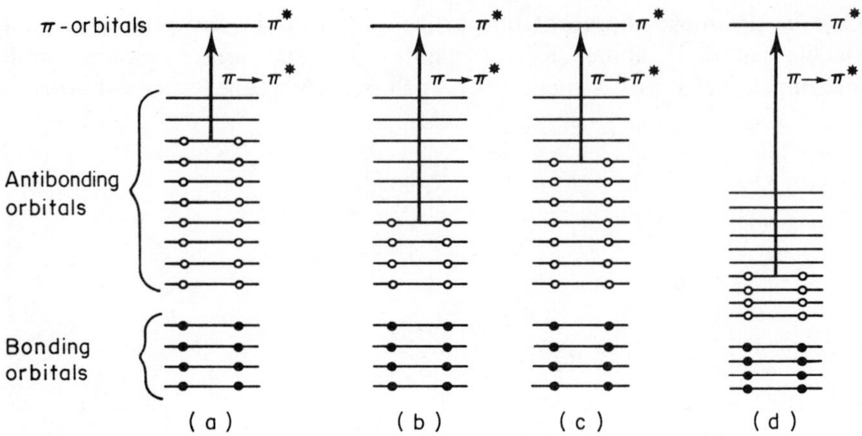

Figure 4.7 Filling of antibonding π orbitals of the phthalocyanine macroring with a heteroatom and metal electrons. (a) H_2Pc; (b) ZnPc; (c) CuPc; (d) $(HSO_4)AlPc$

in acids. The best way to examine the correlation between the coordination properties of a metal and the protonating capacity of a complex is by using metal phthalocyanines as simple models of porphyrins.[289] If the changes in the free energy of a phthalocyanine ligand are taken as a reference, it can be seen from Table 4.17 that formation of only σ bonds in its complexes with metals (ZnPc, HSO_4·GaPc, $(HSO_4)_2$·VPc, Cl_2SiPc) results in a lower basicity and a higher value of ΔG. The values of $\Delta\Delta G_S = \Delta G_S^{MPc} - \Delta G_S^{H_2Pc}$ are positive.

In the many cases where the metal ion has filled d orbitals of π symmetry (d_{xy}, d_{xz}, d_{yz}) it forms backward dative π bonds with the phthalocyanine ligand. The metal serves as a donor of π electrons, and the ligand serves as their acceptor. These bonds are opposite in direction to σ bonds, that is $M \rightarrow N$, and are called backward bonds. The d_π electrons of the metal fill the antibonding π orbitals of phthalocyanine. The filling of the ψ_π^* orbitals increases their energy (see Fig. 4.7), which prevents the n electrons of the peripheral N atoms from entering into a π conjugation as well as enhancing their σ character and capacity for acid protonation. Such a redistribution of electrons is called the *dative π effect of coordination* or, simply, the π effect.[289]

As the metal atom becomes more capable of entering into a π bond, the values of $\Delta\Delta G_S$ become more negative, as can be seen from Table 4.17, hence the capacity for protolytic dissolution in the series CoPc < CuPc < PdPc, PtPc, NiPc $<(HSO_4)IrPc$ < $(HSO_4)RhPc$, $(HSO_4)RuPc$ $<(HSO_4)_2OsPc$ increases. The tabulated data indicate that in the case of CoPc the changes in the acidic properties during coordination of H_2Pc to Co^{2+} are due primarily to the σ effect, while the π effect is secondary. In complex CuPc the π effect on the basic properties has completely compensated for the σ effect. In the rest of the d metals (with the exception of Zn^{2+} whose closed *d* orbital hinders the formation of π bonds) the π effect compensates for the σ effect. This is due to the unusually high capacity of phthalocyanine for π bonding with metals. As was shown in Chapter 3, the same capacity of porphyrins is much lower.

Apart from the backward dative π bond M → N, a forward dative π bond M ← N may occur, coinciding in direction with the dative σ bond. It may appear in complexes whose central atom has vacant d_π orbitals ($(HSO_4)_2VPc$ and Cl_2SiPc). Owing to its minor role the respective *forward dative π effect of coordination* does not exert a tangible influence on the properties of MPc.

In cases where the triply or more charged central ion is coordinated to phthalocyanine or porphyrin and its coordinate bond with extra ligands (HSO_4^-, Cl^-, Ac^-) is weak, a high positive charge is localized at the ion. It must cause a decrease in the energy of ψ_π^* orbitals, promote their filling with *n* electrons of the peripheral nitrogen atoms, and reduce the basicity of phthalocyanines. The present author has proposed[289] that this phenomenon be referred to as the *charge effect of coordination.* This effect accounts for the very low values of the equilibrium constants of protolytic dissolution and their dependence on the H_2SO_4 concentration[37,102,190] in phthalocyanine complexes with th Al^{3+}, Sn^{4+}, and UO^{2+} (see Table 4.17). Such peculiarities cannot be explained in terms of the σ effect since the σ bonds Sn ← N and UO_2 ← N are much weaker than Zn ← N. It may be assumed that in labile phthalocyanine and porphyrin complexes (with Mg^{2+}, Cd^{2+}, Hg^{2+}, Pb^{2+}, Sn^{2+}, and many others) the role of the charge effect is predominant.

In addition to electronic and charge effects of coordination complexes with cyclic π ligands, such as porphyrins, phthalocyanines, and other macroheterocyclic compounds,[292–294] may in some cases display *steric effects of coordination.* If coordination distorts the planar structure of a conjugated cyclic system, the conjugation of *n* electrons of the nitrogen atoms with the macroring π system weakens. As a result the basicity of the molecules increases. If coordination to a metal creates a more coplanar π system than the ligand, the opposite is true. Such ligand changes are not typical of phthalocyanines. It was shown[108] that phthalocyanine complexes with Ni, Rh, Ir, Rn, and Os are characterized by a different steric effect of coordination, associated with the distortion of the valence angles of C—N bonds and manifesting itself only in electronic absorption spectra and not in the basic properties of metal phthalocyanines.

Another factor, apart from the electronic effects, exerting a major influence on the acid–base properties of phthalocyanine molecules is the polymeric state. Polymerization via the central atom and extra ligands is possible in osmium phthalocyanine $(SO_4)OsPc$ and O=OsPc both in the solid phase and in solution, including a sulphuric acid solution, in the form of polymer chains $—OSO_2—OsPc—OSO_2—OsPc$ and —O—OsPc—O—OsPc, respectively.

For such short chains to pass into solution requires a lesser number of protons. If the dissolution of a monomeric complex is characterized by the ratio MPc:H^+ = 1:1; in this case it will be 2:1, 3:1, and even less. It is possible that the dissolution of a chain (—O—OsPc—) requires only one proton rather than *n* protons. Hence the unusual relationship between the solubility of osmium phthalocyanine and the H_2SO_4 concentration (see Table 4.17). A similar phenomenon has been observed in polymeric phthalocyanines with the polymer chain developing via benzene rings.[145–150]

4.6 Kinetics of Formation of Chlorophyll and Its Metal Analogues in Organic Solvents

The formation of metalloporphyrins of structure **XLV** is a relatively slow process and can be measured by methods of classical kinetics. The free energies, enthalpies, and entropies of activation are strongly dependent on the structure of porphyrins and metal salts. Let us begin analysis of the kinetics and mechanism of complex formation with the most important natural porphyrins—chlorophyllic acid and protoporphyrin.

As was mentioned in Chapter 2, magnesium complexes of porphyrins, including chlorophyll, form in solutions from Mg^{2+} salts and porphyrins with great difficulty (for reasons described in the same chapter). On the other hand, salts of most transition metals, as well as Zn^{2+}, Cd^{2+}, and Hg^{2+}, react with chlorophyllic acid (**XXVII** and **XXVIII**) in ethanol at rates suitable for studying the regularities and mechanism of formation of metal analogues of chlorophyll in solution. By metal analogues of chlorophyll are meant complex salts of chlorophyllic acid, in which M^{2+}, M^{3+}, or M^{4+} occupies the same coordination space between the four intracyclic nitrogen atoms of the ligand; consideration of complexes of H_2Chl with Na^+, Li^+, Ag^+ and other metals in which two cations are above and below the molecular plane as metal analogues of chlorophyll is not advised.

The formation of chlorophyllic acid complexes, as opposed to those of other porphyrins, has been studied more completely[295–317] The regularities of their formation were analysed in many works.[318–324] It has been established that chlorophyllic acid *a* reacts in organic solvents with salts as follows:

$$H_2Chl + [MX_2(solv)_{n-2}] \rightarrow MChl + 2HX + (n - 2)solv. \qquad (4.36)$$

The first attempt to study quantitatively the kinetics of formation of a metal analogue of chlorophyll with nickel in methanol was made by Baker and Hodson.[325] They determined only the empirical rate constants and derived an empirical kinetic equation of a fractional order for both reagents, from which no definite conclusion can be drawn as to the reaction mechanism. Berezin and Sosnikova (Volkova)[295–302] examined reaction (4.36) in ethanol with a view to establishing its mechanism and kinetics for many metals: Fe^{2+}, Co^{2+}, Ni^{2+}, Cu^{2+}, Zn^{2+}, Cd^{2+}, Hg^{2+}, Ag^{2+}, and Pd^{2+}. The experiments were conducted under mild conditions at 15–50 °C with the metal salt concentration varying from 1.0×10^{-4} to 2.5×10^{-2} M and always exceeding that of chlorophyllic acid, which was normally equal to

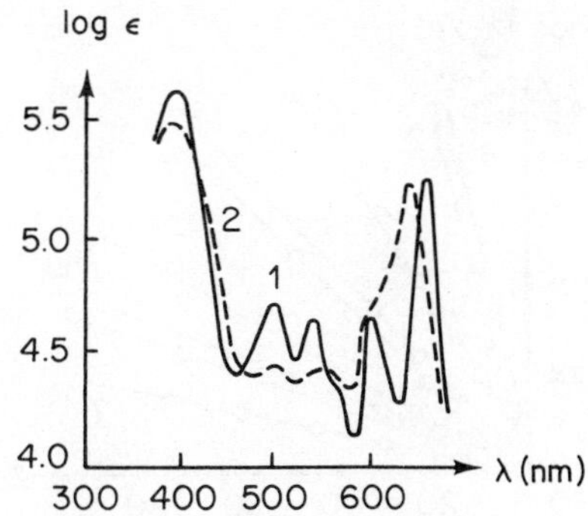

Figure 4.8 Absorption spectra of chlorophyllic acid (curve 1) and its copper complex (curve 2) in ethanol

2.5×10^{-5} M. The wide difference in the electronic absorption spectra of H_2Chl and MChl (Fig. 4.8) permits spectrophotometric determination of the current chlorophyllic acid concentration and the end of the reaction from the equation

$$c_{H_2Chl} = c_0[(D_t - D_\infty)/(D_0 - D_\infty)] \quad (4.37)$$

correlating the initial chlorophyllic acid concentration (c_0) with the optical density at the beginning of the experiment (D_0), at the end (D_∞), and at time t (D_t). The only coloured substances in the solution are H_2Chl at the beginning of the experiment and MChl at the end for reaction (4.36) proceeds all the way in organic solvents.

The time independence of the rate constant and the significant excess of the metal salt give the following simple kinetic equation:

$$-\mathrm{d}c_{H_2Chl}/\mathrm{d}t = k_{\mathrm{eff}}c_{H_2Chl}. \quad (4.38)$$

The effective rate constant k_{eff} depends on the salt concentration. It is correlated with the true rate constant (k_v) of reaction (4.36) by

$$k_{\mathrm{eff}} = k_v c_{MX_2}. \quad (4.39)$$

Thus it was established for the first time that reaction (4.36) in ethanol corresponds to the second-order kinetic equation

$$-\mathrm{d}c_{H_2Chl}/\mathrm{d}t = k_v c_{H_2Chl} c_{MX2}. \quad (4.40)$$

Shown in Fig. 4.9 is the rate of reaction (4.36) versus temperature and salt concentration in a relationship typical of most salts. For most of the salts examined the 'true' rate constant calculated from equation (4.39) does not remain invariable. Formally this is indicative of a fractional order of reaction (4.36) with respect to the salt.

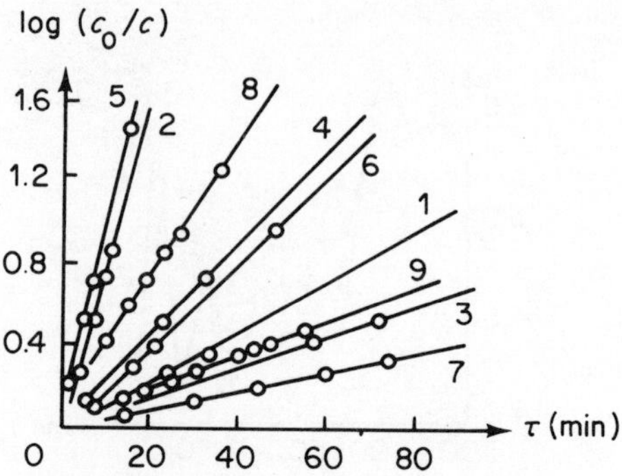

Figure 4.9 Rates of formation of the cadmium analogue of chlorophyll in ethanol. Curves 1, 2 CdI_2 reagent, 30 and 40 °C, 2.5×10^{-3} M; curves 3, 4, $Cd(NO_3)_2$ reagent, 30 and 40 °C, 2.5×10^{-4} M; curve 5, $Cd(NO_3)_2$ reagent, 30 °C, 2.5×10^{-3} M; curves 6, 8, $CdCl_2$ reagent, 30 and 40 °C, 2.5×10^{-3} M; curves 7, 9, $CdCl_2$ reagent, 30 and 40 °C, 2.5×10^{-4} M

However, bearing in mind that the reaction proceeds in organic solvents in which the state of the salts changes with their concentration to a much greater extent than in aqueous solutions and the solvent structure is more variable than that of water, the variability of constant k_v must be ascribed precisely to these factors rather than to changes in the mechanism of reaction (4.36). In this connection it makes no sense to write a fractional-order kinetic equation because it will only obscure the physicochemical phenomena occurring in solutions.

The constants k_v calculated from equation (4.39) at salt concentrations below 2.5×10^{-4} M may be regarded as true constants. At higher salt concentrations constants k_v are 'arbitrary' to a certain degree in cases where they decrease considerably with increasing concentration. The reasons for k_v of reaction (4.36) and similar reactions being concentration-dependent will be discussed below.

It should be noted here that in most cases the concentration dependence of k_v is a qualitative indication of stability of the coordination sphere of salt $MX_2(solv)_{n-2}$ and of changes it undergoes in solution, as well as stability of the solvent structures. The kinetic data obtained in earlier works[295–302] are summarized in Table 4.20.

As can be seen from the latter, in some salts, such as $CuAc_2$, $Cd(NO_3)_2$, CdI_2, and $CdCl_2$, the rate constant does not depend on concentration, that is their structure is stable in ethanol. In other salts, for example $Co(NO_3)_2$, the rate constant is almost independent of concentration. And in such salts as

Table 4.20 Kinetic parameters of reaction (4.36) of formation of metal analogues of chlorophyll in ethanol

Reagent salt	$k_v^{25} \times 10^{-2}$ (l mol^{-1} h^{-1})	E (kcal mol^{-1})	$\Delta S^\ddagger$ (e.u.)
$FeCl_2$	4.48 ± 0.23[a]	13.8 ± 1	16 ± 3
$CoCl_2$	0.018 ± 0.002[b]	31.0	+30
$CoAc_2$	0.011 ± 0.001[a]	23.9	+6.1
$Co(NO_3)_2$	0.106 ± 0.012[a]	24.0	+11
	0.068 ± 0.008[b]	18.0	−10.5
$Ni(NO_3)_2$	0.34 ± 0.04[a]	15.5	−15.4
	1.42 ± 0.28[c]	23.6	+14.4
$CuAc_2$	59.1 ± 0.6	10.6 ± 1.2	−26 ± 4.0
$Cu(NO_3)_2$	91.3 ± 5.5[a]	12.9 ± 1.2	−13 ± 3.0
$Zn(NO_3)_2$	2.30 ± 0.20[a]	14.1 ± 0.5	−16.2 ± 1.6
	15.0 ± 2.0[c]	14.1 ± 0.7	−12.5 ± 2.0
CdI_2	5.25 ± 0.50	19.6 ± 1.7	+4 ± 7.0
$CdCl_2$	18.5 ± 0.8	11.1 ± 0.5	−22 ± 2.0
$CdAc_2$	15.1 ± 1.2[c]	13.4 ± 1.5	−15 ± 5.0
	3.37 ± 0.08[a]	16.5 ± 0.7	−7.5 ± 2.5
$Cd(NO_3)_2$	26.0 ± 1.7	16.2 ± 0.2	−4.5 ± 0.7
HgI_2	23.0 ± 3.0[a]	13.5 ± 2	−14 ± 7.0
$HgCl_2$	21.9 ± 2.4[a]	—	—
$AgNO_3$	2.66 ± 0.15[a]	19.0 ± 0.2	±34 ± 0.5
$PdCl_2$	4.76 ± 0.15[a]	18.0 ± 1	−2 ± 3.0

[a]At a salt concentration of 2.5×10^{-3} M.
[b]At a salt concentration of 2.5×10^{-2} M.
[c]At a salt concentration of 2.5×10^{-1} M.

$Ni(NO_3)_2$, $CdAc_2$, and $Ni(NO_3)_2$ this dependence is great, which is first of all indicative of the salt structures changing in ethanol as their concentration increases. Almost always the equilibrium shifts toward chemically less active forms of salts, for example

$$[Zn(C_2H_5OH)_6]^{2+} \xrightleftharpoons{NO_3^-} [Zn(C_2H_5OH)_5NO_3]^+ \xrightleftharpoons{NO_3^-} [Zn(C_2H_5OH)_4(NO_3)_2].$$

Under comparable conditions the rate of reaction (4.36) is only slightly dependent on the nature of the cation, its charge, radius, and electronic structure. In ethanol only cobalt salts react at slow rates. If in the case of salts of eight different metals the rate constant at 25 °C equals $(9.8 \pm 6.4) \times 10^2$ l mol^{-1} h^{-1} on the average, varying from 59.1×10^2 for $CuAc_2$ to 1.42×10^2 for $Ni(NO_3)_2$, in the case of $CoAc_2$ it is as low as 1.1 l mol^{-1} h^{-1}. The salts of Cu^{2+} react much faster than other salts owing to the strong tetragonal distortion effect (Jahn–Teller effect[326]), while those of Co^{2+} and Ni^{2+} react at an extremely slow rate. The latter must be due to the strong tendency of Co^{2+} and Ni^{2+} to tetrahedral coordination.

The weak dependence of the reaction rate constant on the nature of the cation at the same concentrations and anionic composition permits a tentative and most significant conclusion to be made that an elementary act of bimolecular reaction (4.36) involves removal of two then the remaining solvate molecules from the inner sphere of the salt. Only after two molecules have been removed does the contact between the salt cation and porphyrin molecule (in this case chlorophyllic acid) become possible. This contact, resulting in a coordinate bond M ← N, makes a certain energy contribution and promotes the final and most energy-consuming stages of the MChl formation process.

These stages include stretching and breakage of two N—H bonds and detachment of ions (X^-) of salt MX_2 $(solv)_{n-2}$. This interpretation of the reaction mechanism, proposed by the present author[297] and fully corroborated by the totality of experimental data, was found to apply to all porphyrins as well. The differences in the strength of M^{2+}—solv bonds, depending on the nature of metal M^{2+}, mainly affect the energies and entropies of activation of reaction (4.36). Thus the activation energy varies from 10.6 kcal in the case of $CuAc_2$ to 31 kcal in the case of $CoCl_2$ (see Table 4.20), while the entropy of activation increases from −26 to +30 e.u. for the same salts.

Such a change in entropy is determined by a kinetic compensation effect:

$$E = a + b\Delta S^{\ddagger}. \tag{4.41}$$

The kinetic compensation effect stems from the same factors as the thermodynamic effect (4.13), with the difference that in this case one is dealing with solvation processes in the transition state of the reacting system. The transition state involves ionization of the M—X bonds of the salt and the N—H bonds of the porphyrin (protonation), therefore the transition state is much more polar than the initial state of reaction (4.36). It is not surprising then that in most cases (see Table 4.20) the entropies of activation are negative. Cobalt salts in ethanol make an exception, and their specific properties deserve closer attention. The positive values of entropy in the case of $AgNO_3$ are probably due to changes in the reaction mechanism, whereby the Ag^+ cation is oxidized to Ag^{2+} because the AgChl complex of doubly charged silver is formed.[327]

According to earlier findings[295–302] the rate of formation of metal analogues of chlorophyll depends greatly on the salt anion. For example, in the case of Cd^{2+} and Co^{2+} salts with different anions the order of rates decreases as $NO_3^- > Cl^- > Ac^-$. This is the sequence in which the instability constants of Cd^{2+} and Co^{2+} salts with these anions decrease in solution, that is the covalence of the M—X bond increases.

To determine the effect of the structure of the coordination sphere of the reagent salt on its reactivity measurements were made[299] of the rate of formation of copper phaeophytinate from H_2Chl and various copper salts in absolute and 90% aqueous ethanol. In some cases ethanol was saturated with ammonia for $CuX_2(solv)_{n-2}$ to pass into $Cu(NH_3)_6^{2+}$, just as in the case

Table 4.21 Kinetic parameters of formation of copper analogues of chlorophyll from various copper salts

Reagent salt	$k_v^{25} \times 10^{-3}$ ($l\ mol^{-1}\ h^{-1}$)	E (kcal mol^{-1})	$\Delta S^\ddagger$ (e.u.)	Solvent
Cu(acetate)	5.91	10.6	−26	Absolute ethanol
Cu(nitrate)	4.67[a]	41.7	+82	90% ethanol
Cu(nitrate)	0.445[a]	42.5	+80	90% ethanol + NH_3
Cu(malonate)	0.0273[b]	41.8	+73	90% ethanol + NH_3
Cu(succinate)	0.150[b]	28.4	+31	90% ethanol + NH_3
Cu(citrate)	0.390[b]	21.9	+11	Absolute ethanol
Cu(alaninate)	2.06[b]	6.5	−37	90% ethanol

[a]At a salt concentration of 2.5×10^{-3} M.
[b]At a salt concentration of 2.5×10^{-4} M.

of copper nitrate, or to enhance the solubility of the salt, just as in the case of copper malonate and succinate. The results are presented in Table 4.21.

Analysis of the tabulated data shows that the replacement of the anion in the coordination sphere of the copper salt changes the rate of reaction (4.36) by not more than one order of magnitude, irrespective of the type of coordination of the anion to the metal: linear (acetate), cyclic (malonate), or intercomplex (alaninate, citrate). The strength of the bond in copper citrate, and particularly in alaninate, grows significantly as a result of an increase in its covalence, as compared to copper acetate. This follows from the enormous difference in the instability constants of these salts.[328] However, all salts react with H_2Chl with the same ease. The difference in the anion structure affects the energies and entropies of activation, which diminish considerably in cases where the difference in the polarity of the initial and transition states increases (citrate → acetate → alaninate). These facts attest to reaction (4.36) proceeding with non-dissociated salt $MX_2(solv)_{n-2}$ and to the salt activation stage involving a loss of the solvent molecule rather than the anion.

If reaction (4.36) is conducted in 90% aqueous rather than absolute ethanol, its kinetic parameters are affected. In this case the coordination sphere changes from $[CuAc_2(solv)_4]$ to $Cu(H_2O)_6^{2+}$ or $[Cu(H_2O)_{6-m}solv_m]^{2+}$. The activation energy increases by 31 kcal mol^{-1}, while the entropy of activation increases by 108 e.u., but the reactivity of the salt, that is k_v^{25} (or $\Delta G^\ddagger$) remains the same. The drastic changes in entropy and energy are probably due to the hydrated copper ion having a thick solvation sheath made up of the first, second, and so on coordination spheres. In the transition state they break down as a result of the increasing values of $\Delta S^\ddagger$ and E. Ethanol solvates cannot form stable (second, third, and so on) solvation sheaths via hydrogen bonds.

Saturation of 90% aqueous ethanol with ammonia results in the water molecules being forced out of the inner coordination sphere with formation of cation $Cu(NH_3)_6^{2+}$ in the case of copper nitrate and $CuX_2(NH_3)_{n-2}$ in the case of copper malonate and succinate. The replacement of H_2O by NH_3 practically does not affect the values of E and $\Delta S^{\ddagger}$ (see Table 4.21). However, because of the increasing covalence of the Cu—NH_3 bond as compared to Cu—H_2O, the rate constant of reaction (4.36) in ammoniated ethanol is lower by one order of magnitude than in 90% aqueous ethanol. That the energy and entropy of activation remain the same is explained by the fact that the NH_3 molecules which constitute the first solvation sphere of the copper salt in 90% aqueous ethanol solution saturated with NH_3 are capable of forming the second, third, and so on coordination spheres at the expense of the water molecules. In the transition state these spheres break down with the same effect as is observed in the case of the $Cu(H_2O)_6^{2+}$ ion. Similar phenomena with some variations occur in the case of copper malonate and succinate in 90% aqueous ethanol saturated with ammonia.

A study has been made into the effect of the number of occupation of the coordination sphere of the reagent salt cation by anions.[300,302] Examined for the purpose was the reaction of formation of metal analogues of chlorophyll from chlorophyllic acid with halides of Hg^{2+}, in which the value of m varied from two to four (Table 4.22), with $Fe(CN)_6^{4-}$, $[Fe(CN)_5NO]^{2-}$, $[Co(NH_3)_5Cl]^{2+}$, and $[Co(NH_3)_4(HCOO)_2]^{+}$;

$$MX_m(\text{solv})_{n-m} + H_2Chl \rightleftarrows MChl + (n-2)X^- + 2HX(\text{solv}). \qquad (4.42)$$

It was shown that the reduction in the number of solvent molecules present in the inner coordination sphere of Hg^{2+} from two in HgX_2 to zero in HgX_4^{2-} (see Table 4.22) considerably slows down reaction (4.42) in ethanol, whereas in methanol the rates of formation in the case of salts $HgCl_2$ and $HgCl_4^{2-}$, HgI_2 and HgI_4^{2-} are close. The absence of an outgoing ligand (solv) in the coordination sphere of tetrahedral ions HgX_4^{2-} and their retention of the ability to coordinate the porphyrin molecule indicate that reaction (4.42) may involve not only detachment of two solvent molecules but also displacement of the simple ligands of reagent salt $MX_m(\text{solv})_{n-m}$ from their coordinates. The manner in which the values of E and $\Delta S^{\ddagger}$ vary

Table 4.22 Kinetic parameters of formation of mercury analogues of chlorophyll

Reagent salt	$c^0_{HgHal} \times 10^3$ (M)	$k_v^{25} \times 10^{-3}$ ($l\ mol^{-1}\ h^{-1}$)	E (kcal mol^{-1})	$\Delta S^{\ddagger}$ (e.u.)	Solvent
$HgCl_2$	0.25	28.0	13.9	−5	Ethanol
$HgCl_4^{2-}$	0.25	Very slow	—	—	Ethanol
$HgCl_2$	0.25	0.58	56.4	+101	Methanol
$HgCl_2$	0.50	0.43	48.7	+129	Methanol
$HgCl_4^{2-}$	0.25	0.67	12.7	−16	Methanol
$HgCl_4^{2-}$	2.50	0.046	11.3	−26	Methanol
HgI_2	2.50	0.23	13.4	−14	Ethanol
HgI_4^{2-}	2.50	Very slow	—	—	Ethanol
HgI_2	2.50	0.021	13.1	−22	Methanol
HgI_4^{2-}	0.25	0.068	21.8	+14	Methanol

(see Table 4.22) is fully consistent with a second mechanism being involved, which will be considered at greater length in subsequent chapters.

The predominance in the reacting system of one of these two mechanisms depends largely on the nature of the solvent. A more polar proton-donor solvent (methanol) creates more favourable conditions for the mechanism whereby anions are displaced from their coordinates than a less polar one (ethanol). However, in the case where this mechanism is out of the question for geometrical considerations and no solvating molecules are present in the inner sphere (complex salts $[Fe(CN)_6]^{4-}$, $[Co(NH_3)_5Cl]^{2+}$, $Co(NH_3)_4(HCOO^-)_2$, and the like), reaction (4.42) does not take place no matter what the solvent is.

The effect of the solvent type on the rate of reaction (4.36) and its kinetic parameters was studied in a series of works by Berezin and Smirnova.[305–312] Ethanol was selected as the reference standard and copper and zinc acetates were taken as the reagent salts. Since the variation pattern of the kinetic parameters of reaction (4.36) is the same for both salts, depending on the nature of the solvent, Table 4.23 contains data only for a copper complex.

The rate of reaction (4.36) is at a maximum in glacial acetic acid and rather fast in methanol. In ethanol and acetone the reaction is much slower. The reaction is slowest in dioxane, dimethyl formamide, quinoline, and particularly pyridine. Depending on the organic solvent the rate of the reaction may vary 3000-fold. The solvents listed in Table 4.23 are sufficient to prove that the rate of reaction (4.36) and other kinetic parameters are not at all dependent on the dielectric constant of the medium.

The effect of the solvent resides in that it forms the coordination sphere of reagent salt $MX_n(\text{solv})_{n-m}$, determines the strength of the M—X and M—solv bonds in solution, and decrees the degree of solvation of the reagents, thereby determining the value of the energy and entropy components of the free energy of activation of reaction (4.36). The stability of the coordination sphere of $MX_2(\text{solv})_{n-2}$ and the degree of solvation of

Table 4.23 Kinetic parameters of the reaction $CuAc_2 + H_2Chl \rightarrow CuChl + 2HAc$ in various solvents

Solvent	$c^0_{CuAc_2} \times 10^3$ (M)	$k_v^{25} \times 10^{-3}$ (l mol^{-1} h^{-1})	E (kcal mol^{-1})	$\Delta S^{\ddagger}$ (e.u.)
Ethanol	0.130	5.91	10.6	−26
Methanol	0.150	29.7	14.8	−4
n-Pentanol	0.240	1.99	27.1	31
Acetone	0.280	3.26	11.1	−16
Dioxane	0.300	0.165	17.2	−6
Acetic acid	0.160	95.0	11.2	−14
Dimethyl formamide	0.800	0.131	18.9	−2
Pyridine	0.900	0.0331	22.9	9
Quinoline	0.900	0.101	24.1	15

the reagents are not the only reasons for the strong dependence of the reaction rate on the solvent.

Also studied was the effect of a mixed solvent on the molecular state and reactivity of chlorophyllic acid. As was established from the absorption spectra of H_2Chla solutions,[305] its state does not change perceptibly in the solutions examined. Only in the case of pyridine and quinoline is a small bathochromic shift (5–8 nm) of the first absorption band of H_2Chla observed, which may be caused by the π–π interaction between the aromatic nuclei of heterocyclic porphyrin compounds or their effect on the cyclopentanone ring of porphyrin.

A question arises as to the effect of the solvent on the shift of the keto–enol equilibrium in the cyclopentanone ring of chlorophyllic acid (**XXVII** and **XXVIII**), resulting in changes of the kinetic parameters. Absorption spectra indicate that the equilibrium shifts toward the enol form in ethanol–glycerine (1:1) solutions and in solutions containing more glycerine:[307,312,329]

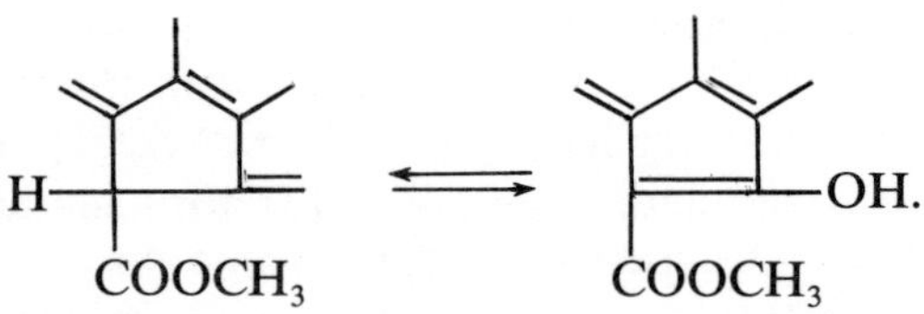

(4.43)

At a ratio of 1:9 association of H_2Chla occurs, accompanied by precipitation of the solid phase under conditions of long-term storage of the solutions.[330] The solvents listed in Table 4.23 do not produce any effect on equilibrium (4.43), which is shifted completely towards the keto form. The effect of the solvent on the electronic absorption spectra and the state of H_2Chlb is still weaker than in the case of H_2Chla.

It was shown[306–311] that the kinetic parameters are determined mainly by the strength of the coordinate bond M—solv and by the steric factor of the solvent molecule's geometry. Molecules of a cyclic structure with a coordinating heteroatom in the cycle (pyridine, quinoline, dioxane) and those exhibiting a similar coordinating capacity (dimethyl formamide) have the lowest rates and the highest energies and entropies of activation of reaction (4.36) (see Table 4.23). The coordinating O or N atom in these molecules is almost totally screened by the rest of the molecule, which prevents the large planar porphyrin molecule from entering into contact with the cation of salt $M(solv)_{n-2}X_2$.

Characteristically, the strengthening of the M—solv bond as one passes from ethanol to methanol) alone does not reduce the rate of reaction (4.36), although the activation energy perceptibly increases. The changing basicity of the solvent (dimethyl formamide → pyridine) and the associated change in the N—H···solv interaction of the pyrrole hydrogen atoms in the porphyrin molecule do not affect the reaction rate, which is also unaffected by the increasing strength of the hydrogen bond (methanol → acetic acid) between the pyrrolenine nitrogen atom and the proton of the solvent, $\geqslant$N···HO— (see Table 4.23). However, we have an entirely different picture if the solvent is not a coordinating one. The expected interaction between the aromatic rings of H_2Chl and nitrous heterocycles does not affect the ability to react with metal salts either. This follows from the same rates of reaction (4.36) in quinoline and in DMFA.

The shift of the keto–enol equilibrium in H_2Chl*a* and the higher viscosity of the medium were shown[312] not to produce any significant effect on the rate of reaction (4.36) (Table 4.24). It was assumed that the enol form has a greater coordinating capacity than the keto form of H_2Chl.

The reaction of formation of Cu^{2+} and Zn^{2+} complexes with chlorophyllic acid was studied in mixed solvents,[313–317,331] and the effect of the nature and amount of the second component of the binary solvent (based on glacial acetic acid) on the reaction rate was established (Figs 4.10–4.13). The second component is an alcohol (methanol, pentanol, ethanol, pyridine, or dioxane). The corresponding energy (activation energy versus solvent composition) and structural (activation entropy versus solvent composition) curves are presented in Figs 4.14–4.16.

All rate-versus-solvent composition curves, except for the case of binary solvent methanol–HAc, feature extrema. In the ethanol–HAc system (see Fig. 4.10) the reaction rate passes through the maximum three times—when small amounts of ethanol are added to HAc, at a percentage ratio of 50:50, and in going from a mixed solvent to pure alcohol. In the pentanol–HAc system the reaction rate grows sharply after 1% of pentanol is added, then gradually decreases without any extrema (see Fig. 4.11). In the pyridine–HAc system extrema appear when 40% of pental are added and in pure pyridine. The change in the complexing reaction rate in the dioxane–HAc system is in many respects similar to that in the pyridine–HAc system, with the difference that the reaction rate diminishes all the way to 100% dioxane.

The extrema result from such phenomena occurring in the solution as breakdown of the pure solvent structure, caused by addition of small amounts of the second component, which substantially alters the properties of the solution and the solvent, as well as resolvation of the inner coordination sphere of the reagent salt, formation of stable mixed structures from both solvent components, changes in the solvating capacity of the mixed solvent, and, as a consequence, changes in the solvation of the starting reagents and the transition state.[250,313–317,331,332]

As has already been mentioned, changes in the dielectric constant of the solvent do not alter the rate of reaction (4.36). Combination of the above phenomena results in the kinetic properties changing non-monotonically. For example, in the ethanol–HAc system the structure of pure HAc breaks down when small amounts (1–10%) of ethanol are added. As a result the solvating capacity of the medium dramatically increases, which is indicated by the diminishing values of $\Delta S^{\ddagger}$ and E (see Fig. 4.14). In this system an

Table 4.24 Kinetic parameters of formation of copper complexes of chlorophyllic acid in the ethanol–glycerine system

Glycerine: ethanol (%)	$c^0_{H_2Chl} \times 10^5$ (M)	$c^0_{CuAc_2} \times 10^4$ (M)	$c^0_{CuAc_2}/c^0_{H_2Chl}$	$k_v^{25} \times 10^3$ (l mol^{-1} h^{-1})	E (kcal mol^{-1})	$\Delta S^\ddagger$ (e.u.)
0:100	1.30	5.00	38	3.42	10.6 ± 0.3	−22.7 ± 1.0
20:80	1.30	5.00	38	1.81	10.8 ± 0.4	−23.4 ± 1.4
50:50	1.30	5.00	38	1.46	11.9 ± 1.3	−20.0 ± 4.3
80:20	1.30	1.30	10	0.44	12.8 ± 0.2	−19.4 ± 0.6

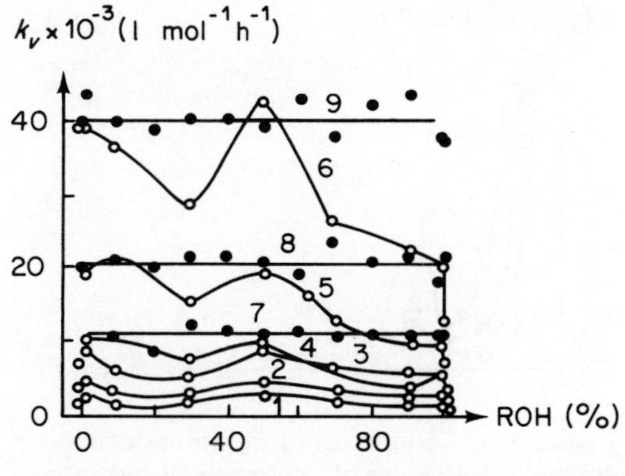

Figure 4.10 Rates of formation of the copper analogue of chlorophyll in the systems ethanol–HAc (curves 1–6) and methanol–HAc (curves 7–9). Temperature: curves 1, 4, 7, 15 °C; curves 2, 5, 8, 25 °C; curves 3, 6, 9, 35 °C, $CuAc_2$ concentration: curves 1–3, 2.5×10^{-3} M; curves 4–6, 1.5×10^{-4} M; curves 7–9, 2.5×10^{-4} M

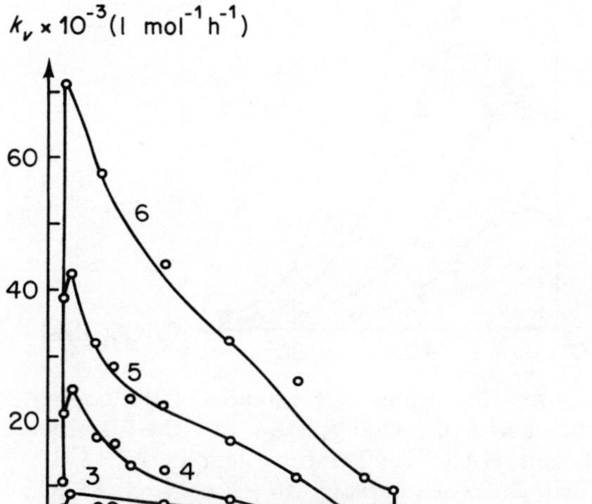

Figure 4.11 Rates of formation of the copper analogue of chlorophyll in the system pentanol–HAc. Temperature: curves 1, 4, 15 °C; curves 2, 5, 25 °C; curves 3, 6, 35 °C. $CuAc_2$ concentration: curves 1–3, 2.5×10^{-3} M; curves 4–6, 2.5×10^{-4} M.

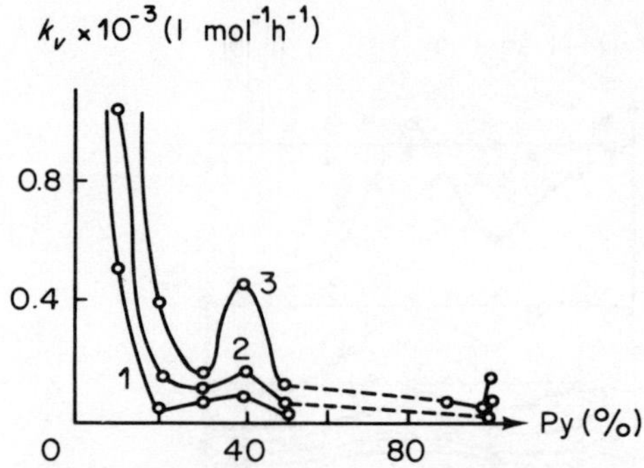

Figure 4.12 Rates of formation of the copper analogue of chlorophyll in the system pyridine–HAc. Temperature: curve 1, 25 °C; curve 2, 35 °C; curve 3, 45 °C. $CuAc_2$ concentration: 2.5×10^{-3} M

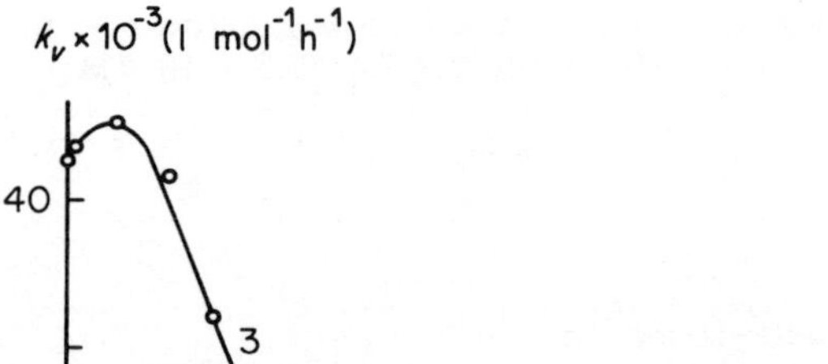

Figure 4.13 Rates of formation of the copper analogue of chlorophyll in the system dioxan–HAc. Temperature: curve 1, 15 °C; curve 2, 25 °C; curve 3, 35 °C; curve 4, 45 °C. $CuAc_2$ concentration: 2.5×10^{-4} M

extremum appears at 5–10% of ethanol, which may be caused by partial resolvation of the first coordination sphere of $[Cu(HAc)_{n-2}Ac_2]$ with formation of $[Cu(HAc)_{n-2-m}\cdot(C_2H_5OH)_mAc_2]$ particles that are possibly more reactive. In the case of the pentanol–HAc system the effect of breakdown of the structure is highly pronounced. Addition of 1% of

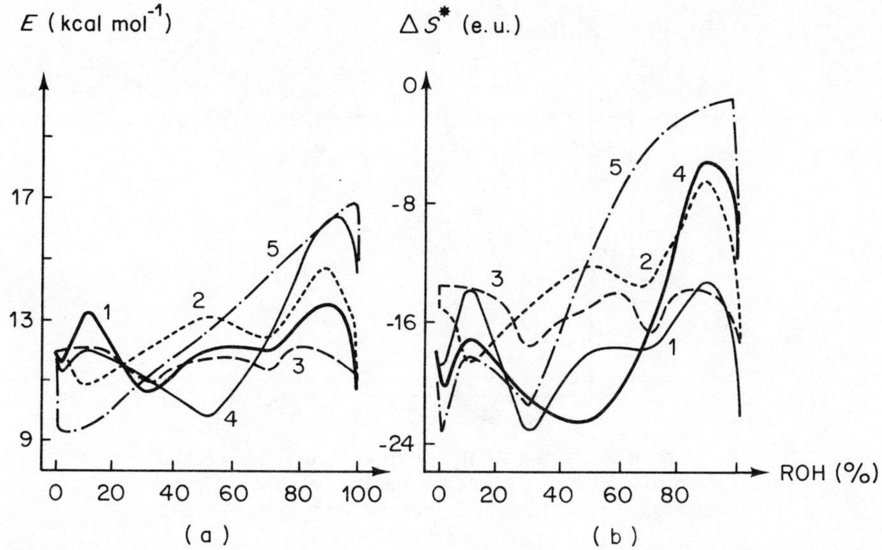

Figure 4.14 Activation energy (a) and entropy (b) of formation of the copper analogue of chlorophyll versus solvent composition. Curves 1, 2, ethanol–HAc; curve 3, methanol–HAc; curves 4, 5, pentanol–HAc. $CuAc_2$ concentration: curves 1, 4, 2.5×10^{-3} M; curves 2, 3, 5, 2.5×10^{-4} M

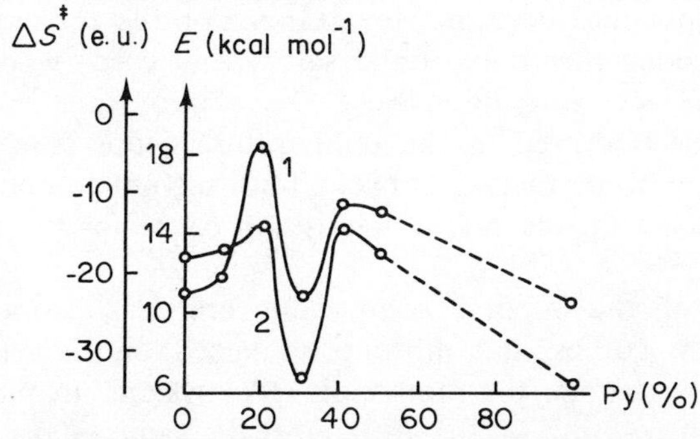

Figure 4.15 Activation energy (curve 1) and entropy (curve 2) of formation of the copper analogue of chlorophyll in a mixed Py–HAc solvent

pentanol doubles the reaction rate. When such negligibly small amounts are added, no rearrangement of the inner sphere of the copper salt may take place.

The curves presented indicate that the effect of breakdown of the solvent structure depends on the reagent salt concentration. It is pronounced at a concentration of 2.5×10^{-4} M and in more dilute solutions of the copper salt when the properties of the solution approach those of an ideal one. At salt concentrations of 3.5×10^{-3} M and above the salt changes the solvent

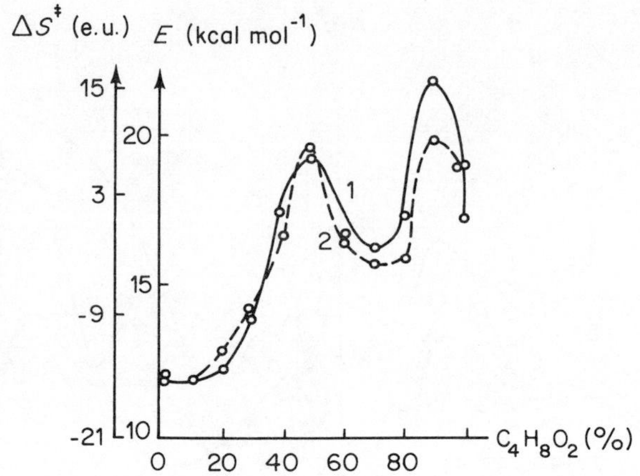

Figure 4.16 Activation energy (curve 1) and entropy (curve 2) of formation of the copper analogue of chlorophyll in a mixed dioxan–HAc solvent

structure to such an extent that the effect of its breakdown as a result of addition of the second component on the composition–rate curves is in most cases insignificant.

High salt concentrations overshadow all of the above-mentioned effects of the second component; therefore, the extrema on the curves stand out more sharply at low concentrations of the salt. In all cases where the solution components interact more intensively, the extrema are smoothed out. A similar picture is observed as the temperature drops (see Fig. 4.10) with increasing energy of interaction between both solvent components (note the absence of extrema on the rate–composition curve for the methanol–HAc system).

The effects of the second component are less pronounced on the rate–composition curves and more pronounced on the structure–energy ones. Take, for example, the methanol–HAc system. In pure solvents the rate constants of reaction (4.36) are almost the same, and the energies and entropies of activation are close in value. Nevertheless, on the structure–energy curves (see Fig. 4.14) one can see the same but less clearly defined extrema as in the ethanol–HAc system when the salt concentration in the latter is 2.5×10^{-3} M.

The influence of the salt concentration on the structure of pure solvents depends on the salt species and can be clearly seen on the rate–composition curves. The structure of HAc is so changed by the presence of salt $CuAc_2$ at a concentration of 2.5×10^{-3} M and more that it becomes almost insensitive to addition of 1% of pentanol (see Fig. 4.11). Pure pentanol is very sensitive to addition of salts, which is why even at a $CuAc_2$ concentration of 2.5×10^{-4} M it practically does not respond to small (1%) amounts of added HAc. Pure pyridine is, on the contrary, less sensitive to addition of

2.5×10^{-3} M $CuAc_2$, for under such conditions its structure and properties change significantly as a result of addition of 1% HAc. In this case the rate constant becomes two or three times as low, the activation energy becomes twice as low, and the entropy is reduced by 36 e.u.[317]

The kinetic method based on reaction (4.36), which is highly sensitive to the solvent structure, may be employed as an effective means for rapid and reliable qualitative analysis of different aspects of structural stability of pure and mixed solvents, formation of new structures, resolvation of the reagents, and similar phenomena. These changes in solutions are manifest both on the rate–solvent composition and structure–energy curves. The $\Delta S^{\ddagger}$–composition curves permit tracing changes in the solvating capacity of the solvent. Owing to the clearly defined compensation effect in reactions (4.36) and (4.42) these curves are almost identical with the E–solvent composition ones (see Figs 4.14–4.16).

In recent years new concepts, such as 'structural stability',[333] have started to be used for interpretation of the processes and phenomena occurring in solutions. The structural $\Delta S^{\ddagger}$–solvent composition curves may become instrumental in estimating the relative stability of structures. Obviously stabilization must lead to a higher value of $\Delta S^{\ddagger}$ for reactions (4.36), while destabilization, or breakdown of the solvent structure, must result in a lower value of $\Delta S^{\ddagger}$, that is in higher solvating capacity of the solvent. Changes in the solvent structure manifest themselves not only in the kinetic but also in other characteristics of the solution, provided that the dissolved substances or products of their interaction are sensitive to solvation.[252–254] These changes must be taken into account in studying and analysing all reactions and processes occurring in solutions. Unfortunately, in most works on porphyrins published outside the USSR, the effects of the solvent are neglected.

In discussing the effect of the solvent type on reaction (4.36) one cannot ignore such an important factor as the supporting electrolyte. The supporting electrolyte alters the structure and solvating capacity of the solvent, as well as influencing the coordination sphere of the salt. It may be coordinating or non-coordinating (inert). A coordinating supporting electrolyte changes the structure of the coordination sphere of reagent salt $M(solv)_{n-2}X_2$ and shifts the equilibrium between various coordination forms (e.g. tetrahedron $\rightleftarrows$ octahedron, *cis*-octahedron $\rightleftarrows$ *trans*-octahedron), thereby promoting the ligand exchange of the X^- ions and solvent. An inert supporting electrolyte enables these changes to occur only via the solvent, altering its structure and the molecular linkage. The supporting electrolyte permits one to stabilize particular coordination forms of the reagent salt.[301,334] The effect of supporting electrolytes on reaction (4.36) is illustrated by Table 4.25, which presents the parameters of formation of a chlorophyll analogue with Co^{2+} in alcohols.[334]

The introduction into the reacting system $H_2Chl + Co(NO_3)_2$ of an inert supporting electrolyte $LiNO_3$ does not bring about any changes in the

Table 4.25 Kinetic parameters of formation of chlorophyllic acid complex with cobalt in alcohols

Reagent salt	Supporting electrolyte	Solvent	k_v^{25} ($l\ mol^{-1}\ h^{-1}$)	E ($kcal\ mol^{-1}$)	$\Delta S^{\ddagger}$ (e.u.)
$Co(NO_3)_2$	—	Methanol	9.2	10.3	−39
	$LiNO_3$	Methanol	86.0	13.4	−21
	—	*n*-Pentanol	18.8	20.0	−1
	$LiNO_3$	*n*-Pentanol	38.3	18.6	−5
$CoCl_2$	—	Methanol	6.4	15.5	−24
	LiCl	Methanol	17.6	12.3	−27
	—	*n*-Pentanol	30.3	12.9	−24
	LiCl	*n*-Pentanol	6.1	8.4	−42
$CoAc_2$	—	Methanol	47.6	23.7	+12
	NaAc	Methanol	4.3	—	—
	—	*n*Pentanol	1.2	25.0	+10
	LiAc	*n*-Pentanol	4.2	12.0	−31

Note: The concentration of the reagent salt is 2.5×10^{-3} M and that of the supporting electrolyte is 1.31–3.28 M. The dash indicates that the latter is absent.

octahedral structure of reagent salt $[Co(solv)_6]^{2+}$ in methanol and $[Co(solv)_4(NO_3)_2]$ in pentanol, as can be seen from the electronic spectra. However, the reaction rate increases two- to 10-fold. Therewith, small energy and structural changes in methanol (on the increasing side) and in pentanol (on the decreasing side) are opposite. Hence $LiNO_3$ primarily affects the solvent structure. It was shown that up to a concentration of 0.1 M $LiNO_3$ the effect of the supporting electrolyte is not noticeable in the kinetics of reaction (4.36). At high concentrations $LiNO_3$ affects the linkage of the solvent molecules. If in methanol it is stronger while the solvating capacity decreases (with the result that $\Delta S^{\ddagger}$ and E increase, too), in pentanol the effect of the supporting electrolyte is different (both kinetic parameters slightly increase).

The effect of the coordinating supporting electrolyte LiCl stabilizes the tetrahedral form of the salt, shifting the equilibrium $CoCl_2(solv)_4 \rightleftarrows CoCl_4^{2-}$ to the right. This is corroborated by absorption spectra.[334] As a result of stabilization of the salt the concentration dependence of k_v in methanol decreases approximately three times. In pentanol $CoCl_2$ yields a stable form predominant in the solution, probably $CoCl_2(solv)_2$, therefore all kinetic parameters at $CoCl_2$ concentrations of 2.5×10^{-3} and 2.5×10^{-2} M are practically the same.[301] Judging from the absorption spectra, LiCl in pentanol promotes formation of $CoCl_4^{2-}$ predominantly. However, LiCl also affects the solvent structure. Pentanol is more susceptible to LiCl than $CoCl_4^{2-}$. LiCl doubles the reaction rate in methanol owing to E going down for the most part ($\Delta S^{\ddagger}$ decreases insignificantly); in pentanol, on the other hand, the reaction rate is reduced five times because of the predominant decrease in $\Delta S^{\ddagger}$ (E decreases by 4.5 $kcal\ mol^{-1}$).

Thus pentanol saturated with LiCl or $LiNO_3$ is a more solvating solvent than pure ethanol. It behaves similarly in the presence of the coordinating supporting electrolyte LiAc. In the latter case the rate of formation of CoChl increases approximately 3.5-fold, E decreases by 13 kcal mol^{-1}, and $\Delta S^{\ddagger}$ decreases by 41 e.u. (see Table 4.25).

In methanol reaction (4.36) with supporting electrolyte LiAc proceeds at a 10-fold slower rate. LiAc in methanol is probably conducive to an equilibrium shift toward the tetrahedral form of cobalt acetate with the result that the strong concentration dependence of constant k_v in methanol is completely eliminated[301] by its addition. In pentanol $CoAc_2$ prevails in a tetrahedral form even without the supporting electrolyte, and additions of LiAc do not alter the spectrum.[334]

These data clearly suggest a highly pronounced specific effect of the nature of the solvent on the stabilization of individual coordination forms of the salt in the presence of a supporting electrolyte, which, as will be shown later, may change the reaction mechanism. It is assumed[322] that in methanol the conditions under which reaction (4.42) proceeds in accordance with the ligand shift mechanism are more favourable than in pentanol. A large ligand (pentanol) and a less polar medium favour, in the latter case, a mechanism involving detachment of two solvent molecules from the coordination sphere.

As was shown earlier,[334] different salts of the same reacting cation, for example cobalt, are kinetically dissimilar in solvents of the same nature. A different cation of the reagent salt at the same anionic composition of the medium may also alter the trend of the kinetic parameters of reaction (4.36), depending on the solvent. For example, the rates of formation of complexes CuChl and ZnChl from their acetates in reaction (4.36) in various solvents decrease as follows: pentanol $<$ethanol $<$ methanol. In the case of CoChl we have: pentanol $=$ ethanol $<$ methanol (cobalt acetate), methanol $=$ ethanol $<$ pentanol (cobalt nitrate), ethanol $<$ methanol $<$ pentanol (cobalt chloride). Pentanol is found to be the most favourable solvent for synthesis of a metal analogue of chlorophyll with cobalt.

The literature contains scant data on the effect of the solvent on the reaction of formation of metal analogues of chlorophyll, depending on the metal and the salts it forms, and whatever data are available pertain, mainly, to salts of Cu^{2+}, Zn^{2+}, and Co^{2+}.

Unfortunately the appropriate conditions for formation of chlorophyll from H_2Chl and Mg^{2+} salts in solutions in reaction (4.36) are yet to be established. In this chapter the reaction will be described in the context of structural analogues of chlorophyll.

Finally, I should like to dwell upon different aspects of formation of metal analogues of chlorophyll *b* in solution. All the previous kinetic data were given for metal analogues of chlorophyll *a*.

The complexing of Cu^{2+} and Zn^{2+} salts with H_2Chlb was studied in ethanol[245] and in glacial acetic acid.[247] The kinetic parameters of the

Table 4.26 Kinetic parameters of formation of complexes of chlorophyllic acids *a* and *b* under comparable conditions

Complex	$k_v \times 10^{-2}$ ($l\ mol^{-1}\ h^{-1}$)	E (kcal mol^{-1})	$\Delta S^{\ddagger}$ (e.u.)
Ethanol			
CuChl*b*	1.80	14.0	−17
CuChl*a*	12.9	14.5	−12
Glacial acetic acid			
CuChl*b*	23.4	18.7	+4
CuChl*a*	191	10.5	−19
ZnChl*b*	1.25	15.2	−14
ZnChl*a*	133	11.6	−17

reactions yielding CuChl*b* and ZnChl*b* are summarized in Table 4.26. Chlorophyllic acid of the *b* series is less likely to coordinate to metal salts than that of the *a* series. Complex CuChl*b* in ethanol (the salt-to-H_2Chl ratio being 10:1) forms at a rate seven times slower and eight times slower in glacial acetic acid than complex CuChl*a*.[245] The difference in the reaction rates is still greater in the case of zinc acetate, which reacts 106 times more slowly with H_2Chl*b* than with H_2Chl*a*. The differences in energy and entropy between the two chlorophyllic acids are determined by the solvent. In ethanol this difference is smaller, and the rate of reaction (4.36) decreases in the case of H_2Chl*b* due to the decrease in entropy, that is due to a greater degree of solvation in the transition state. In glacial acetic acid, on the other hand, the entropies of activation increase in the case of H_2Chl*b* but the predominant increase in the activation energy (see Table 4.26) slows down the coordination reaction.

The different behaviours of such closely related porphyrin molecules as chlorophyllic acids *a* **(XXVII)** and *b* **(XXVIII)** is not surprising. As was shown in the beginning of this chapter, the *b* acid dissociates less readily than the *a* acid because of the lesser degree of protonation of the N—H bond as a result of the higher electron density at the N atoms (+C effect of the formyl group). The lesser degree of protonation, that is the greater covalence of the N—H bond, seems to be the main reason for the lesser tendency of H_2Chl*b* to coordinate to metals. The increase in the electron density at the pyrrolenine nitrogen atoms (⋛N) does not affect the coordination rate for the M···N coordination and is not a decisive factor at the activation stage. The weak dependence of the rates of formation of metal analogues on the coordinating capacity of the metal (see Table 4.20) proves this point quite well. The wide differences in energy, involved in the formation of complexes of both chlorophyllic acids in glacial acetic acid, are probably due to the greater degree of specific solvation of the nitrogen atoms in H_2Chl*b*. The overall effect of desolvation of these atoms and stretching of the more covalent N—H bond in H_2Chl*b* is responsible for the greater values of activation energies as well as entropies of activation.

4.7 Kinetics of Formation of Haemin and Its Analogues in Solution

Porphyrin of the blood, or protoporphyrin DME (**XXVI**), enters into a reaction with $FeCl_2$ yielding haemin as a result of the Fe^{2+} complex known as protohaeme, which forms in the reaction

$$H_2PP + MX_2(solv)_{n-2} \rightarrow MPP + 2HX + (n-2)solv, \tag{4.44}$$

immediately oxidizing to the triply charged iron complex (Cl)FePP known as haemin. The kinetics of the reaction between Fe^{2+} salts and protoporphyrin have not been studied.

Salt $FeCl_3$ reacts with protoporphyrin or its water-soluble amide derivatives, yielding haemin.[75] Studies of the reaction in acetone and chloroform have shown that an intermediate unstable complex is formed first, having structure **XLVI** in which only the reactive site of porphyrin, including four nitrogen atoms, can be seen.

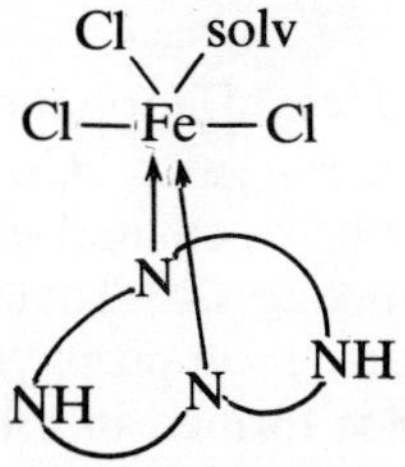

XLVI

This intermediate, called 'sitting-atop' (SAPP) by Fleischer and Wang,[75] is a mixed complex of trivalent iron, less stable than solvate complexes with water, alcohols, pyridine, and other polar molecules. Addition of a small amount of any one of these solvents to the solution of the intermediate immediately destroys it, and spectra reveal only H_2PP.

Hence protoporphyrin may form an intermediate only in weakly polar or weakly coordinating solvents. This factor is most essential in understanding the mechanism of the metalloporphyrin-yielding reactions, which will be discussed in this chapter. Fleischer and Wang[75] observed the formation of intermediates of type **XLVI** in non-polar media with salts of Fe^{2+}, Fe^{3+}, Co^{2+}, Zn^{2+}, Sn^{2+}, Pt^{IV}, and other metals. No protoporphyrin intermediates with salts of Ni^{2+}, Cu^{2+}, Hg^{2+}, and Cd^{2+} have been detected as they yield metalloporphyrins directly in reaction (4.44).

In cases where intermediates are formed, heating of the solutions to a temperature above 60 °C gradually produces true stable protoporphyrin complexes. In the intermediate examined in detail by Fleischer and Wang[75] the N—H bonds are considerably stretched, which is why the frequencies at 3.05 and 6.15 μm are shifted to 3.11 and 6.64 μm, respectively.

The stability constant of intermediate $FeCl_3$-SAPP forming in the reaction

$$FeCl_3 + H_2PP \rightleftarrows FeCl_3 \cdot H_2PP \tag{4.45}$$

was found to equal $(3.9 \pm 0.2) \times 10^3$ in chloroform plus 2.5% of methanol.

It becomes quite evident that this constant will be much greater in a pure non-polar solvent. In this case the formation of the intermediate results in almost complete bonding of porphyrin, and because of this reaction (4.44) practically stops at room temperature.

Fleischer and Wang[75] had determined the composition of the intermediate ($FeCl_3$:H_2PP = 1:1) and showed that its electronic absorption spectrum is similar to that of a diprotonated $H_2PP_2^{2+}$ particle. All these properties corroborate structure **XLVI**. A detailed thermodynamic study of the $FeCl_3$–H_2PP system in chloroform with added amounts of methanol has shown that addition of methanol in an amount of 1% and less yields a binuclear intermediate by the reaction

$$Fe_2Cl_6 + 2H_2PP \rightleftarrows Fe_2Cl_6 \cdot 2H_2PP, \tag{4.46}$$

with a stability constant of 1.4×10^8. Heating of the intermediate-containing solutions leads to formation of haemin. The kinetics of haemin formation are yet to be studied.

The literature contains the rates of formation of metal analogues of haemin—protoporphyrin complexes with Cu^{2+}, Zn^{2+}, and Cd^{2+}—in ethanol,[335] glacial acetic acid,[336] binary ethanol–benzene mixtures,[337,338] and *tert*-butanol–trichloracetic acid mixtures.[339] Table 4.27 gives the formation rate constants of copper and zinc protoporphyrin complexes in absolute ethanol. Copper protoporphyrin is formed in ethanol in the presence of a five- to 25-fold excess of $CuAc_2$ at a rate approximately three times faster than complex CuChl*a*. In this case k_v is not dependent on concentration.

Zinc protoporphyrin forms in ethanol at rate constant $k_v^{25} = 1.65 \times 10^3$ l mol^{-1} h^{-1}, activation energy $E = 9.2$ kcal mol^{-1}, and entropy $\Delta S^{\ddagger} = -28$ e.u. Thus the formation of ZnPP, just as that of CuPP, involves lower energy and entropy of activation as compared to H_2Chl*a* complexes. The most probable reason for the sharp decrease in the values of E and $\Delta S^{\ddagger}$ is the absence of the long phytol residue that is present in chlorophyllic acid and is a source of considerable steric hindrance to coordination.

Cadmium protoporphyrin is formed in ethanol, at 25 °C, at a rate constant of 5.3×10^3 l mol^{-1} h^{-1}, $E = 13.7$ kcal mol^{-1} and $\Delta S^{\ddagger} = -14$ e.u.,[338] that is at a rate four times faster than its analogue with H_2Chl*a* under similar conditions but with roughly the same activation parameters.

The formation of metal protoporphyrins in glacial acetic acid is much faster than in ethanol. For example, in the case of ZnPP, constant $k_v^{25} = 7.5 \times 10^5$ l mol^{-1} h^{-1}, that is 450 times greater than in ethanol.† From the available data on the formation of metal analogues of haemin one can predict the rate of formation of haemin itself from $FeCl_2$ and H_2PP—it must exceed that of a chlorophyll analogue by one to two orders of magnitude.

†In all metalloporphyrin-yielding reactions metal acetates are used unless otherwise specified.

Table 4.27 Rates of formation of copper and zinc protoporphyrin complexes in ethanol

t (°C)	$c^0_{CuAc_2} \times 10^3$ (M)	k^{25}_{eff} (h^{-1})	$k^{25}_v \times 10^{-3}$ (l mol^{-1} h^{-1})	$c^0_{ZnAc_2} \times 10^3$ (M)	k^{25}_{eff} (h^{-1})	$k^{25}_v \times 10^{-3}$ (l mol^{-1} h^{-1})
25	0.25	0.912 ± 0.09	3.6 ± 0.36	0.50	0.825 ± 0.02	1.65 ± 0.04
25	0.625	2.10 ± 0.25	3.36 ± 0.40	1.25	2.65 ± 0.10	2.12 ± 0.08
25	1.25	3.42 ± 0.33	2.74 ± 0.26			
35	0.25	1.20 ± 0.13	4.80 ± 0.50	0.50	1.45 ± 0.05	2.90 ± 0.10
35	0.625	3.42 ± 0.25	5.50 ± 0.40	1.25	4.56 ± 0.30	3.65 ± 0.24
35	1.25	5.85 ± 0.32	4.68 ± 0.25			
45	0.25	2.13 ± 0.08	8.53 ± 0.32	0.50	2.15 ± 0.10	4.3 ± 0.20
45	0.625	5.4 ± 0.3	8.65 ± 0.48	1.25	7.0 ± 0.6	5.6 ± 0.5
45	1.25	11.1 ± 0.8	8.90 ± 0.64			

Note: Protoporphyrin concentration is 5×10^{-5} M.

4.8 Reactivity of Other Porphyrins in Coordination Reactions

Over the past two decades a great deal of interest has been aroused in the kinetics of formation and other properties of metal complexes with other porphyrins. The sufficiently high stability of porphyrin complexes and the limitless possibilities of varying the molecular structure by functional substitution make metalloporphyrins one of the most promising classes of compounds from the point of view of development of industrial and biological catalysts on their basis. Particularly promising is combination of metalloporphyrins into extra complexes with polypeptides and proteins. Such complexes have been synthesized and studied in recent years.[26,340]

The specific structural features of a porphyrin molecule manifest themselves primarily in the rates and mechanism of formation of metalloporphyrins. In this connection it becomes a task of paramount importance to accumulate data on the rates and activation parameters of coordination reactions involving a wide range of metals with major porphyrin molecules, both natural and synthetic. Unfortunately no such data are available for most porphyrins except for chlorophyllic acid *a*.

Fleischer *et al.*[341,342] have examined the formation of complexes of *meso*-tetrapyridylporphin with salts of Cu^{2+}, Co^{2+}, Ni^{2+}, and Mn^{2+} in glacial and aqueous acetic acids. Tetrapyridylporphin is structurally close to tetraphenylporphin (**XXX**). The difference between the two is that in the former pyridine residues replace the four phenyl residues in the meso position. This is why H_2TPP and its complex salts become soluble in aqueous acid solutions. Fleischer *et al.*[341] have studied the rate of the complexing reaction

$$H_2TPP + M^{2+}_{solv} \rightarrow MTPP + 2H^{+}_{solv}. \qquad (4.47)$$

The reaction was found to be of first order with respect to porphyrin and the metal. Unfortunately the reaction rate was studied within narrow salt concentration limits (*c*. 4×10^{-3} M). An unfounded assumption was made that in glacial acetic acid and its aqueous solutions reagent salt $CuAc_2$ is in the form of the Cu^{2+} ion.

The rate constants of reaction (4.47) calculated from a second-order equation at 25 °C are equal to 1.40×10^3, 0.31×10^3, 0.021×10^3, and 0.087×10^3 $l\,mol^{-1}\,h^{-1}$, respectively, for $CuAc_2$, $CoAc_2$, $MnAc_2$, and $NiAc_2$ in acetic acid.

The reaction is much slower in glacial acetic acid than with chlorophyllic acid *a*, which can be seen from Table 4.23. It is important that the rate of reaction (4.47) yielding tetrapyridylporphin complexes with Cu^{2+} and Zn^{2+} from perchlorates in 8.3% aqueous acetic acid with 0.01 M $HClO_4$ is characterized by rate constants of 0.63×10^3 and 0.022×10^3 $l\,mol^{-1}\,h^{-1}$. Under these conditions H_2TPP is protonated at the pyridine residues yielding $C_5H_4\overset{+}{N}H$ fragments which make porphyrin soluble in water. This acidity is not sufficient for protonation of the reactive site of its molecule, therefore the reaction rate decreases only two to three times as a result of H_2O substituting for HAc in the coordination sphere of the salt cation.

It was found[341,342] that the rate of reaction (4.47) in the pH range 1.44–2.16 varies inversely with the proton concentration, and in H_2O–HAc mixtures it passes through a maximum in a 90% acetic acid solution and through a minimum in a 50% acetic acid solution. These facts could not be explained earlier[341] because of the lack of data on the state of the porphyrin, salt, and solvent under these conditions.

The concepts set forth in this chapter regarding the reactions yielding metal analogues of chlorophyll provide an easy explanation to the above facts. This will be done in part a bit later. At this point it should be noted that the appearance of extrema (Fig. 4.17) on the rate–HAc–H_2O binary solvent composition curve in the neighbourhood of 90% aqueous HAc is due either to protonation of H_2TPP at the pyridine nuclei, which leads to polarization of the N—H bonds of porphyrin and promotes the reaction, or to breakdown of the acetic acid structure and the resulting greater solvating capacity of the medium. Both assumptions are equally possible, and both factors are probably involved.

In the region of 90–50% HAc the reaction rate diminishes as a result of rearrangement of the coordination sphere of the copper salt from $[Cu(HAc)_4Ac_2]$ to $[Cu(H_2O)_4Ac_2]$, whereby it is substantially stabilized. The water molecule is much more difficult to remove than the weakly coordinating acetic acid molecule. Nor could Fleischer *et al.*[341,342] explain why H_2TPP does not enter into a coordination reaction with salts of Cd^{2+}, Pb^{2+}, and Hg^{2+}. The impossibility of this coordination in the selected proton-donor media[341] stems from the extremely high lability (see Chapter 5) of complexes HgP, CdP, and PbP.

A most interesting communication was presented by Baum and Plane[65] concerning direct introduction of Mg^{2+} ions into deuteroporphyrin DME **(XLVII)** from a solution of Mg^{2+} salts in ordinary organic solvents (acetone, methanol, etc.).

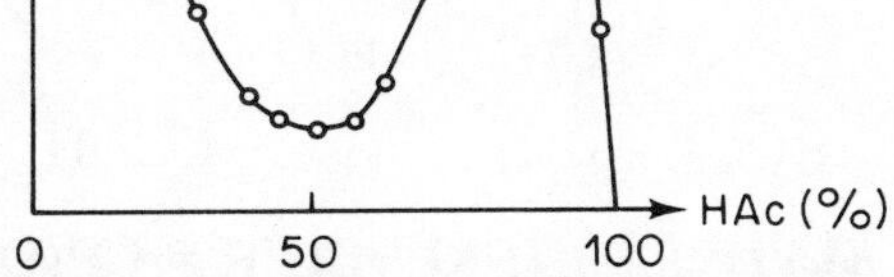

Figure 4.17 Rate of formation of copper tetrapyridylporphin versus HAc concentration

The rate of incorporation of cation Mg^{2+} from its perchlorate in acetone into deuteroporphyrin in the presence of catalysing pyridine additives is described by the following kinetic equation:

$$d[MP]/dt = k_1([H_2P]/[H^+]) + k_2[H_2P][Mg^{2+}][Py]/[H^+]), \qquad (4.48)$$

in which $k_1 = 4.7 \times 10^{-4}\ \mathrm{l\,mol^{-1}\,h^{-1}}$ corresponds to a non-catalytic reaction and $k_2 = 3.6\ \mathrm{l\,mol^{-1}\,h^{-1}}$ corresponds to a reaction catalysed by pyridine and other nitrogen-containing solvents. The activation energy of the non-catalytic reaction is 36.5 kcal $\mathrm{mol^{-1}}$, and that of the catalytic reaction is 15.8 kcal $\mathrm{mol^{-1}}$.

The catalytic action of pyridine is due to the replacement of the oxygen-containing molecules stably coordinated to magnesium in the coordination sphere of reagent salt $Mg(solv)_6^{2+}$ by weakly coordinated pyridine molecules. Unfortunately no entropy data have been produced. It may be assumed though that the reaction yielding predominantly ionic complex MgDP under catalytic conditions involves easy detachment of pyridine molecules from the coordination sphere of magnesium. The catalytic action is so strong that the rate of incorporation of magnesium into deuteroporphyrin approaches those of coordination of salts $CoCl_2$ and $CoAc_2$ to chlorophyllic acid in ethanol solutions (see Table 4.20). The data presented by Baum and Plane[65] give every reason to believe that the Mg^{2+} ion can be incorporated directly into chlorophyllic acid and provide a new insight into the formation of chlorophyll[3] in the living cell.

In contrast to most work published outside the USSR on the kinetics of formation of metalloporphyrins, in which no attention is paid to the entropy characteristics of the reactions, in their work[343] Longo *et al.* summarized valuable data on the kinetic parameters of formation of complexes of various metals with haematoporphyrin **(XLVIII)**, tetrapyridylporphin, tetraphenylporphin, and its numerous substituents in the benzene rings.

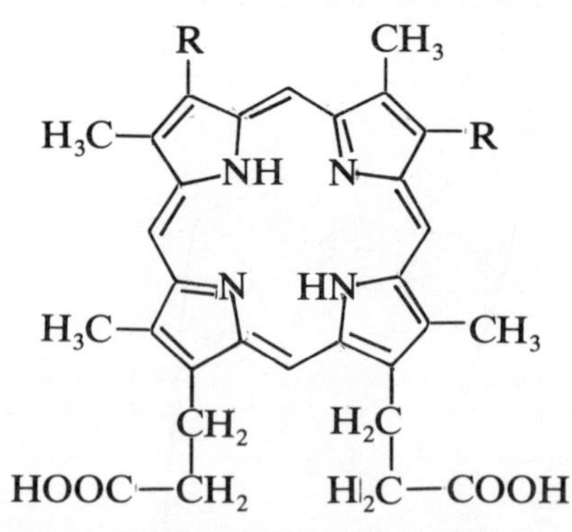

XLVII: R = H; **XLVIII**: R = $CH(OH)CH_3$

Let us analyse the data and findings of this work more closely. Of particular interest is the reaction yielding an analogue of protohaeme, which is a complex of Fe^{2+} with haematoporphyrin (**XLVIII**). It proceeds in accordance with the following equation, written schematically by Longo *et al.*:[343]

$$MX_2 + H_2P \rightleftarrows MP + 2HX, \quad (4.49)$$

the activation energy being 15.2 kcal mol^{-1}, the negative entropy of activation being −12 e.u., and the reaction rate being 5.5×10^3 l mol^{-1} h^{-1}. These parameters are given for salt $FeAc_2$ in glacial acetic acid and are practically the same as those of formation of the Fe^{2+} complex with chlorophyll *a* (see Table 4.20). Under similar conditions the complexes of haematoporphyrin with ions Mn^{2+} and Ni^{2+} are characterized by the following parameters:

	k_v^{50} (l mol^{-1} h^{-1})	E (kcal mol^{-1})	$\Delta S^{\ddagger}$ (e.u.)
MnHP	9×10^2	15.7	−12
NiHP	58.0	14.3	−20

Low rates of formation are typical of the nickel complex. Longo *et al.*[343] leave this fact without explanation. Interestingly, in the case of tetrapyridylporphin an entirely different picture is observed: in glacial acetic acid complex NiTPP forms, at 25 °C, at a rate of 83 l mol^{-1} h^{-1}, whereas the rate of formation of MnTPP is 21 l mol^{-1} h^{-1}. Hence the determining effect of the porphyrin molecule structure on the order and rates of formation of metalloporphyrins.

Slight variations in the geometry and electronic structure of the main porphyrin coordinating centre and even another molecular fragment may produce drastic changes in the kinetics. This is vividly illustrated by the data of Table 4.28, obtained from a reaction between salt $CuCl_2$ and tetraphenylporphin, its *o*-, *m*- and *n*-phenyl substituents, porphin, and etioporphyrin, in DMFA at 54 °C.[343]

In the series of molecules of tetraphenylporphin derivatives it is the state of their peripheral fragment, namely the phenyl residue, that changes and not that of the coordination centre. As a rule, an alkyl substituent occupying the *n*- and *m*-positions (a total of four substituents according to the number of phenyl residues) practically does not affect the reaction rates since the state of the porphyrin nitrogens and N—H bonds does not change. The introduction of electron-acceptor substituents (*p*-CN, *p*-NO_2) slows down the reaction four times. This has nothing to do with the changing state of the nitrogen atoms due to a C effect for the phenyl groups are not conjugated with the macroring. Nor has this anything to do with the −I effect of the above groups, bearing in mind that the latter effect would have reduced the degree of covalence of the N—H bond and sped up reaction (4.49).

Table 4.28 Kinetic parameters of formation of copper metalloporphyrins in DMFA at 54 °C

Porphyrin	E (kcal mol^{-1})	$\Delta S^{\ddagger}$ (e.u.)	$\Delta G^{\ddagger}$ (kcal mol^{-1})
Tetraphenylporphin	16.5	−10	19.6
Substituted tetraphenylporphins			
p-Cl	14.7	−15.6	19.8
p-CN	12.0	−24.6	20.3
p-NO_2	13.5	−20.4	20.2
p-OH	8.5	−30.6	18.8
p-OCH_3	14.7	−12.2	18.7
o-OC_2H_5	23.8	11.2	20.2
p-CH_3	12.1	−22.6	19.5
m-CH_3	15.8	−11.9	19.7
o-CH_3	24.9	13.1	20.6
Porphin	19.6	−2.6	19.8
Etioporphyrin	16.6	−8.6	19.4

The entropy of activation decreasing by 10–15 e.u. is indicative of a substantial increase in the transition state solvation. The activation energy also decreases in this case. The introduction of strong electron-donor substituents (OH, OCH_3), on the other hand, makes the reaction 10 times faster. In this case, the OCH_3 group alters the value of $\Delta S^{\ddagger}$ only slightly and that of E is modified to a greater extent, while the phenol hydroxyl reduces both parameters significantly. If the influence of these groups had resulted from the +I effect, it would have been exactly opposite, that is the reaction rate would have diminished, and the effect of the OCH_3 group would have been much stronger than that of the OH group.

Hence, the effect of these substituents is not related to the electronic effects in the macroring either but is due to changes in the transition state solvation. The solvation pattern may change not only when the phenyl nucleus is replaced by groups affecting its electronic state but also by groups changing its orientation with respect to the macroring plane. Such groups include *o*-CH_3 and *o*-OCH_3. Ortho substitution by any groups considerably slows down the reaction (eight to 10 times), significantly increases the activation energy (by 7–8 kcal mol^{-1}), and the entropy of activation becomes positive, changing by not more than 20 e.u. The last-mentioned change is also due to the porphyrin solvation pattern changing as a result of variations in its geometrical configuration. Possibly the structure of the solvent surrounding the entire molecule also changes significantly (cell effect).

The data of Table 4.28 suggest an important conclusion that complexes of Cu^{2+} with porphin and tetraphenylporphin form at practically the same rates, while the activation parameters differ widely in DMFA.

Longo *et al.*[343] point out a rigorous kinetic compensation effect in reaction

(4.49), typical, as has been mentioned frequently in work by the present author,[318–324] of all reactions involving porphyrin molecules.

Presented in Tables 4.29 and 4.30 are the kinetic parameters[343] of formation of complexes of metals from their chlorides with tetraphenylporphin in DMFA and the effect of the solvent on the formation of CuTPP. In accordance with the reaction rates depending on the metal involved, the following series can be written $Cu^{2+} > Cd^{2+} > Zn^{2+} > Pd^{2+} > Co^{2+}, Ni^{2+} \gg Mg^{2+}$.

The rates of formation of Co and Ni tetraphenylporphins are almost three orders of magnitude lower than that of CuTPP. Longo *et al.*[343] provide no clues as to the above rate series. On the basis of the data presented in this chapter concerning chlorophyllic acid complexes it may be assumed that the lower rates correspond to octahedral complexes of Co^{2+} and Ni^{2+}. Therefore, in the transition state, two solvent molecules are lost from the coordination sphere. More positive values of $\Delta S^{\ddagger}$ for these metals corroborate this well (see Table 4.29). The compensation effect applies to the entire series of metal salts.

The activation energy is closely associated with the transition state solvation: wherever additional solvation of the transition state is less, more energy is involved in activation.

The activation energy is extremely high in the formation of Mg^{2+} tetraphenylporphin, that is a structural analogue of chlorophyll, which is probably due to the extremely strong solvation of salt $MgCl_2$, high stability of the solvation sheath of $Mg(solv)_6^{2+}$, and an insignificant gain in energy via the coordinate Mg—N bond of the porphyrin molecule. The weak

Table 4.29 Kinetic parameters of formation of tetraphenylporphin complexes from metal chlorides in DMFA

Parameter	$CuCl_2$	$ZnCl_2$	$CdCl_2$	$NiCl_2$	$PdCl_2$	$CoCl_2$	$MgCl_2$
$\Delta G^{\ddagger}$ (kcal mol^{-1})	20.4	21.6	20.8	23.2	21.8	23.1	—
E (kcal mol^{-1})	16.3	15.3	12.1	23.4	14.8	24.9	*c.* 60
$\Delta S^{\ddagger}$ (e.u.)	−9.8	−18.3	−25.4	0.6	−20.4	5.2	—

Table 4.30 Kinetic parameters of formation of copper tetraphenylporphin complex in various solvents from $CuCl_2$ at 40 °C

Parameter	DMFA	THF	DMSO	Cyclohexane	HMFA
$\Delta G^{\ddagger}$ (kcal mol^{-1})	19.5	17.8	20.3	16.4	24.0
E (kcal mol^{-1})	16.4	19.9	19.4	15.6	17.7
$\Delta S^{\ddagger}$ (e.u.)	−9.9	6.7	−2.9	−2.6	−20.1
ε	36.1	7.6	45.0	*c.* 20.0	30.0
DN[a]	26.6	20.0	29.8	*c.* 17.0	38.8

[a]Donor number of the solvent.

coordination affinity between the Mg^{2+} and Cl^- ions does not promote the reaction, which is why Mg^{2+} coordinates perceptibly to tetraphenylporphin in DMFA only at temperatures above 80 °C.

The rate of formation of complex CuTPP (see Table 4.30) depends strongly on the nature of the solvent, which is essential in understanding the mechanism of the reaction whose rate is fastest in ketones and ethers exhibiting minimal σ-donor capacity and creating coordination spheres $[CuCl(solv)_5]^+$ of a minimal stability. As the coordinating capacity of the solvent increases (DMFA < DMSO < HMFA), the rate decreases by four orders of magnitude. The rate of reaction (4.49) is not dependent on the dielectric constant (ε) (see Table 4.30). All solvents listed in Table 4.30 have a similar configuration and the same coordinating oxygen atom, therefore the steric hindrances to reaction (4.49) are virtually the same.

Among other interesting results of the work under consideration[343] should be pointed out the independence of the rate constant of the reaction between salt $CuCl_2$ and tetraphenylporphin in DMFA of the concentration of the inert supporting electrolyte ($NaNO_3$, $LiClO_4$), a slower reaction rate in the case of a coordinating supporting electrolyte (NaCl, LiCl), and its becoming much slower when a sufficient amount of acid is added. The introduction of 6×10^{-2} M HCl makes the reaction 350 times slower. Extremely wide differences are observed in the rates of formation of CuTPP in glacial acetic and propionic acids. In the latter the reaction is 47 times faster. Longo *et al.*[343] attribute this difference to partial protonation of porphyrin in glacial acetic acid. This conclusion seems to be premature because protonation of such weak bases as porphyrins in the weakly solvating and protonating acetic acid is hardly possible. In any event one should not ascribe it to the formation of cationic products of the $H_2TPPH_2^{2+}$ type. Protonation discontinues at a stage where a more or less strong hydrogen bond is formed. Longo *et al.*[343] make no mention of the stoicheiometric and intimate mechanism of reaction (4.49).

The inference as to the second order of the metalloporphyrin-yielding reactions, activity of the H_2P and inertness of the H_4P^{2+} particles, strong dependence of the kinetic parameters on the nature and coordinating capacity of the solvents, and the effect of the supporting electrolyte is in full agreement with the results of earlier works conducted in the present author's laboratory[320–342] and considered in this chapter.

From the work by Longo *et al.*[343] and some other investigators no definite conclusions can be drawn about the effect of individual structural units of porphyrin molecules on the kinetics of the metalloporphyrin-yielding reactions. Systematic studies of this problem have been described in many works.[245,246,335,245–350]

The porphyrins examined differ from one another in the nature of the side substituents (alkyl, alkylene, hydrogen, formyl group, formic or propionic acid residue), their position (pyrrole and meso substitution), the presence of an additional fifth ring (a so-called cyclopentanone ring), and a hydrogenated double bond in one of the pyrrole rings. They also differ in the distribution of the π-electron density in the macroring, the energy of the bonding and antibonding π orbitals, the solubility, and hence the degree of solvation in the ground state, and the protonation of the N—H bonds, that is the acid ionization constants.

Consider now the effect of the structural units of porphyrins using the formation of copper complexes in ethanol as an example. The reaction proceeds as follows:

$$H_2P + MX_2(solv)_{n-2} \rightarrow MP + 2HX + (n-2)solv. \quad (4.50)$$

The main kinetic parameters of this reaction have been obtained under standard conditions and are summarized in Table 4.31. The formulae of some porphyrins were presented earlier. Here the formulae of phylloporphyrin (**XLIX**) and pyrroporphyrin (**L**) are given.

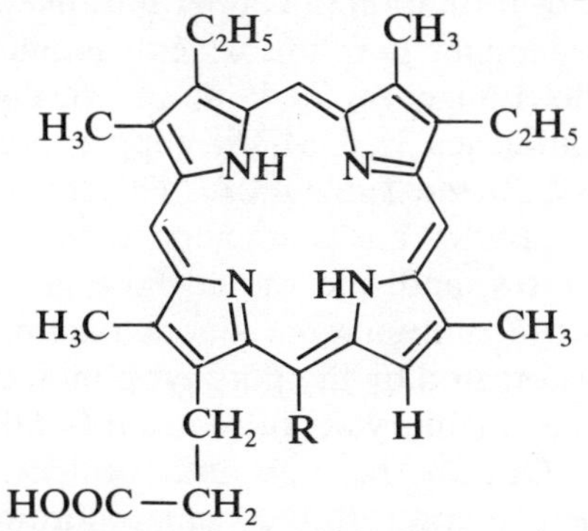

XLIX: R = CH_3; **L**: R = H

Table 4.31 Kinetic parameters of formation of copper complexes of natural and synthetic porphyrins in ethanol

Porphyrin	$c^0_{CuAc_2} \times 10^4$ (mol l^{-1})	$k_v^{25} \times 10^{-3}$ (l mol^{-1} h^{-1})	E (kcal mol^{-1})	$\Delta S^‡$ (e.u.)
Chlorophyllic acid *a*	3.75	1.29	14.5	−12
Chlorophyllic acid *b*	4.50	0.18	14.0	−17
Phaeophorbide *a*	3.75	4.75	7.5	−28
Chlorin e_6	4.50	8.20	6.5	−34
Rhodine g_7	2.70	1.70	6.4	−38
Synthetic chlorin	5.50	2.20	11.2	−21
Porphin	4.45	2.00	13.2	−16
Mesoporphyrin	2.50	6.70	7.5	−31
Protoporphyrin	2.50	3.60	8.8	−29
Pyrroporphyrin	2.64	15.50	7.0	−32
Phylloporphyrin	0.90	48.50	10.0	−18
Etioporphyrin I	2.30	7.00	13.4	−12
Etioporphyrin II	2.30	6.90	12.8	−14
Etiochlorin I	2.30	7.50	11.3	−18
Etiochlorin II	2.30	7.50	11.3	−18
Tetraphenylporphin	3.30	7.40	14.0	−10

Copper acetate reacts in ethanol with all of the 15 porphyrins examined at relatively slow rates, low activation energies, and negative entropies of activation. As can be seen from Table 4.31, when significant changes occur in the porphyrin structure, that is in going from H_2Chl*a* to H_2Chl*b* or from H_2Chl*b* to phylloporphyrin, and so on, the kinetic parameters change drastically despite the fact that the main structural unit of the porphyrin molecule, that is the macroring with a continuous 16-membered π system (see Chapter 3), remains unchanged. The limiting values of the rate constants changing by a factor of 270, the activation energy becoming twice as great, and the entropy of activation changing by 26 e.u. can be explained by the different steric conditions under which the salt attacks porphyrin and by the different solvation of the reactive site, polar substituents (specific solvation), and peripheral portions of the molecule (universal solvation).

Note the higher rate of formation of copper tetraphenylporphin (3.5 times higher) as compared to copper porphin, which has much lower values of E and $\Delta S^{\ddagger}$. According to Longo *et al.*,[343] at practically the same rates of formation of these complexes in DMFA, copper porphin features much higher values of E and $\Delta S^{\ddagger}$ (see Table 4.28). This means that, depending on the solvent, various porphyrins are characterized by quite different relative coordination reaction rates, and one should take into account the specific features of each solvent in analysing the kinetic parameters.

The steric hindrances created by the porphyrin molecule and coordination sphere of the salt in an elementary act of reaction (4.50) depend strongly on the porphyrin species. Due to the high and specific sensitivity of various porphyrins to rearrangement of the salt coordination spheres, the concentration dependence on the reaction rate constant is different for each porphyrin (Table 4.32). Such high specificity and selectivity of porphyrins with respect to different coordination spheres of salt $MX_2(\text{solv})_{n-2}$ may be caused by the insignificant changes in the geometrical parameters of the porphyrin molecule centre, which are not detectable radiographically.

The concentration dependence of the reaction rate constant can be taken into account quantitatively from the standpoint of the theory of relative activity coefficients,[351] in which case the relationship between the rate constant and concentration will be written as

$$k_v^c = k_v^{c\to 0}\,\Sigma\,{}^0\gamma_i^c, \tag{4.51}$$

where $k_v^{c\to 0}$ corresponds to a low salt concentration (below 5×10^{-4} M), k_v^c corresponds to any finite salt concentration, $\Sigma\,{}^0\gamma_i^c$ is the ratio of the activity coefficients of reagents MX_2 and H_2P as well as activated complex $[MX_2{\cdot}H_2P]^{\ddagger}$ or, for brevity, $X^{\ddagger}$:

$$\Sigma\,{}^0\gamma^c = {}^0\gamma^c_{MX_2} \times {}^0\gamma^c_{H_2P}/{}^0\gamma^c_{X^{\ddagger}}. \tag{4.52}$$

Relative activity coefficients ${}^0\gamma^c_A$ represent changes in standard chemical potential μ_i of dissolved substance A as one passes from a solvent with a low salt concentration ($c_{MX_2} \to 0$) to a solvent with any finite salt concentration:

Table 4.32 Concentration dependence of the constant of the rate of formation of metalloporphyrins in ethanol at 25 °C

Porphyrin	Reagent salt	$c^0_{MAc_2} \times 10^3$ (M)	$k^{25}_v \times 10^{-3}$ (l mol^{-1} h^{-1})
Chlorophyllic acid *a*	$CuAc_2$	0.375	1.29
		1.50	1.42
Chlorophyllic acid *b*	$CuAc_2$	0.45	0.177
		1.50	0.185
Phaeophorbide *a*	$CuAc_2$	0.375	4.75
		1.50	3.00
Rhodine g_7	$CuAc_2$	0.265	1.70
		1.60	1.56
	$ZnAc_2$	0.28	3.80
		0.85	1.75
Chlorin e_6	$CuAc_2$	0.45	8.20
		1.85	5.70
	$ZnAc_2$	0.28	9.54
		0.55	4.60
Synthetic chlorin	$CuAc_2$	0.247	2.36
		2.475	1.80
	$ZnAc_2$	0.191	1.33
		1.91	0.28
Porphin	$CuAc_2$	0.178	2.12
		0.89	1.59
	$ZnAc_2$	0.191	1.10
		1.91	0.48
Mesoporphyrin	$CuAc_2$	0.250	6.70
		1.25	6.25
	$ZnAc_2$	0.50	5.16
		1.25	6.20
Protoporphyrin	$CuAc_2$	0.250	3.60
		1.25	2.74
	$ZnAc_2$	0.50	1.65
		1.25	2.12

$$\bar{\mu}^c_A = \mu^0_A + RT \ln {}^0\gamma^c_A. \tag{4.53}$$

The methods for measuring the relative activity coefficients are outlined elsewhere.[315] However, such an approach is experimentally difficult to apply to porphyrins. Besides, it does not take into account the chemical activity of the solvent itself changing with the concentration of the reagents, nor is it applicable to reactions involving an intermediate. Another important factor is that this approach disregards the physical picture of the changes undergone by the reagent particles in the solution. A quantitative approach that neglects structural changes in the reagents and solvent can hardly be fruitful. A more fruitful approach involves investigation of the structure of the reacting particles and the solvent at different concentrations and temperatures.[322]

Coming back to Table 4.31, I should like to point out again that the factors responsible for a slower reaction rate in passing from H_2Chl*a* to H_2Chl*b* and from chlorin e_6 to rhodine g_7 were discussed in conjunction with the data presented in Table 4.26 and explained by the +C-electron effect of the formyl group in porphyrin systems.

The detachment of the phytol residue as one passes from H_2Chl*a* to phaeophorbide *a* promotes coordination. The rate of the complexing reaction increases 3.5 times, the activation energy sharply decreases, and favourable conditions are created for the transition state solvation. Thus the phytol residue creates formidable hindrances to the reagent salt along the path of reaction (4.50). Transitions H_2Chl*a* → chlorin e_6 and H_2Chl*b* → rhodine g_7 involving opening of the cyclopentanone ring, appearance of a carboxyl group at C6, and detachment of the phytol residue speed up the reaction six to 10 times and significantly reduce the energy and entropy of activation as compared to detachment of the phytol residue alone. Hence the cyclopentanone ring considerably hinders the coordination when it forms from chlorin e_6 (**XXXVII**) or rhodine g_7 (**XXXV**) as a result of detachment of water units from the COOH group with C6 and the CH_2 group being in the γ position.

The fact that the reaction yielding a complex with chlorin e_6 is almost twice as fast as that yielding a complex with phaeophrobide *a* is also indicative of the hindrances to coordination of the metal salt, created by the cyclopentanone ring. Chlorin e_6 and rhodine g_7 are structurally related in the same manner as H_2Chl*a* and H_2Chl*b*. Chlorin e_6 has a CH_3 group at C3, while rhodine g_7 has a CHO group at the same atom. Table 4.31 suggests that the transition from chlorin to rhodine is associated with the reaction rate being reduced almost five times due to the decrease in the value of $\Delta S^{\ddagger}$, that is due to additional solvation of the N—H bonds of rhodine g_7 in the transition state. These data agree well, just as in the case of formation of H_2Chl*a* and H_2Chl*b* complexes, with the assumption that the covalence of the N—H bonds grows under the effect of the CHO group.

Synthetic chlorin and porphin are characterized by practically the same rates of coordination of copper acetate in ethanol but considerably different activation parameters, which is probably due to the structural and electronic differences in the coordination centres of these porphyrins. Naturally, the same rates of formation of CuP and CuChl complexes in ethanol give no reason to believe that these differences are non-existent. Interestingly, the ratio between the kinetic parameters of etioporphyrin II (**XXIX**) and synthetic etiochlorin II (**LI**) is exactly the same as in the case of chlorin and porphin, that is the hydrogenation of the double bond in one of the pyrrole rings produces the same effect regardless of the degree of alkylation of the pyrrole nuclei.

It goes without saying that etioporphyrin I (**LII**) and etioporphyrin II feature the same coordination rates, energies, and entropies of activation since a change in the relative position of the CH_3 and C_2H_5 groups in the pyrrole nuclei without a change in their number can hardly alter the state of their molecules. It was shown[335] that the transition chlorin $e_6 \rightarrow$ mesoporphyrin during which one carboxyl group conjugated with the macroring at the C6 atom disappears along with a vinyl group at C2 affects the kinetic parameters but slightly. This is probably due to the fact that substituents —COOH and —CH=CH$_2$ of the same electronic nature are also replaced by the same alkyl and pseudo-alkyl substituents.

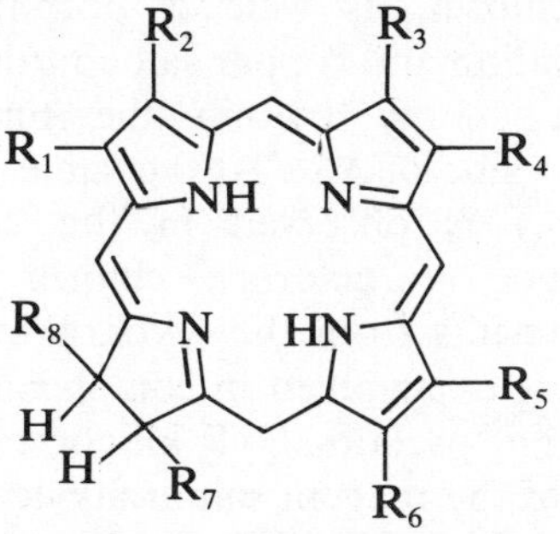

LI: $R_1, R_4, R_5, R_8 = CH_3$; $R_2, R_3, R_6, R_7 = C_2H_5$
LII: $R_1, R_3, R_5, R_7 = CH_3$; $R_2, R_4, R_6, R_8 = C_2H_5$

These substituents at C2 and C6 seem to act in opposite directions, with the result that the kinetics remain the same. Unfortunately the electronic effects of substitution in the aromatic molecule of porphyrin have not been studied experimentally or theoretically. The effect of substituents in it is much more complex than in the benzene molecule, which can be readily seen from the spectral manifestations of these effects.[8] The effect of substituents on the chemical reactivity is considerably more complicated than the spectral picture as a result of the complex solvation pattern in porphyrin molecules substituted in different positions. All this requires experimental proof in the near future.

In protoporphyrin the rate of coordination of copper salts was found[335] to be half as fast as in the case of mesoporphyrin. The transition mesoporphyrin **(XXV)** → protoporphyrin **(XXVI)** is structurally related to the replacement of two ethyl substituents at C2 and C4 by vinyl ones which are expected to exhibit a similar but somewhat weaker +C effect than in the case of the formyl group. The +C effect must lead to a lesser degree of protonation of the N—H bond and a slower reaction rate. The electronic effect of the CH=CH group can best be described as follows (in accordance with the concepts set forth in Chapter 4).

The lower-energy antibonding π-electron orbitals (ψ_π^*) are filled by the π electrons of the ethylene bond to a small degree but sufficient to displace partially the *n* electrons of the nitrogen atoms from the lowest ψ_π^* orbitals of porphyrin. The conversion of the *n* electrons from π to σ states increases the σ-electron density at the nitrogen atoms and strengthens the N—H bonds, with the result that the coordinating capacity of porphyrin is reduced, that is the formation of metalloporphyrins slows down.

The marked differences in the rates of formation of complexes of chlorin

e_6 and synthetic chlorin, porphin, and etioporphyrin can be explained only by the different solvation patterns of these molecules in the initial and transition states. In both cases alkylation of the pyrrole nuclei (including pseudo-alkylation with $CH_2—CH_2—COOH$ groups) speeds up the reaction three to four times (see Table 4.31). The changes in other kinetic parameters depend on the substituents (purely alkyl or pseudo-alkyl) included in the pyrrrole nuclei.

Although etiochlorin and synthetic chlorin, as well as porphin and etioporphyrin, differ in the degree of alkylation and copper salt coordination rates, they have the same activation energy (within the limits of experimental error) and a slightly greater value of $\Delta S^{\ddagger}$. At the same time, a sharp decrease in the activation energy is observed in the case of pseudo-alkyl substitution (synthetic chlorin → chlorin e_6, etioporphyrin → mesoporphyrin), which results from the favourable effect of the propionic acid residues on the energies involved in complexing with metal salts. As was assumed earlier,[335] the 'pseudo-alkyl' kinetic effect is associated with the primary coordination of the reagent salt at alkylcarboxyl groups, proceeding as follows: $[Ac_2(solv)_{n-3}]^M \leftarrow (CH_3OOC)PH_2$.

Owing to the favourable arrangement of the methylcarboxyl groups with respect to the main coordination centre $[N_2(NH)_2]$, the primary coordination of the salt may facilitate incorporation of both reactive sites into the 'cell' of the solvent, whereby the activation energy of the process is reduced. At the same time the entropy of activation also decreases to a considerable degree (by 13–17 e.u.). Hence primary coordination promotes the preliminary desolvation of the reactive site of a pseudo-alkyl-substituted porphyrin. A similar effect could be produced by the steric hindrances created by the $(CH_2)_2COOCH_3$ groups during solvation of the main coordination centre with the solvent molecules, that is the influence of the field of these groups on the reactive site. To solve the problem finally requires studies into the kinetics of coordination and thermodynamics of solvation of the above porphyrin pairs in different solvents.

Another extremely interesting fact is the slower complexing rates in the case of unsubstituted porphyrin as compared to etioporphyrin (see Table 4.31). The alkyl substituents in the porphyrin nucleus exhibit a +I effect and increase the electron density on the orbitals of the ethylene bonds of the pyrrole nuclei, hence on the orbitals of the macroring. The +I effect is probably associated with the increase in the frequency of N—H vibrations during alkylation of porphyrins (see Table 4.12) as a result of strengthening of the N—H bonds. However, the coordination rate of alkyl substituents in porphyrins increases rather than decreases. Consequently, along with the energy factors an important role in the coordination reaction is played by the solvation factor which determines the reaction rate in the case of alkyl substitution.

Despite the little knowledge we have about this phenomenon there is irrefutable proof of its major role. Listed in Table 4.33 are the rates of

Table 4.33 Rate constants of formation of cadmium complexes of various porphyrins in the ethanol–benzene system

$C_2H_5OH:C_6H_6$	$k_v^{25} \times 10^{-2}$ ($l\ mol^{-1}\ h^{-1}$)					
	H_2P	H_2TPP	H_2EP	H_2PP	H_2MP	H_2Chla
100:0	8.36 ± 0.25	57.9 ± 2.4	405 ± 7	52.7 ± 2.7	186 ± 10	7.46 ± 0.17
90:10	6.38 ± 0.42	51.6 ± 1.9	283 ± 4	35.4 ± 1.3	120 ± 5	5.47 ± 0.24
80:20	—	29.1 ± 2.2	190 ± 4	22.2 ± 1.3	73.2 ± 3.0	3.42 ± 0.10
70:30	3.37 ± 0.16	20.6 ± 0.4	118 ± 5	13.2 ± 0.6	43.9 ± 1.0	2.04 ± 0.12
60:40	—	14.1 ± 0.5	67.6 ± 1.6	7.48 ± 0.9	26.2 ± 0.6	1.52 ± 0.09
50:50	1.96 ± 0.07	10.1 ± 0.3	33.7 ± 1.0	4.77 ± 0.11	15.1 ± 0.2	1.14 ± 0.07
40:60	1.65 ± 0.06	9.29 ± 0.2	17.6 ± 0.5	3.07 ± 0.16	8.17 ± 0.12	0.99 ± 0.03
30:70	1.64 ± 0.09	9.00 ± 0.15	9.59 ± 0.14	2.45 ± 0.06	4.82 ± 0.09	1.15 ± 0.08
	5.1	6.4	42.2	21.5	38.6	6.5

Note: Given in the last line is the rate constant ratio in ethanol and in 30:70 ethanol–benzene.

formation of cadmium complexes with chlorophyllic acid and some porphyrins in an ethanol–benzene medium of gradually changing composition.[338]

The rates of formation of porphyrin complexes with cadmium are minimal in the case of porphin and maximal in the case of etioporphyrin at all binary solvent ratios. In ethanol they differ by a factor of 48, while in a 70:30 benzene–ethanol mixture they differ only by a factor of six. In this range of solvent compositions the rate constant of CdP decreases five times, whereas that of CdEP decreases 42 times. Such a sharp difference in the kinetic behaviour of porphyrins is due to the widely different solvation pattern of porphyrins in the initial and transition states and the different structure of the solvent near their molecules. Porphin, tetraphenylporphin, and chlorophyllic acid presumably have a greater tendency to specific solvation, whereas alkyl and pseudo-alkyl porphyrins tend to universal solvation.

The different relationship between porphyrins and solvents manifests itself in different changes in the energy and entropy of activation of complex formation in an ethanol–benzene medium. An interesting case of changing solvation of a porphyrin molecule occurs when structural changes are caused by asymmetrical meso substitution, such as the transition pyrroporphyrin **(L)** $\rightarrow$ phylloporphyrin **(XLIX)**. As a result of a single CH_3 group occupying the γ position in the π system of the macroring, the π system is strongly polarized (which is indicated by the bathochromic shift of the first absorption band), whereby the solvation of the entire molecule is promoted. Consequently, the transition state in phylloporphyrin becomes more probable than in pyrroporphyrin (the value of $\Delta S^{\ddagger}$ of the former exceeds that of the latter by 14 e.u.). The compensation effect increases the activation energy by 3 kcal mol^{-1} in the case of phylloporphyrin as opposed to pyrroporphyrin.

Symmetrical alkylation of the meso or pyrrole positions of porphyrin brings about changes in solvation of a different nature having nothing to do with the polarization of the macroring as a whole. In spite of the enormous interest in the solvation of porphyrins it has not been studied completely enough.

Thus, at present, not only do we know that the structural features of a porphyrin molecule produce a tangible effect on its reactivity in coordination reactions but also that the role of individual functional substituents and molecular fragments in the kinetics has been established. The effect of energy factors (changing the state of the N—H bond), solvation factors, and primary coordination on the kinetic parameters has also been elucidated.

4.9 Mechanism of Formation of Metalloporphyrins in Solution

The reactions yielding complexes with large aromatic ligands which are in a continuous conjugation along the periphery of porphyrin molecules are characterized by a number of features distinguishing them from other reactions in which complexes with simple ligands are formed.

These features are determined by:

(1) the low mobility of the ligand in solution, which is a target for the more mobile reagent, namely the salt molecule (or ion);

(2) the weak acidity of the ligands whose ionization, $H_2P \rightleftarrows 2H^+ + P^{2-}$, is extremely difficult;

(3) the presence of a dense π cloud of the conjugated system of the macroring, which forms a thick screen around the reactive site of porphyrin, including four intracyclic nitrogen atoms;

(4) the complexity of solvation of the porphyrin ligand which is a carrier of various solvation centres of a widely different nature.

A salient feature of the metalloporphyrin-yielding reactions is their slow rate as compared to reactions involving other ligands.[352–354] It stems from the above characteristics of the porphyrin ligand. Summarizing the regularities presented in this chapter one may write the following general stoicheiometric equation for the metalloporphyrin-yielding reactions:

$$M(\text{solv})_{n-m}X_m + H_2P(\text{solv})_p \rightarrow MP(\text{solv})_q + 2HX(\text{solv})_r + (n - m + p - q - 2r)\text{solv}. \qquad (4.54)$$

Equations (4.36), (4.42), (4.44), (4.47), (4.49), and (4.50) are particular cases of this general equation. In equation (4.54) n is the maximum coordination number of the cation of a salt of any charge, under given conditions of the medium, m is the number of anions, which may vary from zero (the salt is completely ionized) to n (pure anionic acid), and $(p - q)$ is the number of solvent molecules lost by the porphyrin molecule as a result of coordination.

In organic solvents considered in this monograph the reagent salt is, in most cases, in a non-ionized state. The salt may be dimeric. The porphyrin molecule is always monomeric unless otherwise specified.

At low reagent salt concentrations reaction (4.54) is in strict compliance with the second-order kinetic equation—first-order with respect to the porphyrin and first-order with respect to the salt. Analysis of the kinetic parameters of similar reactions in which the structure of one of the reagents is altered provides valuable information on the mechanism of the reaction and the activation phenomena involved.

Let us now examine the roles of the outgoing, incoming, and accompanying ligands in reaction (4.54). By the *outgoing ligand* in reaction (4.54) in which simple molecules are exchanged for porphyrin is here meant a ligand that practically loses its association with the metal atom as the reaction proceeds toward the top of the potential barrier and is not included in the transition state configuration. This definition is different in principle from that applied to reactions in which the ligands in the complex are replaced by simple ones.[352] In this case by an outgoing ligand is implied that which is absent in the end product of the reaction. From this point of view all ligands of salt $M(solv)_{n-m}X_m$ would have been outgoing. Such a definition is certainly not correct bearing in mind that different solvent and salt anion molecules in reaction (4.54) behave in a substantially different fashion in the course of activation.

Analysis of the kinetic data of a reaction yielding metal analogues of chlorophyll (see Table 4.20 and others that follow) has shown that in the case of salts of octahedral structure $[M(solv)_4X_2]$, which include most of the salts examined, the outgoing ligands are two solvent molecules. Their removal at the first stage of activation of reacting system (4.54) ensures contact and coordinate bonding between the metal and porphyrin. This path of the reaction and transition state scheme **(LIII)**, proposed

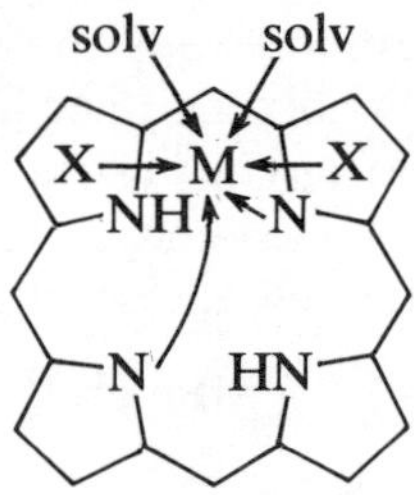

LIII

earlier,[297–299] adequately explain the strong dependence of the rate and activation parameters of the metalloporphyrin-yielding reaction on the nature of the solvent, both individual and mixed, as well as the weak dependence of the reaction rate on the covalence of the M—X bond and on the nature of the metal.

The remaining $(n - m - 2)$ solvent molecules, as well as the anions of X, are *accompanying ligands*, for they remain in the transition state configuration and leave the inner coordination sphere of the cation beyond the top of the potential barrier. Naturally, the activation parameters will be affected first of all by the outgoing ligands. Depending on the nature of the outgoing group, that is the solvent molecule, the rate of formation of CuChl (see Table 4.23) changes by a factor of 500, the value of E by a factor of 2.5, and the value of $\Delta S^{\ddagger}$ by 54 e.u. A similar ratio was observed in the case of ZnChl. The outgoing group (solvent) strongly influences the activation energy in the first place since it participates directly in the activation of the reacting system. This is only natural for the ease of detachment of two solvent molecules and the stretching of the remaining M—solv bonds are determined by the strength of their coordinate bond with the salt cation. This strength can be evaluated qualitatively from the heats of solution of the salts in various solvents[250,274,332,355] and from the donor numbers.[344]

However, the activation energy is not always strictly proportional to the strength of the M—solv bond. The point is that in the case of such specific incoming ligands as porphyrins an important role is probably played by the steric hindrances created by the solvent molecules in the coordination sphere of the salt as it attacks the porphyrin molecule. It may be assumed that the activation energy increases with the volume occupied in the inner sphere of the salt by the solvent molecule in the free rotation state. In this case the reaction rate is generally slower. Such sterically hindering molecules examined by us include quinoline, pyridine, DMFA, pentanol, ethyl acetate, and others. No perceptible hindrances are created by such solvents as methanol, ethanol, acetone, and acetic acid, which can be seen from their Stuart models. They exhibit much greater coordination rates and low activation energies (see Table 4.23).

Unfortunately the role of the solvent in reaction (4.54) is not restricted only to performing the function of an outgoing group. The solvent also acts as a medium actively solvating the protons of the N—H bonds of porphyrin and anions of the M—X salt. However, the above data suggest that the X^- and H^+ ions are not outgoing ligands and that the role of the solvent as a medium is but secondary. The latter clearly follows from comparison of solvent pairs different in nature (Table 4.23). For example, if one takes the acetic acid–pyridine pair, the first component solvates protons weakly and has a low dielectric constant, while the behaviour of the second one is opposite. Nevertheless, the activation energy in the case of acetic acid is low and the coordination rate is high, whereas in the case of pyridine the opposite is true. In this and other examples the predominant role of the solvent as a structural unit of the inner coordination sphere of the reagent salt is clearly evident. The acetic acid molecules are weakly coordinated by the copper salt and easily leave the sphere, while pyridine forms sufficiently stable complexes with copper, and their detachment is difficult.

Analysis of all available kinetic data indicates that the mechanism of reaction (4.54) can be best described as follows. As the porphyrin and salt molecules come closer together within a single solvent cell, the two adjacent M—solv bonds are cleaved (if the metal salt is not a pure anionic acid). Then valence angle solv—M—solv increases, and both bonds are stretched. The two protons of the main coordination centre of the porphyrin molecule leave the molecular plane under the effect of the field of the metal and occupy a position on one side of this plane as is shown schematically at the top of page 124 (see over).

Therewith the protonation of the N—H bond, or its polarity, increases with the strength of the bond between the proton and solvating solvent molecule. The number of proton-solvating molecules remains the same. As the reagents come still closer together under the effect of the force field of porphyrin, the deformation of the two adjacent M—solv bonds continues, and two solvent molecules disassociate themselves from the salt cation and are removed from the inner coordination sphere. Their place is occupied by two tertiary nitrogen atoms of porphyrin. The emerging donor–acceptor σ bond $M \leftarrow N$ is weak but partially compensates for the energy expended in

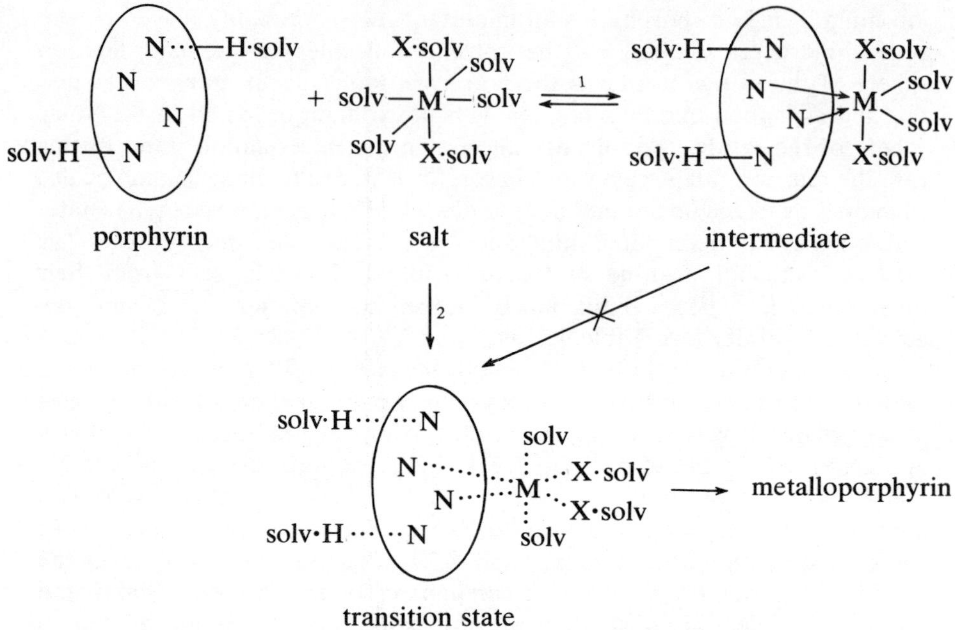

the detachment of two solvent molecules. At this stage the interaction may proceed along one of the following two possible paths:

Path one: On the above scheme this path is indicated by paired arrows. If the M—solv bond is not sufficiently strong, just as the emerging H^+···solv and X^-···solv bonds (weakly polar solvents HAc, $CHCl_3$, CCl_4), the detachment of two solvent molecules will involve small energy expenditures without affecting the rest of the coordination sphere of the salt. At the same time the remaining M—solv and M—X bonds are practically unchanged. In this case the reaction may discontinue at the stage of formation of a mixed complex which was termed 'intermediate' in earlier works by the present author[32] or 'SAPP complex' in non-Soviet literature.[75] It is quite obvious that the metalloporphyrin intermediate does not reach the top of the potential barrier in reaction (4.54) because the formation of a porphyrin complex requires a considerable amount of energy to be expended in the stretching of the M—solv, M—X, and N—H bonds which are in a state similar to that of the starting reagents.

Hence the energy profile of this path of reaction (4.54) will take the form shown in Fig. 4.18 for the intermediate. The latter is in a dynamic equilibrium with the starting reagents. Its stability is low and will depend on the extent to which the energy of the system increased during its rise toward the top of the potential barrier.

Path two: If the bond between the cation and solvent molecule is sufficiently strong and the solvation interactions between the N—H···solv and M—X···solv bonds are significant (as in the case of polar, easily

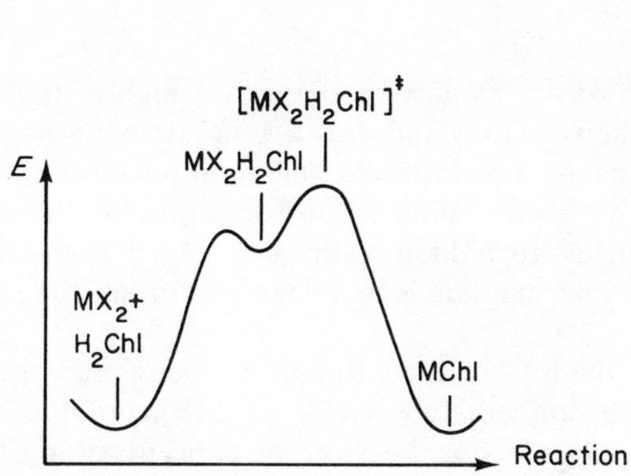

Figure 4.18 Energy diagram of a metalloporphyrin-yielding reaction

coordinated solvents which represent the majority of solvents in Table 4.23), the detachment of two solvent molecules will be difficult, involve high activation energy expenditures, and may occur only near the top of the potential barrier of reaction (4.54), while the remaining M—solv, M—X, and N—H bonds will be strongly deformed and stretched. In this case the salt–porphyrin system will react bimolecularly in accordance with a self-consistent activation mechanism (indicated by a single arrow in the scheme) and no intermediate can form. It takes a single elementary step for the reacting system to transform via the transition state to a stable intercomplex compound (metalloporphyrin) in which the metal atom lies in the molecular plane between four nitrogen atoms.

Both paths of reaction (4.54) are easily traced in experiments. It should be noted that in a reaction with chlorophyllic acid the copper salts do not yield intermediates in any one of the examined solvents, including 70% ethanol solution of benzene. However, intermediates have been detected both spectrally and kinetically in a reaction between chlorophyllic acid and salts $ZnCl_2$, $CoCl_2$, $FeCl_2$, $PdCl_2$, and $CoAc_2$, conducted in glacial acetic acid.[71,264,306] Salts $CoCl_2$ and $CoAc_2$ with chlorophyllic acid form intermediates in acetone,[303] while $Co(NO_3)_2$ does not under the same conditions. At the beginning of this chapter I mentioned protoporphyrin intermediates formed by salts of some metals in chloroform and acetone.[75]

The intermediates of porphyrins are unstable in media containing even small amounts of polar strongly coordinating solvents (e.g. water, alcohols, pyridine). This means that intermediates $MX_2(solv)_{n-4}{\cdot}H_2P$ are much weaker than solvate complexes $MX_2(solv)_{n-2}$ as far as their thermodynamic stability is concerned. Polar solvent molecules immediately displace the porphyrin from the intermediate, which is indicated by absorption spectra as well as changing coloration of the solution. These experiments suggest that, firstly, in polar strongly coordinating solutions reaction (4.54) cannot proceed via an intermediate and, secondly, the potential well on the energy curve (see Fig. 4.18), occupied by the intermediate, is not deep. Under the examined conditions[71,75,264] intermediates are highly stable. If intermediate solutions are allowed to stand for a long time at normal temperature, metalloporphyrins are not formed—this requires strong heating. As can be inferred from the above, direct transformation of the intermediate to a metalloporphyrin is forbidden (indicated by the crossed arrow in the scheme) and the intermediate is in a dead end rather than on the pathway of reaction (4.54).

When heated, the intermediate dissociates perceptibly back to the salt and porphyrin, which immediately yield an MP complex in a bimolecular reaction. The reaction is slow because the concentration of the free reagents is low. Here one is dealing with a situation precisely like the one which occurs in an acidic medium where H_2P transforms to $H_2PH_2^{2+}$. Protonation also takes the reagents off the main path,[264,321] The activation energy of metalloporphyrin formation will be high under these conditions for it will also include the energy expended in the dissociation of the $MX_2 \cdot H_2P$ or $H_2PH_2^{2+}$ particles.

The logical conclusion from the foregoing is that solvate complexes of metals with a purely solvate or mixed structure interact with porphyrins by an S_E2 mechanism according to the classification by Ingold and Hughes or $I_{a,d}$ according to the more perfect classification by Langford and Gray.[352] In this case a metal salt which is a more mobile molecule than porphyrin should preferably be selected as the reagent. In some works H_2P is selected as the reagent and the mechanism in S_E2.[23]

The role of the incoming ligand in metalloporphyrin-yielding reactions is as important as that of the outgoing ligand. Table 4.31 suggests that, depending on the porphyrin molecule structure, the rate of the metalloporphyrin-yielding reaction changes by a factor of 270, the activation energy by a factor of two, and the entropy of activation by 26 e.u. Hence the activation stage involves significant changes not only in outgoing but also in incoming ligands. Such changes include, first of all, stretching of the N—H bonds with their hybridization changing in the transition state, resolvation of the reactive site of porphyrin, its environment, and polar side substituents. All kinetic features involved in the substitution of the relatively simple ligands considered elsewhere[352] were in accordance with a self-consistent mechanism with associative (I_a) or dissociative (I_d) activation. Reactions with porphyrins are based on a self-consistent mechanism with associative–dissociative activation ($I_{a,d}$).

In cases where the reagent salt does not contain solvent molecules in the inner sphere, that is the salt is an anionic acid, reaction (4.54) may proceed in accordance with a displacement mechanism. It has long been noticed[356,357] that the substitution of the solvent molecules in the coordination sphere by an anion may initially speed up the metalloporphyrin-yielding reaction, then slow it down, and, finally, bring it to a complete halt. This is precisely what results from addition of hydroxyquinoline (HOx) to a solution containing a copper salt and mesoporphyrin.

Lowe and Phillips[357] could not explain this phenomenon because the reaction mechanism was unknown. The explanation follows naturally from the above-described reaction mechanisms. As one goes from ion $[Cu(solv)_n]^{2+}$ to $[Cu(solv)_{n-2}Ox]^+$ then to $CuOx_2(solv)_{n-4}$ and $[CuOx_3]^-$, the reaction with porphyrin is accelerated first owing to the Cu—solv bond weakening as a result of the lower effective charge at the Cu atom in the case of covalent bonding of the Ox^- anion. The weakening of the Cu—solv bond promotes the reaction, increasing its rate and reducing the activation energy. A decrease in the number of solvent molecules in the inner sphere acts in the opposite direction and may lead to a complete inertness of the salt when the number of solvent molecules equals zero.

Any molecules capable of strong coordination to the central atom behave similarly to protons. Such behaviour has been established in the case of formation of a cobalt complex with H_2Chl in the presence of urea, thiourea, and imidazole.[358] It is not always possible to distinguish quantitatively between the effect of solvent substitution in the inner sphere and that of addition of a proton or molecular compounds on the solvent structure. In many cases though this can be done.[358]

The fact that reaction (4.54) is possible with tetraacid complexes ($HgCl_4^{2-}$, HgI_4^{2-}, $CoCl_4^{2-}$, $CoAc_4^{2-}$) and impossible with hexacoordinated complexes ($Co(NH_3)_6^{3+}$, $Co(NH_3)_5Cl^{2+}$, $Co(NH_3)_4(HCOO)_2^+$, $Fe(CN)_6^{4-}$) indicates that apart from the mechanism involving detachment of the solvent molecules one is also dealing with displacement of ligands from their coordinates. The displacement of ligands along with packing of the coordination sphere is possible, in the transition state, in the case of particles MA_4^{2-}, ML_4^{2+} (L is a covalently bound ligand, A is an acid ligand (proton)) but impossible in the case of particles MA_6^{m-}, ML_6^{m+}. In the course of packing of the coordination sphere two coordinate places are freed for contact with the porphyrin molecule.[300] The larger the ligand, the more difficult the reaction conditions and the higher the activation energy which must be expended to counter the forces of mutual repulsion of the ligands and stretching of the M—A bonds. For example, in a reaction yielding complex HgChl in methanol the activation energy is higher by 9 kcal mol^{-1} with ion HgI_4^{2-} than with $HgCl_4^{2-}$, which is due to the difficult displacement and packing of the larger iodine atoms. Characteristically, the reaction with particles HgA_4^{2-} requires a highly polar medium. It proceeds in methanol and is almost unnoticeable in ethanol (see Table 4.22).

Methanol provides for a stronger solvation of the iodine atoms in HgI_4^{2-}, promotes stretching of the Hg—I bonds in the course of activation, and leads to rearrangement of the tetrahedral structure of the HgI_4^{2-} ion into an octahedral one with two vacant vertices. In this case one is dealing with an $I_{a,d}$ displacement mechanism; however, the outgoing ligand as such is absent. Its function is performed by accompanying ligands A^- or L whose state undergoes considerable changes during activation of the metal salt–porphyrin system. No intermediates can form under such conditions.

From the viewpoint of structure and energies the displacement mechanism must involve lower entropies of activation, which is precisely what has been observed.[334] Depending on the nature of the solvent and the outgoing or accompanying ligand the activation energy may vary in any direction. More often than not it decreases. To elucidate the problem further more experimental data are required.

The above regularities and mechanism of metalloporphyrin formation, established from a wealth of available data, are common not only to all porphyrins and phthalocyanines but also to many other planar macrocyclic ligands such as annulenes and cyclic polyesters (so-called 'crowns').[359]

To conclude this discussion of the mechanism of reaction (4.54) I should like to dwell on certain concepts developed in the non-Russian literature on the subject.[341,342,360–367] Studies involving porphyrins as well as tetrasulphophthalocyanine[255,257] have been conducted in aqueous solutions, that is under conditions conducive to association of porphyrins and phthalocyanines. This factor often complicates the simple kinetic picture typical of metalloporphyrin formation, and investigators tend to introduce a great number of equilibrium constants which are purely conjectural and experimentally corroborated.[362] The major drawbacks of this approach include complete disregard of the state of the solvent and changes in its concentration, the structure and state of the coordination sphere of the reagent salt not being duly taken into account, and the lack of data on energies and entropies of activation, which are absolutely necessary for discussing different aspects of the reaction mechanism. Since a review of the relevant works does not throw new light on the mechanism of the coordination reaction, I shall briefly outline only a few principal concepts.

On the basis of studies into the tetrapyridylporphyrin-yielding reaction which is of second order with respect to the cation concentration, due either to primary coordination of M^{2+} at the pyridine nitrogen atoms or to the existence of porphyrin molecule associates $(H_2P)_i$, Fleischer *et al.*[342] put forward the idea of preliminary formation of outer-sphere (OS) complexes:

$$H_2P + M^{2+} \rightleftarrows M^{2+}\text{—OS—}H_2P, \qquad (4.55)$$

$$M^{2+}\text{—OS—}H_2P \rightleftarrows M\text{—OS—}HP^+ + H^+, \qquad (4.56)$$

$$M^{2+}\text{—OS—}HP^- + M^{2+} \xrightarrow{\text{slowly}} MP + H^+ + M^{2+}. \qquad (4.57)$$

According to Fleischer *et al.*[342] and Hambright[362] outer-sphere complex $M(H_2O)_6^+ \cdot H_2P$ promotes the metalloporphyrin-yielding reaction, facilitating the proton detachment step. In their work[342] concerned with speeding up of a metalloporphyrin-yielding reaction in the presence of salts of 17 different cations Fleischer *et al.* calculated the outer-sphere complex formation constants. These constants will not be cited here because they are random, purely empirical values having nothing to do with the structural and energy characteristics of ions. For example, the values of K_{OS} for ions $(CH_3)_4N^+$, NH_4^+, and $(C_2H_5)_4N^+$ are given as being equal to 4.4, 6.2, and 26.3, respectively.

It should be noted that the values of K_{OS} were determined in an acidic medium at pH = 2, that is under conditions in which outer-sphere complexes of the $(H_2O)_{n-1}M^{n+}\cdots OH—H\cdots N(NH)_2N(P)$ type cannot form. Besides, not only do they fail to promote formation of metalloporphyrins (as would have been the case if the medium had been less acidic) but, on the contrary, they hinder it by taking the reagents off the reaction path.

Hambright[362] examined the speeding up of formation of Cu^{2+} complexes with porphyrins in the presence of the salt $LiNO_3$. It has been concluded that outer-sphere complexes Li—OS—H_2P are formed, exhibiting a particular activity in reaction (4.57). Proceeding from the porphyrin complex formation mechanism outlined in the beginning of this chapter, it is quite obvious that outer-sphere complexes feature a porphyrin reactive site bound as in an acid. The accelerating effect of the salt $LiNO_3$ on the formation of CoChl (see Table 4.25) is observed in alcohols in which outer-sphere complexes of the above type cannot form. A similar effect of the supporting electrolyte is due to changes in the structure and solvating capacity of the solvent, as well as changes in the structure of the coordination sphere of the salt and solvation of the initial and transition states of the system.

It has been proposed[364,365] that the porphyrin molecule is deformed in the course of a metalloporphyrin-yielding reaction as a result of formation of intermediates (SAPP complexes), and a 'general' mechanism of metalloporphyrin formation is set forth:

$$H_2P + MX_2 \rightleftarrows MX_2\text{—sat-}H_2P, \tag{4.58}$$

$$MX_2\text{—sat-}H_2P \rightleftarrows MX_2\text{—sat-}H_2P_{(\text{deformed})}, \tag{4.59}$$

$$MX_2\text{—sat-}H_2P_{(\text{deformed})} + M'X_2 \rightarrow M'P + MX_2 + 2HX. \tag{4.60}$$

In this description the idea of the conjugated system of the porphyrin macroring being broken seems to be fallacious. The conjugation energy is estimated to be *c*. 400 kcal mol^{-1},[368] and even a slight displacement of two pyrrole nuclei from the molecular plane may involve expenditure of tens of kilocalories which cannot be compensated by the formation of intermediates because of the weak coordinate interaction in them. This would have resulted in colossal activation energies. The impossibility of reducing them by solvation discredits this idea. Hambright and other investigators[360–367] have not determined even roughtly the values of E and $\Delta S^{\ddagger}$ for the systems examined by them.

The idea of deformation of porphyrins stems from the finding that the reaction involving tetrapyridylporphyrin in aqueous solutions is of second order with respect to the cation. However, in the system with *meso*-tetra(*N*-methylpyridyl)porphin, studied by the same investigators[361] in an aqueous solution, the reaction had already been found to be of first order with respect to Cu^{2+}. If one takes this porphyrin as an example, in which primary coordination of the Cu^{2+} ions at the N atoms of the pyridyl residues is impossible and the probability of formation of associates $(H_2P)_i$

is very low, the 'general' mechanism of reactions (4.58)–(4.60) does not make sense any more. Nor is the 'general' mechanism applicable to all porphyrins examined in non-aqueous solutions.[296–324]

The proponents of the idea that the porphyrin molecule is deformed try to substantiate it by the fact that the rate of formation of *N*-methyletioporphyrin complexes with Zn^{2+} and Cd^{2+} is 10^5 times greater than that of similar etioporphyrin complexes.[364] The high degree of non-coplanarity of the $N{=}CH_3$ derivatives of etioporphyrin in the solid state and also, judging from the high protonation constants, in solution is not sufficient for concluding that the transition state of reaction (4.54) must or may involve deformation of the planar porphyrin molecule. Incidentally, most other porphyrins are planar in solution.[5] As one goes from H_2EP to its $CH_3{-}N$ derivative, not only deformation occurs but also the structure of the reactive site changes drastically from $N_2(NH)_2$ to $N_3(NH)$. These porphyrins are characterized by entirely different transition states owing to the changing number of broken N—H and M—X bonds.

The values assigned by Shah *et al.*[364] to the energies and entropies of activation of formation of Zn–etioporphyrin complexes (13.1 kcal mol^{-1}; −27 e.u.) and its $CH_3{-}N$ analogue (8.0 kcal mol^{-1}; −21 e.u.) differ so insignificantly that the absence of deformation is beyond any doubt. As has already been mentioned, if the macroring had been broken, the values of E would have been so great and those of $\Delta S^{\ddagger}$ so small that reaction (4.54) could not have taken place in weakly polar organic solvents. Nevertheless its rate is faster in acetic acid than in DMFA, faster in dioxane than in quinoline (see Table 4.23), and so on.

In addition to the arguments against deformation of the porphyrin macroring, put forward above and examined in detail earlier,[324] there is yet another, the most indisputable one. A deformation of the phthalocyanine macroring in the solid phase is impossible in a reaction with salts of Cu^{2+} and other metals, and the reaction seems to be out of question. However, as was established earlier,[369,370] even with this exceptionally rigid conjugated ligand the reaction proceeds at a surprisingly fast rate in the solid phase (much faster than with porphyrins in solution) at room temperature. With the hydrated copper complex $[Cu(H_2O)_6]^{2+}$ the reaction is complete within a few minutes. Initially it yields an intermediate exhibiting a number of extremely interesting properties.[369]

And one final remark. The formation of SAPP complexes is assumed to be the most energy-consuming or, in other words, slowest process. As was shown at the beginning of this chapter, they form instantaneously, are easy to detect spectroscopically, and slow down rather than speed up reaction (4.54) by altering its path. Thus the idea of the accelerating effect of formation of outer-sphere complexes of the above type or intermediates of the SAPP-complex type and deformation of the porphyrin macroring at the activation stage has no theoretical ground and is without any experimental proof.

4.10 The Role of Acid–Base Interaction in Complexing Reactions

It has long been known that the complexing reaction

$$M^{n+} + nL \rightleftarrows ML_n^{n+} \tag{4.61}$$

and the ligand protonation reaction

$$L + H^+ \rightleftarrows LH^+ \tag{4.62}$$

have much in common. According to the early concepts advanced by Bjerrum[371] the capacities of a ligand for protonation and complexing are equivalent. Departures from this rule had been difficult to understand until various types of donor–acceptor bonds (σ bond, forward π bond, backward π bond, δ bond, and others) were discovered. It became obvious that the similarity between the complexing and acid–base properties of ligands exists until π orbitals occupied by a pair of electrons and unoccupied π orbitals appear in metal ions in addition to vacant σ orbitals. This always happens in metal cations with unfilled *p* or *d* shells. In this case (see the beginning of this chapter) a forward or backward dative bond is formed depending on the presence of a vacant or filled π orbital at the ligand.

Such orbitals can be found in porphyrin and particularly phthalocyanine molecules, which is why one cannot expect corresponding changes in acidity (basicity) and coordinating capacity. This is an extremely important consideration in view of the fact that one of the motive forces in the metalloporphyrin- and metal–phthalocyanine-yielding reactions is the emergence of a coordinate M → N bond with simultaneous rupture of the N—H bond. Hence the processes of metal coordination and acid–base interaction are competitive in reaction (4.54). The picture becomes still more complicated when reaction (4.54) is conducted in an acidic medium. This complication is due to equilibria (4.3) and (4.4) of protonation of the porphyrin molecules. The situation in an acidic medium is extremely complicated because questions arise as to the forms in which the ligand is present in these solutions, their reactivity in reactions (4.54), and the concentration of these forms.

Therefore, if the reaction rate is measured in acidic media with respect to the total ligand concentration but without taking into account the different reactivity of the ligand forms, the results may be misleading in so far as the kinetics and mechanism of reaction (4.54) are concerned. Still, many investigators who started studying the kinetics and mechanism of metalloporphyrin formation used aqueous acid or aqueous organic solutions.[255,257,341,343,361,362] It is natural to assume that the mechanism of reaction (4.54) should be investigated in such media in which porphyrin is present in the molecular form. The best from this point of view is a neutral medium with pH $\simeq$ 7. An acidic or basic medium only makes the task more difficult and does not add anything new to our knowledge of the reaction mechanism.

In the consensus of most investigators porphyrins reacting with metal salts are in the molecular form or, as many call it, in the form of a free base (H_2P). However, as was wrongly assumed in the earliest works concerned with the formation of tetrasulphophthalocyanine complexes,[255,257] even at pH = 2.5–6.5 the ligand reacts in the form of monoanion $[HPc(SO_2O^-)_4]^-$. Calculations show that in a 10^{-5} M phthalocyanine solution at pH = 6 only 0.04% of the substance is in the monoanionic form, that is the ligand participates in the reaction in the form of $H_2Pc(SO_3^-)_4$. There is no reason to believe that this form is not reactive, for assuming otherwise would contradict the extant experimental data on metalloporphyrins. It is also possible that due to the instability of the ionic forms of porphyrins and their extremely intensive solvation (see Chapter 4) they will not differ considerably from H_2P in coordination reactions. The data of Table 4.21 indicate that when reaction (4.54) is conducted in 90% ethanol saturated with ammonia, that is under conditions highly favourable to porphyrin ionization, the reaction rate decreases by one order of magnitude rather than increases.

Schiller *et al.*[255] have established the following order of rates of coordination of cations to tetrasulphophthalocyanine: $Cu^{2+} \gg Zn^{2+} > Co^{2+} > Mn^{2+} \gg Ni^{2+}$. This series does not correspond to the coordinating capacity of the metals, that is their tendency to accept electron pairs of ligands or give away their electrons to the π orbital of the ligand. Hence it is not the electronic factors that are decisive in reaction (4.54), as is contended by Schiller *et al.*,[255] but a combination of structural and energy factors.

The unsubstantiated conclusion[255] as to the low reactivity of the H_2Pc forms was then referred to as an indisputable fact in some other works. However, Choi and Fleischer,[341] who studied the formation of CuTPP in an acidic medium and had established a linear relation between rate constant k_v and $1/a_{H^+}$ dismissed the idea put forward by Schiller *et al.*[255] as inconsistent with their finding. At the same time the simple assumption that an acidic medium yields the porphyrin monoprotonation product H_2PH^+, which is not capable of coordinating to metal salts, would have easily explained the dependence of the formation rate of MP on pH. It can readily be shown that the concentration of the only reactive form is

$$[H_2P] = K_R[(c_0 - c_{H_2P})/a_{H^+}]. \qquad (4.63)$$

As can be inferred from equation (4.63), at high acidities $[H_2P] = \text{const.}/a_{H^+}$, which gives a linear relation between the reaction rate and the hydrogen ion concentration, found by Choi and Fleischer.[341]

The first proof of H_2PH^+ and $H_2PH_2^{2+}$ lacking coordination properties was provided by the present author in an earlier work.[264] There it was also demonstrated that the CuChl-yielding reaction practically discontinues in 95% formic acid and 90% acetic acid. Under these conditions both media protonate chlorophyllic acid transforming it to H_3Chl^+ or H_4Chl^{2+} forms.

The protonating capacity of 100% formic and acetic acids is not sufficient to transform H_2Chl to cationic forms. The protonation is clearly seen in absorption spectra. Table 4.34 lists the effective and true rate constants in 95% formic acid.

The constant rate decreases, as compared to glacial HAc, by 3.5 orders of magnitude, E increases threefold to 29 kcal mol^{-1}, and $\Delta S^{\ddagger}$ increases from -19 to $+36$ e.u.

The protons of the cationic forms of porphyrins bind the electron pairs of nitrogen atoms, thereby preventing their coordination to the metal. The coordination reaction proceeds at an elevated temperature as a result of dissociation of cations H_3P^+ and H_4P^{2+} to H_2P and H^+. The activation energy of the MP-yielding reaction includes the energy expended for ionization of the cationic forms of porphyrins. (The inertness of H_4P^{2+} was established later.[343])

An interesting phenomenon is observed in the coordination reactions between metal salts and phthalocyanines. As is known, H_2Pc is monoprotonated by concentrated H_2SO_4 at a peripheral rather than a central nitrogen atom. The H_2PcH^+ has an unbound coordination centre. This is why phthalocyanine readily forms complexes when the Cu^{2+} salt is brought in contact with its sulphuric acid solution. These facts were established for the first time in the laboratory of the Ivanovo Chemical Technology Institute.

In cases where the complexing reaction is conducted on solid phthalocyanine or its polymers,[161,162,369,370] the presence of a diluted acid (H_2SO_4, HCl, HAc) considerably slows down the reaction probably due to the central nitrogen atoms being bound by the H_3O^+ ions which form a hydrogen bond with them. As was shown earlier (Chapter 4), H_3O^+ cannot protonate phthalocyanines because of their extremely weak basicity. Probably the central nitrogen atoms are not bound in protonated phthalocyanine H_2PcH^+. The fractional positive charge appearing on them rules out the possibility of formation of a hydrogen bond with H_3O^+. Therefore, concentrated H_2SO_4 promotes formation of stable metal

Table 4.34 Kinetic parameters of formation of copper complex of chlorophyllic acid in 95% formic acid

t (°C)	$c^0_{CuAc_2} \times 10^4$ (M)	k_{eff} (h^{-1})	$k_v \times 10^{-3}$ (l mol^{-1} h^{-1})
50	7.20	0.218 ± 0.02	0.320
60	6.60	0.716 ± 0.043	1.08
63	4.39	0.708 ± 0.008	1.61
70	4.66	1.99 ± 0.12	4.27
70	5.07	1.90 ± 0.06	3.75

Note: Initial H_2Chl concentration is 1.7×10^{-5} M.

phthalocyanines. Naturally, labile phthalocyanines cannot form under similar conditions.

The facts presented in this section should be taken into account in selecting appropriate experimental conditions for studying coordination processes on metalloporphyrins. For example, one cannot take sufficiently protogenic media (aqueous HAc, pyridine + H_2SO_4, dimethyl formamide + HCl) for studying the formation of CdP, HgP, PbP, and similar complexes because they are highly labile and form only in neutral solvents.

Chapter 5

The kinetics and mechanism of protolytic dissociation of porphyrin and phthalocyanine complexes

The first among the metalloporphyrin molecules to be studied kinetically for stability in solution was chlorophyll.[372–374] Later, additional studies were conducted,[375] but again without definitive results. The kinetics and mechanism of dissociation of chlorophyll,[73,375] its labile and stable metal analogues,[376–379] haemin,[76] and some other metalloporphyrins[247,336,339,380–384] were elucidated only in recent years. Model systems such as metal phthalocyanines were used in studies into the mechanism of their dissociation.[37,102,108–110,113,114,191,284,385–389] It is advisable to start discussing the kinetics and mechanism of dissociation with simpler model systems.

5.1 Dissociation of Phthalocyanine Complexes and Their Polymers in Proton-donor Media

Phthalocyanine complexes of the general formula **LIV** are characterized by an extremely high stability which has been demonstrated on phthalocyanine complexes with Cu^{2+}, Ni^{2+}, Co^{2+}, Pd^{2+}, and Al^{3+} in hot concentrated H_2SO_4.

No data are available on their thermodynamic stability from

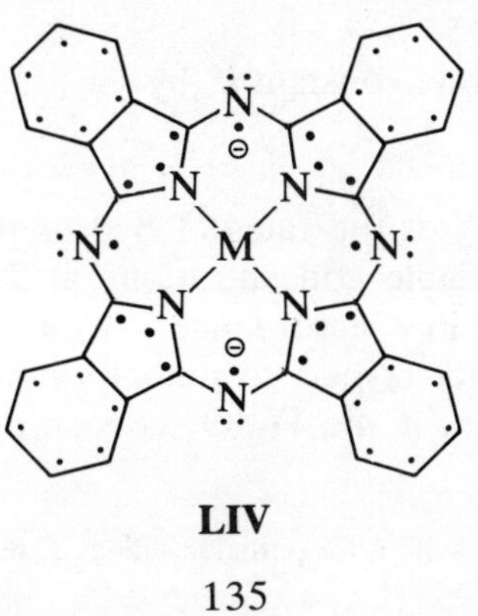

LIV

measurements of the equilibria of the reactions in which they are formed:

$$MX_2(\text{solv})_{n-2} + H_2Pc \rightleftarrows MPc + 2HX + (n-2)\text{solv}. \tag{5.1}$$

This is due to the extremely strong M—N bonds and very low complex dissociation rates. Nobody has succeeded so far in finding a solvent in which equilibrium (5.1) would not be in one of its extreme positions. Under conditions of phthalocyanine complex formation (see Chapter 2) their dissociation is not perceptible.

The pronounced unidirectionality of the processes of formation and dissociation of phthalocyanine complexes was one of the factors that made it necessary to find a kinetic criterion for evaluating their stability,[191] as well as to use a spectral stability criterion.

All phthalocyanine complexes and the H_2Pc ligand itself are liable to dissociate in sulphuric acid solutions, with the result that the intensely coloured green solution becomes completely colourless. The decomposition reaction† includes two steps:

$$MPcH^+ + 2H_3O^+ \xrightarrow[k_1]{} M^{2+}_{aq} + H_2PcH^+, \tag{5.2}$$

$$H_2PcH^+ \xrightarrow[k_2]{H_2O,\ H^+} \text{colourless products}. \tag{5.3}$$

The ratio between the rates of these steps is determined by the electronic properties of the cation.

Characteristically, if the cation is not capable of forming a sufficiently strong σ bond because of the weak electron affinity (Li^+, Na^+, Mg^{2+}, Ca^{2+}, Ag^+, Mn^{2+}, Sn^{2+}, Pb^{2+}) or because of its not being capable of dsp^2 hybridization, that is planar arrangement of the σ bonds (most of the above-mentioned cations plus Be^{2+}, Cd^{2+}, Hg^{2+}), then complexes of the labile group (labile phthalocyanines) are formed. They feature, primarily, an ionic type of bonding and detach the cation in the course of passage into the sulphuric acid solution. For these complexes $k_1 \geqslant k_2$, therefore the rates of discoloration of their sulphuric acid solutions and H_2Pc solutions in sulphuric acid are the same. Given in Table 5.1 are effective rate constants k_{eff} calculated from the equation

$$-dc_{H_2Pc}/dt = k_{eff}c_{H_2Pc} \tag{5.4}$$

and related to the true rate constant k_v by

$$k_{eff} = k_v c^4_{H_3O^+}. \tag{5.5}$$

A preliminary study[130] of the rates of hydrolytic dissociation of H_2Pc in H_2SO_4 solutions of variable concentrations at 25–60 °C has shown that H_2PcH^+ dissociates in accordance with a fifth-order kinetic equation—first-order with respect to phthalocyanine and fourth-order with respect to the H_3O^+ ion, if the H_2SO_4 concentration is within the Brand

†For simplicity the reaction is written for phthalocyanines of doubly charged cations.

Table 5.1 Rates of dissociation of phthalocyanine and its labile complexes in H_2SO_4 solutions at 25 °C

Complex	$c^0_{H_2SO_4}$ (M)	$c_{H_3O^+}$ (M)	k_{eff} (h^{-1})	$\tau_{1/2}$ (h)
H_2Pc	17.65	6.18	0.137 ± 0.008	5.1
	15.40	12.90	5.30 ± 0.14	0.131
MgPc	17.65	6.18	0.113 ± 0.011	6.14
	15.40	12.90	3.40 ± 0.26	0.204
Ag_2Pc	17.65	6.18	0.170 ± 0.010	4.08
	15.40	12.90	4.78 ± 0.20	0.145
PbPc	17.65	6.18	0.109 ± 0.008	6.36
	15.40	12.90	5.08 ± 0.13	0.136
SnPc	17.65	6.18	0.145 ± 0.024	4.79
	15.40	12.90	2.26 ± 0.42	0.306
$(HSO_4)FePc$	17.65	6.18	0.198 ± 0.014	3.50
	15.40	12.90	0.550 ± 0.025	1.26
$(HSO_4)_2SnPc$	17.65	6.18	0.0155 ± 0.0014	45
	15.40	12.90	0.160 ± 0.018	4.3
BePc	17.65	6.18	0.095 ± 0.008	7.30
	15.40	12.90	1.40 ± 0.06	0.50
GaPc	17.65	6.18	0.104 ± 0.02	6.65
	15.40	12.90	5.18 ± 0.23	0.13

region (15–18.8 M). The dissociation mechanism of H_2Pc has not yet been elucidated and is probably complicated. The true constant of H_2PcH^+ dissociation at 25 °C is equal to 1.12×10^{-4} $1^4 \, mol^{-4} \, h^{-1}$, and the activation energy is equal to 19.5 kcal mol^{-1}. The fourth order with respect to the H_3O^+ ion is responsible for the extremely high dissociation rates of H_2Pc in H_2SO_4 solutions of medium concentrations and its high stability in the monohydrate.

As can be seen from Table 5.2, complexes MgPc, Ag_2Pc, PbPc, SnPc, GaPc, and BePc in H_2SO_4 solutions easily lose the metal during their dissolution in H_2SO_4. Only protonated phthalocyanine ligand H_2PcH^+ is practically present in these solutions. However, $(HSO_4)FePc$ in 15 M H_2SO_4 already dissociates 10 times more slowly than H_2Pc. The phthalocyanine complex with Sn(IV) of composition $(Cl_2)SnPc$ or $(HSO_4)_2SnPc$ is even more stable than all of the above complexes. In this case $k_1 \ll k_2$, and the limiting step of decomposition in H_2SO_4 solutions is the dissociation in accordance with equation (5.2).

The stability of phthalocyanine complexes increases sharply when the cation is capable of forming σ bonds, as a result of its greater electron affinity and the possibility of the 3*d* or 4*d* levels being filled with the ligand electrons (Sn^{4+}, Ga^{3+}, Al^{3+}, VO^{2+}, Zn^{2+}). The decomposition rate of these complexes in H_2SO_4 solutions is extremely low at room temperature. It is limited by reaction (5.2), that is dissociation of the complex, and may be studied at temperatures of about 100 °C and higher. These are stable phthalocyanines.

Table 5.2 Rate constants of dissociation of protonated phthalocyanine complexes in accordance with equation (5.6)

Complex	t (°C)	$c^0_{H_2SO_4}$ (M)	$c_{H_3O^+}$ (M)	k_{eff} (h^{-1})	$k_v \times 10^2$ ($l^2\ mol^{-2}\ h^{-1}$)
$CuPcH^+$	100	18.20	3.52	0.0104 ± 0.0004	0.0843
		18.20	3.52	0.0109 ± 0.0007	0.0880
		16.20	10.40	0.100 ± 0.005	0.0925
		16.18	10.40	0.090 ± 0.005	0.0843
	120	18.20	3.94	0.185 ± 0.012	1.19
		18.00	4.88	0.230 ± 0.010	0.966
		17.55	6.24	0.435 ± 0.023	1.15
		16.70	9.40	0.735 ± 0.014	0.912
		16.25	10.40	1.19 ± 0.02	0.830
	136	17.27	7.65	3.57 ± 0.30	6.10
$PtPcH^+$	100	16.27	10.15	0.0054 ± 0.0005	0.0053
	120	18.20	3.94	0.0075 ± 0.0005	0.048
		17.10	8.00	0.028 ± 0.002	0.044
		16.64	9.50	0.038 ± 0.001	0.042
		15.73	11.47	0.056 ± 0.001	0.042
		14.82	13.00	0.079 ± 0.004	0.046
	155	18.20	5.10	0.295 ± 0.036	1.13
$PdPcH^+$	100	16.27	10.15	0.0055 ± 0.0003	0.0054
	120	17.10	8.00	0.029 ± 0.002	0.045
		16.61	9.50	0.037 ± 0.004	0.041
		16.18	10.50	0.047 ± 0.004	0.043
		16.78	11.47	0.066 ± 0.002	0.050
	155	18.00	5.65	0.450 ± 0.002	1.40
		16.37	10.63	1.39 ± 0.020	1.23
		16.35	10.63	1.53 ± 0.07	1.36
$ZnPcH^+$	65	16.00	11.00	0.085 ± 0.07	0.070
	100	17.15	7.50	0.28 ± 0.02	0.50
		16.70	8.95	0.41 ± 0.02	0.52
		16.27	10.15	0.56 ± 0.02	0.54
		16.20	10.40	0.52 ± 0.05	0.50
	120	18.20	3.94	0.31 ± 0.03	2.0
		16.37	10.10	1.76 ± 0.16	1.7
		16.25	10.40	2.09 ± 0.18	1.9
$CoPcH^+$	100	18.20	3.52	0.0147 ± 0.0003	0.118
		16.38	9.75	0.101 ± 0.009	0.106
	120	18.20	3.94	0.085 ± 0.007	0.549
		16.20	10.50	0.474 ± 0.028	0.430
	140	16.37	10.35	1.29 ± 0.23	1.20
$NiPcH^+$	100	16.20	10.40	0.0499 ± 0.0034	0.0463
	120	18.20	3.94	0.023 ± 0.002	0.150
	155	17.17	8.37	0.780 ± 0.050	1.11
		17.15	8.37	0.960 ± 0.078	1.37
$(HSO_4)AlPcH^+$	100	16.30	10.09	0.036 ± 0.004	0.036
	108	16.36	9.50	0.049 ± 0.002	0.055
	120	14.37	13.23	0.217 ± 0.016	0.124

Table 5.2 *continued*

Complex	t (°C)	$c^0_{H_2SO_4}$ (M)	$c_{H_3O^+}$ (M)	k_{eff} (h^{-1})	$k_v \times 10^2$ ($l^2\ mol^{-2}\ h^{-1}$)
		15.75	11.47	0.146 ± 0.007	0.110
		16.20	10.50	0.119 ± 0.007	0.108
		17.07	7.96	0.063 ± 0.007	0.099
		17.53	6.14	0.045 ± 0.004	0.119
	138	15.70	11.65	0.406 ± 0.008	0.298
		17.47	6.85	0.119 ± 0.016	0.254
$(HSO_4)GaPcH^+$	100	14.82	13.00	0.405 ± 0.030	0.240
		15.27	12.40	0.351 ± 0.002	0.228
		16.18	10.40	0.287 ± 0.037	0.266
		16.62	9.17	0.242 ± 0.040	0.288
	120	17.08	8.00	0.403 ± 0.009	0.631
		18.00	4.88	0.130 ± 0.012	0.544
	138	17.47	6.85	0.550 ± 0.011	1.17
$(HSO_4)RhPcH^+$	120	15.72	11.47	0.039 ± 0.002	0.0292
		16.64	9.50	0.0258 ± 0.0007	0.0287
		17.50	6.28	0.0127 ± 0.0005	0.0322
	138	17.27	7.65	0.060 ± 0.006	0.102
	155	16.35	10.63	0.838 ± 0.023	0.202
$(HSO_4)_2SnPcH^+$	25	17.18	8.00	0.0201 ± 0.0013	0.0314
	35	17.18	7.80	0.0425 ± 0.0059	0.0695
	45	17.18	7.60	0.119 ± 0.011	0.205
	55	17.18	7.40	0.324 ± 0.009	0.590
	65	17.18	7.13	0.902 ± 0.018	1.78
	66	17.70	5.20	0.484 ± 0.030	1.79
	40	15.08	13.00	0.326 ± 0.018	1.93
		17.00	8.30	0.160 ± 0.008	2.32
		17.70	5.55	0.063 ± 0.004	2.04
$(HSO_4)_2VPcH^+$	60	16.20	10.50	0.0125 ± 0.0005	0.0114
	100	18.20	3.52	0.057 ± 0.003	0.46
		18.00	4.33	0.080 ± 0.001	0.43
		17.40	6.45	0.201 ± 0.005	0.48
		16.40	9.78	0.58 ± 0.02	0.61
	120	18.20	3.94	0.38 ± 0.01	2.44
$(HSO_4)_2OsPcH^+$	120	16.35	10.10	0.0139 ± 0.0012	0.0136
	138	16.23	10.53	0.0477 ± 0.0021	0.0434
	157	10.95	10.95	0.181 ± 0.012	0.151
$(HSO_4)RuPcH^+$	135	17.07	8.35	0.0335 ± 0.0014	0.048
		16.70	9.30	0.0435 ± 0.0010	0.051
		16.63	9.30	0.0463 ± 0.0043	0.048
		16.20	10.70	0.0542 ± 0.0012	0.0475
	155	17.58	6.62	0.0803 ± 0.0040	0.182
		16.20	10.70	0.234 ± 0.020	0.205
		16.20	10.70	0.235 ± 0.020	0.200
$(HSO_4)IrPcH^+$	119	17.58	6.07	0.011 ± 0.001	0.0297
	135	17.10	8.17	0.041 ± 0.008	0.0612
		15.50	11.80	0.080 ± 0.010	0.0576
	155	17.58	6.62	0.0415 ± 0.0021	0.117

Finally, the stability of MPc increases to the maximum when the cation becomes capable of forming σ or π bonds (e.g. Co^{2+}, Ni^{2+}, Cu^{2+}, Pt^{2+}, Pd^{2+}, Ru^{3+}, Rh^{3+}, Ir^{3+}, Os^{4+}, Si^{4+}). The rate of their decomposition in an H_2SO_4 solution even at 100 °C is extremely low and difficult to measure. This is why the dissociation constant of these stable phthalocyanines was studied at higher temperatures (100–155 °C).

It has been established[37,386] that the rate of reaction (5.2) decreases considerably as the H_2SO_4 concentration grows and remains in the Brand region proportional to the square of the H_3O^+ concentration. Thus the following universal kinetic equation of dissociation of all stable metal phthalocyanines has been derived:

$$-\mathrm{d}c_{\mathrm{MPcH}^+}/\mathrm{d}t = k_v c_{\mathrm{MPcH}^+} c^2_{\mathrm{H_3O}^+}, \tag{5.6}$$

applicable over the entire range of the temperatures examined. Given in Table 5.2 are the values of effective dissociation rate constants of stable phthalocyanines ($k_{\mathrm{eff}} = k_v c_{\mathrm{H_3O}^+}$) and true constants of equation (5.6).

In recent years studies have been made into the kinetic stability of phthalocyanine complexes with titanium, $(Cl_2)TiPc$ and $(HO)_2TiPc$;[390] germanium, $(HO)_2GePc$;[391] silicon, $(Cl_2)SiPc$;[116] uranyl, UO_2Pc,[113] and molybdenyl.[114] The Ti, Ge, and Si complexes have been found to be stable, while UO_2Pc turned out to be a typically labile compound; MoO_2Pc occupies an intermediate position between stable and labile phthalocyanine complexes.

Rate constants k_v of all stable phthalocyanine complexes, including $(HSO_4)MPc$ and $(HSO_4)_2MPc$, are invariable over a broad range of temperatures and H_2SO_4 concentrations. This is possible only in the case of complete detachment of the HSO_4^- counter-ion at high temperatures, that is when their dissociation is measured, and may serve as a demonstration of the existence of coordinatively unsaturated $MPcH^{2+}$ and $MPcH^{3+}$ ions (M is a triply or quadruply charged cation) in solutions at 100–150 °C. It also proves that no sulphonation of the benzene rings takes place under experimental conditions.

The rate constants are strongly dependent on the metal atom, its electronic structure, charge, radius, and degree of coordinative saturation. The values of the rate constants correlate well with the capacity of the central atom to form σ and π bonds of various strength, that is with the degree of covalence of the M—N bonds.[37,88,387]

The kinetic equation of protolytic dissociation being independent of the nature of the metal atom can be explained in terms of a common mechanism for this process. The concepts of trimolecular dissociation mechanism will be set forth later. The structural similarity of the coordination centres of phthalocyanine and porphyrins, as well as the aromatic portion of their molecules, has enabled the present author[37] to predict that both metal phthalocyanines and metalloporphyrins of the stable group are described by the same kinetic equations and characterized by the same dissociation mechanisms; this prediction was corroborated experimentally at a later date. The rate constants were used in determining the energies, free energies, and entropies of activation of MPc, XMPc, and X_2MPc, listed in Table 5.3. The rate constants are accurate to within 5–10%, the values of $\Delta S^{\ddagger}$ to within 2–4 e.u., and the values of E to within 0.5–1 kcal mol^{-1}.

A most significant finding from the data obtained in the earlier work[37] is the high kinetic activity of the H_3O^+ ions during dissociation and inactivity of the more protogenic H_2SO_4 molecules. To the best of my knowledge no examples could be found in the literature, before publication of my own works[102,191] on protolytic complex dissociation reactions with a limiting step involving the least protogenic of the two particles. In the above-cited works this phenomenon was explained as follows.

In the case of stable covalent complexes, such as metal phthalocyanines and metalloporphyrins, breakage of strong covalent M—N bonds is the most energy-demanding stage of activation of the molecules in the course of dissociation. The energy demand may be a minimum if the attacking particle carries both the proton necessary for binding electron pairs of the N atom and the strongly solvating molecule capable of occupying the coordinate place which becomes vacant in the inner sphere of the cation and lowering its effective charge. Two such ideal attacking particles are H_3O^+ in aqueous media and H^+_{solv} in non-aqueous media.

The molecules of acids even as strong as H_2SO_4 do not have this advantage. They are good proton transferrers but their carrier (HSO_4^-) is a much weaker solvating agent (in the most general sense) than water. Besides, H_2SO_4 differs from H_3O^+ by poorer mobility in solution and a less favourable geometric configuration. There is no reason to doubt that the

Table 5.3 Kinetic parameters of dissociation of stable phthalocyanine complexes in accordance with reaction (5.2)

Complex	$k_v \times 10^5$ (l^2 mol^{-2} h^{-1})	E (kcal mol^{-1})	$\Delta G^{\ddagger}$ (kcal mol^{-1})	$\Delta S^{\ddagger}$ (e.u.)
$ZnPcH^+$	1.43	17.6	29.0	−38
$CoPcH^+$	0.080	21.4	30.7	−31
$NiPcH^+$	0.066	19.0	30.8	−40
$CuPcH^+$	0.00049	35.7	33.8	+7.1
$PdPcH^+$	0.000084	32.4	34.8	−7.7
$PtPcH^+$	0.000094	32.1	34.7	−8.5
$(HSO_4)GaPcH^+$	5.75	11.9	28.2	−55
$(HSO_4)IrPcH^+$	0.198	12.5	30.1	−59
$(HSO_4)AlPcH^+$	0.142	16.3	30.4	−47
$(HSO_4)RhPcH^+$	0.0032	22.4	32.6	−34
$(HSO_4)RuPcH^+$	0.00077	24.3	33.3	−30
$(HSO_4)_2SnPcH^+$	31.2	21.7	27.2	−17
$(HSO_4)_2VPcH^+$	0.240	23.1	30.1	−24
$(HSO_4)_2OsPcH^+$	0.0019	21.8	32.9	−37

solvating proton is the dissociation-initiating particle in most stable complexes with an anion in acidic media. A similar interpretation of the protolytic dissociation of porphyrin complexes, assuming simultaneous protonation and solvation, was offered elsewhere.[364,392]

A striking feature is the highly negative values of the entropies of activation (see Table 5.3), which vary for most metal phthalocyanines from −20 to −50 e.u. on the average. Only in the case of Cu, Pd, and Pt complexes are they greater. The values of $\Delta S^{\ddagger}$ suggest that the transition state of the reacting system MPc + $2H_3O^+$ is much more polar than the initial state. This is due to the strong polarization of the M—N bond, which is also responsible for high values of the activation energies. They range from 12 to 35 kcal mol^{-1}. The Cu, Pd, and Pt complexes whose central atom is coordinatively saturated and forms the strongest bonds are characterized by the highest activation energies. In cases where the covalence of the M—N bonds decreases (Zn, Ga) and the central atom becomes coordinatively unsaturated (Ga, Zn, Al, Ir), the activation energy is lower. These regularities will be discussed at greater length in Section 5.4.

In conclusion here are some results of studies into the kinetics of dissociation of polymeric phthalocyanines. In earlier works[147,148] the present author investigated the kinetic stability, in H_2SO_4 solutions, of Zn, Cu, Ni, Ga, and Os complexes with polymeric phthalocyanine in which the conjugation chain passes through the benzene rings (see Fig. 2.2). By analogy with metal phthalocyanines the overall dissociation process may be written for labile polymeric phthalocyanine complexes as

$$(\mathrm{MPc})_i\mathrm{H}_m^{m+} + 2i\mathrm{H_3O^+} \rightleftarrows (\mathrm{H_2Pc})_i\mathrm{H}_m^{m+} + i\mathrm{M}^{2+}_{\mathrm{aq}}. \quad (5.7)$$

A polymer molecule with polymerization degree i, monoprotonated in a sulphuric acid solution at m units (it is shown that $m \leqslant i$), undergoes protolytic dissociation at all units and becomes an m-protonated ligand. The mechanism of dissociation at a chain unit must be identical with that in the case of MPc. The dissociation of stable polymeric complexes takes place under the same stringent conditions as that of a monomer (highly concentrated sulphuric acid solution; temperature between 50 and 150 °C).

As can be seen from the kinetic curves (Fig 5.1), the dissociation of stable complexes does not correspond to general equation (5.7) but follows a sequential dissociation–decomposition pattern (stoicheiometric 'link-after-link' mechanism):

$$(\mathrm{MPc})_i\mathrm{H}_m^{m+} + 2\mathrm{H_3O^+} \xrightarrow{k_1} \{(\mathrm{MPc})_{i-1}\mathrm{H_2Pc}\}\mathrm{H}_m^{m+} + \mathrm{M}^{2+}_{\mathrm{aq}}. \quad (5.8)$$

The slow dissociation in the first link is then followed by a fast disintegration of the now metal-free macroring into colourless X fragments (phthalimide and pyromellitimide):

$$\{(\mathrm{MPc})_{i-1}\mathrm{H_2Pc}\}_m^{m+} \xrightarrow{k_{\mathrm{hydr}}} (\mathrm{MPc})_{i-1}\mathrm{H}_{m-1}^{(m-1)+} + 3\mathrm{X}, \text{ etc.} \quad (5.9)$$

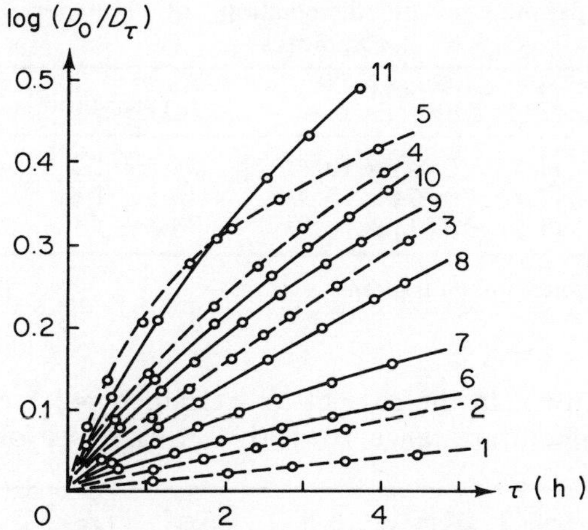

Figure 5.1 Kinetic curves of dissociation of copper (curves 1–5) and gallium (curves 6–11) polyphthalocyanines Temperature: curves 1, 2, 100 °C; curves 3, 4, 120 °C; curve 5, 130 °C; curves 6, 7, 70 °C; curves 8, 9, 70 °C; curves 10, 11, 90 °C. H_2SO_4 concentration: curves 1, 3, 6, 8, 10, 17 °C; curves 2, 4, 5, 7, 9, 11, 16 °C

The final stage of dissociation is process (5.2), characterized by final-stage dissociation rate constant k_i. In the general case one may expect that the numerical values of the first- (k_1) and final- (k_i) stage constants are different. In earlier works[147,148] all of the polymeric complexes examined were shown to be inferior in stability to monomers. The first-stage dissociation constants of polymers are several times greater than constants k_v of monomers, while final-stage constants k_i practically coincide with the latter, which corroborates the sequential pattern of polymer decomposition (5.8) and (5.9).

Table 5.4 lists kinetic dissociation parameters calculated from the first-stage dissociation constants at the beginning of the kinetic curves (see Fig. 5.1). The tabulated data indicate that polymerization drastically alters all kinetic parameters. The point is that the polymeric phthalocyanines of this type (ribbon polymers with conjugation over the entire macromolecule) feature an average degree of polymerization not greater than 6–8.[393] This is probably why $(MPc)_i$ complexes with polymerization degrees of several tens and hundreds will differ significantly from monomers. The difference in the stability of MPc and $(MPc)_i$ complexes was ascribed[148] to some distortions of the planar structure of the macroring in polymers and to mutual influence of the metal atoms in the same macromolecule. This influence increases with the chain length. The order (n) of the dissociation reaction of polymers with respect to the H_3O^+ ion, calculated from the graphical relation $\log k_{eff} = \log k_i + n \log C_{H_3O^+}$, was

Table 5.4 Kinetic parameters of dissociation of polymeric phthalocyanine complexes

Polymer	k_v^{25} (l^2 mol^{-2} h^{-1})	E (kcal mol^{-1})	$\Delta S^{\ddagger}$ (e.u.)
$(CuPc)_i$	$(1.58 \pm 0.04) \times 10^{-8}$	30.3 ± 1.3	−9.7
$[(HSO_4)GaPc]_i$	$(7.43 \pm 0.07) \times 10^{-5}$	16.9 ± 1.6	−36.6
$[(SO_4)OsPc]_i$	$(1.72 \pm 0.01) \times 10^{-6}$	15.6 ± 1.1	−48.4

Note: The first-stage rate constant is given.

found to approach two. In the case of all examined polymers this relation is linear. In the temperature range 70–150 °C n varies as follows:

Polymer	$(CuPc)_i$	$[(HSO_4)GaPc]_i$	$(NiPc)_i$	$[(SO_4)OsPc]_i$	$(ZnPc)_i$
n range	1.96–2.12	1.67–1.70	1.93	1.63–1.89	*c.* 2

The present author also studied[394–396] the kinetic stability of polymeric Co^{2+} phthalocyanines of different structures.

In Co^{2+} complexes with a ligand structure (see Fig. 2.2) first-stage rate constant k_1^{25} is three orders of magnitude greater than in a monomer, which is indicative of a strong distortion of the macroring in the polymer chain. The reaction was found to be of second order with respect to the H_3O^+ ion, and the polymerization degree was found to be 6–8.

Polymeric phthalocyanine complexes with a cross-linked structure (see Fig. 2.2 between lines AA and BB) may be more stable than the monomer. This is due to the absence of distortion of the planar macroring structure in polymers of this kind. However, the stability of such polymeric complexes is virtually unknown. In a previous work[396] a study was made into the kinetic stability of a polymeric complex of phthalocyanine with silicon in which the polymer chain passes through the central atom (bridged polymers). Their properties differ in principle from those of the polymers described above. They exhibit a higher stability than monomeric $(HO)_2SiPc$, and their decomposition involves a different stoicheiometric mechanism for which I proposed[396] the term 'skeleton mechanism'. A more detailed description of these compounds will be given in Chapter 6.

5.2 Dissociation of Chlorophyll, Haemin, and Their Analogues

Porphyrin complexes with most metals are highly stable in solution, but their stability is much inferior to that of the corresponding phthalocyanine complexes. In a concentrated H_2SO_4 solution even the most stable metalloporphyrins (Fe^{3+}, Ni^{2+}, Cu^{2+}, and Pt^{2+} complexes) dissociate instantaneously in spite of the fact that they have an advantage over metal phthalocyanines. The point is that in a highly protogenic medium metal phthalocyanines are present in a protonated form ($MPcH^+$), and the peripheral proton seems to affect the M—N bond strongly, weakening the latter. In metalloporphyrins no such thing occurs because most of them lack heteroatoms in the conjugated system, with the exception of four complexed N atoms.

As compared to phthalocyanines, porphyrins and their complexes have the advantage of being more readily soluble in organic solvents of different types. Moreover, about 10 water-soluble porphyrins have already been obtained. All this offers good possibilities for the study of the effect of the nature of the solvent on their stability in solution. Most studies into the kinetic stability and mechanism of dissociation of porphyrin complexes have been conducted in organic solvents or their mixtures (glacial acetic acid, its mixtures with ethanol, monohydrate of H_2SO_4, or water, butanol–trichloroacetic acid, acetone–water, and other systems).

Like metal phthalocyanines porphyrin complexes are divided, depending on the metal species and irrespective of the structure of the side branches of the conjugated portion of the molecule, into two clearly defined groups—labile and stable complexes. The former include complexes of Li, Na, Ca, Be, Hg, and Pb (their dissociation is caused by H_2O) and Mg^{2+} and Cd^{2+} (dissociation in diluted acid solutions). The latter are complexes of Cu^{2+}, Ni^{2+}, Co^{2+}, Pd^{2+}, Ag^{2+}, Fe^{3+}, Mn^{3+}, and Zn^{2+} (dissociation in organic solvents in the presence of monohydrate of H_2SO_4). In some cases the rates of their dissociation can be measured in glacial acetic acid at temperatures in the neighbourhood of 100 °C.

The overwhelming majority of studies into the dissociation of metalloporphyrins have been carried out over the past decade. Characteristically, in his monograph[23] published in 1964, Falk presented no quantitative data on the stability of metalloporphyrins.

In order to have a general idea of the kinetic stability of metalloporphyrins, depending on the structural factors as well as metal and porphyrin species, it is now appropriate to review the results of systematic studies into the protolytic dissociation of chlorophyll *a*, its metal analogues, and complexes of protoporphyrin, mesoporphyrin, and tetraphenylporphin.

The first attempt to study the dissociation kinetics of chlorophyll qualitatively was made by Mackinney and Joslyn.[372–274] The reaction

$$MgChl + 2HX \rightarrow Mg^{2+} + H_2Chl + 2X^- \qquad (5.10)$$

was investigated by them in 90% aqueous acetone in the presence of one of the following acids: $H_2C_2O_4$, H_2SO_3, H_2SO_4, and HCl. The reaction was tentatively found to be of first order with respect to the acid, but its order with respect to chlorophyll was not established. Reliable kinetic parameters could not be obtained either. The unsuccessful results stem from improper selection of the proton-donor solvent in which partial protonation of H_2Chl to $H_2ChlH_2^{2+}$, aggregation and other phenomena could be observed in an elaborate spectral picture permitting measurement of the dissociation rates.

Among the interesting results of this work[373–374] is the finding that chlorophyll *a* dissociates at a rate seven times faster than chlorophyll *b*.

Studies into the dissociation kinetics of chlorophyll and some of its structural derivatives were repeated by Schanderl, Chichester, and Marsh[397] in 80% aqueous acetone in the presence of 10^{-4} M HCl. Under these conditions chlorophyll *a* was found to dissociate 5.5 times faster than chlorophyll *b* at the same activation energies, equal to 10.5 kcal mol^{-1}. It was also established that the substitution of C_2H_5, CH_3, and H for the phytol residue, that is transition from chlorophylls to ethylphaeophorbides, methylphaeophorbides, and phaeophorbides, accelerates the dissociation only slightly (the rate constants at 25 °C for compounds of the *a* series equal, respectively, 0.0103, 0.0127, 0.0156, and 0.0171 min^{-1}). The factors responsible for a similar effect of the alkyl size on the dissociation rate were not explained but, judging from the invariable values of the activation energy, it may be assumed that they are purely structural and involve changes in entropy. This study has provided no information on the mechanism of reaction (5.10) and the nature of bonding between Mg^{2+} and porphyrins.

The dissociation of chlorophyll and its numerous metal analogues was examined systematically in other works[71,73,247,375–379,398] under comparable conditions in adequately selected solvents creating no difficulties. It is advisable to consider the dissociation of chlorophyll and its labile metal analogues separately from stable complexes. The best medium for studying the dissociation of chlorophyll and its labile analogues is glacial acetic acid or its mixtures with ethanol. Stable chlorophyll analogues were also studied in glacial acetic acid.

Presented in Fig. 5.2 are changes in the absorption spectrum of MgChl*a* in ethanol with 2.0 M HAc. In the course of dissociation MgChl transforms to H_2Chl (curve 7). A similar spectral change is observed during dissociation of CdChl. However, the dissociation of CdChl in an ethanol solution requires the addition of acetic acid in amounts 100 times smaller than in the case of MgChl. Complex HgChl dissociates instantaneously after a trace amount of HAc is added to its alcohol solution. These data give a better insight into the difference in the covalence of the M—N bonds in Mg^{2+}, Cd^{2+}, and Hg^{2+} metalloporphyrins.

It is quite obvious that HgChl is in fact a purely ionic complex and the interaction in it is purely electrostatic, which is why its dissociation in a polar proton-donor medium is fast. CdChl is another predominantly ionic complex. However, a certain degree of covalence in the Cd—N bonds renders the ionization non-instantaneous. MgChl is already an essentially covalent complex which is an absolutely stable compound from the kinetic standpoint in extremely weak proton-donor media (e.g. ethanol–0.2 M HAc).

The dissociation of MgChl and CdChl is characterized by being first order with respect to the complex, which can be seen from the kinetic

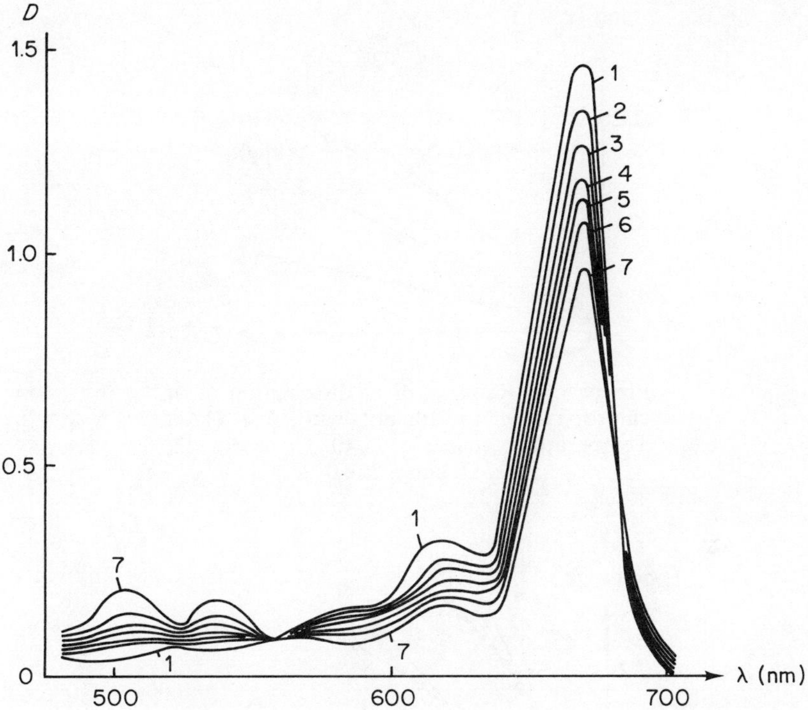

Figure 5.2 Changes in the spectrum of chlorophyll *a* in the course of dissociation in ethanol with 2 M HAc at 60 °C Time: curve 1, 0 min; curve 2, 15 min; curve 3, 45 min; curve 4, 75 min; curve 5, 120 min; curve 6, 180 min; curve 7, 600 min

curves of Figs 5.3 and 5.4. The linearity in the case of MgChl persists for several hours, and the dissociation under experimental conditions is irreversible. The dissociation of CdChl in ethanol plus 0.02–0.06 M HAc is perceptibly reversible and can be described by the following kinetic relation:

$$-\mathrm{d}c_{\mathrm{CdChl}}/\mathrm{d}t = k_{\mathrm{diss}}[\mathrm{CdChl}]\,[\mathrm{H}^{+}_{\mathrm{solv}}]^{n} - k_{\mathrm{rev}}[\mathrm{H_2Chl}][\mathrm{CdAc_2}]. \qquad (5.11)$$

Initially, when the concentration of $CdAc_2$ and H_2Chl is small, the rate of formation of CdChl can be ignored. At the same time (see Fig. 5.4) the rate constants remain invariable. The rates are the same in the temperature range 25–45 °C (curve 1), which is indicative of zero activation energy. This fact, along with the second order of the dissociation with respect to HAc, definitively prove that CdChl is an ionic complex and that the dissociation involves not only H_3O^+ or H^+_{solv} ions but also predominantly undissociated acetic acid molecules. Table 5.5 lists the values of effective rate constants of the MgChl and CdChl dissociation reaction that can be

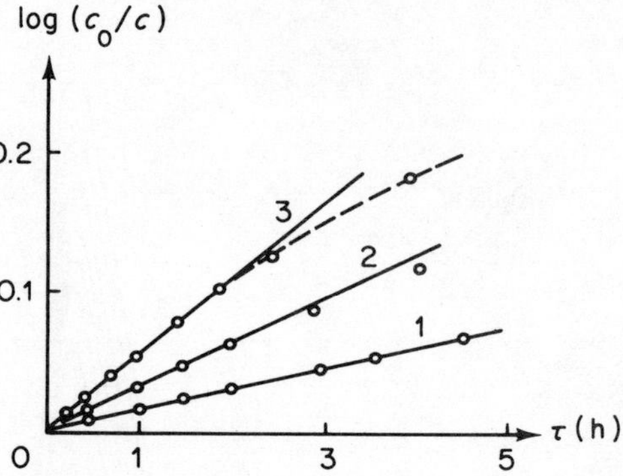

Figure 5.3 Rates of dissociation of chlorophyll *a* in ethanol with 2 M HAc. Temperature: curve 1, 40 °C; curve 2, 50 °C; curve 3, 60 °C

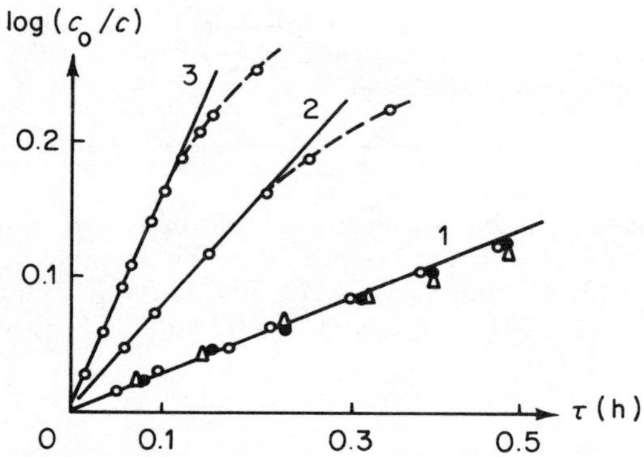

Figure 5.4 Rates of dissociation of the cadmium analogue of chlorophyll *a* in ethanol with HAc. HAc concentration: curve 1, 0.02 M at 25, 35, and 45 °C; curve 2, 0.04 M at 25 °C; curve 3, 0.06 M at 25 °C

written, in the general form, as follows:

$$MChl + 2H^+_{solv} + 2Ac^- \rightarrow MAc_{2solv} + 2H_2Chl. \qquad (5.12)$$

The constants have been calculated with predominant H^+_{solv} ion concentration from the following equation of pseudo-first-order reactions:

$$-dc_{MChl}/dt = k_{eff}c_{MChl}. \qquad (5.13)$$

The dissociation rate is measured with reference to the porphyrin complex concentration, which is calculated from the optical density values using the

Table 5.5 Rate constants of dissociation of chlorophyll *a* and its cadmium and mercury analogues in ethanol

Complex	HAc addition (M)	t (°C)	k_{eff} (h^{-1})
MgChl	2.0	40	0.038 ± 0.003
		50	0.076 ± 0.006
		60	0.142 ± 0.010
		25	0.013[a]
	4.0	25	0.34 ± 0.04
	6.0	25	8.3 ± 1.2
	10.0	25	41.5 ± 5.0
CdChl	0.02	25	0.68 ± 0.05
		35	0.64 ± 0.06
		45	0.62 ± 0.06
	0.04	25	1.86 ± 0.17
	0.06	25	3.8 ± 0.3
	0.1	25	13.8 ± 2.0
	0.2	25	83 ± 10
HgChl	0.02	25	Very fast

[a]Calculated from the Arrhenius equation: $E = 13.7 \pm 0.5$ kcal mol^{-1}, $\Delta S^{\ddagger} = -37 \pm 1$ e.u.

following formula for a mixture of two coloured substances:

$$c_{\text{MChl}} = c^0_{\text{MChl}}\ [(D_\tau - D_\infty)/(D_0 - D_\infty)] \tag{5.14}$$

and taking into account that during the reaction the system may include not more than two coloured compounds (c_{MAc_2} is negligibly small, while H_2Chl either is stable under the experimental conditions or decomposes to colourless compounds). The validity of equations (5.13) and (5.14) has been rigorously proved in experiments with all metalloporphyrin systems.

Table 5.5 suggests a strong dependence of the dissociation rate of complexes MgChl and CdChl on the HAc concentration. In the case of CdChl the dissociation is of second order with respect to HAc. This implies that the dissociation involves not only H^+_{solv} ions whose concentration is very small under the experimental conditions but also undissociated acetic acid molecules which determine the kinetics and mechanism of the reaction in the case of ionic complex CdChl. The situation with MgChl is more complicated. Large amounts of acetic acid added to ethanol do not permit establishing the dependence of the dissociation rate on its concentration for the amount of the added acid affects the structure of the solvent and the state of associates $C_2H_5OH{\cdots}HAc$, thereby significantly altering the reaction rate. The effect of the solvent composition and structure on the rates of metalloporphyrin formation was discussed at length in Chapter 4.

In order to establish the order of the chlorophyll dissociation reaction with respect to the acid and solvated proton the dissociation of chlorophyll was studied[375] in *tert*-butanol with 0.1–0.35 M CCl_3COOH. The acidity of the medium was characterized by the concentration of the protonated form of an *m*-nitroaniline indicator, which was calculated from the indicator relation

$$\log I = \log[(\varepsilon_I - \varepsilon)/(\varepsilon - \varepsilon_{IH^+})] \tag{5.15}$$

The logarithm of the ratio between the concentrations of the ionized and non-ionized forms of the indicator (log I) was found from the molar extinction coefficients of the indicator (ε_I), its protonated form (ε_{IH^+}), and their mixtures under actual experimental conditions (ε).

At relatively low acidities of the solutions the concentration of solvated proton $C_4H_9OH_2^+$ which takes an active part in the protolytic dissociation of metalloporphyrins and the protonated form of the indicator (IH^+) changes concommitantly with increasing amount of trichloroacetic acid in *tert*-butanol.[375] Therefore, the dependence of rate constant k_{eff} on the IH^+ ion concentration permits one to establish the order of the dissociation reaction with respect to the solvated proton. A similar relationship for MgChl and ZnChl is shown in Fig. 5.5. In the case of chlorophyll in the *tert*-butanol–trichloroacetic acid system the order of the reaction with respect to H^+_{solv} is clearly two. Similar relationships have been derived for structural analogues of chlorophyll, such as magnesium protoporphyrin[382] and magnesium mesoporphyrin.[339]

The dissociation of ZnChl is also of second order with respect to the solvated proton (see Fig. 5.5), however, the straight line deflects at a point corresponding to the reaction becoming faster in the acidity range where

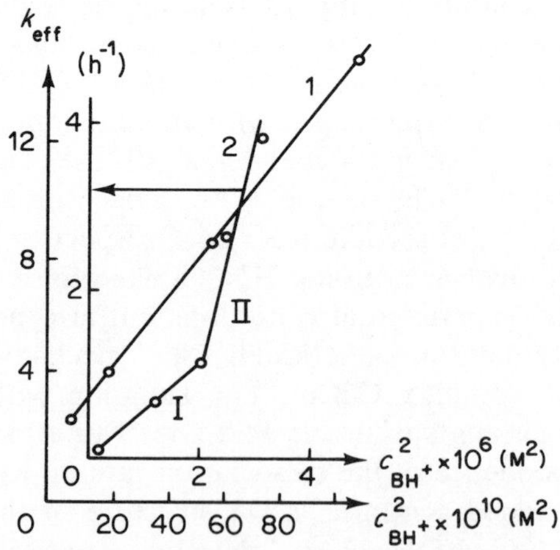

Figure 5.5 Effective dissociation rate constant of chlorophyll (curve 1) and its zinc analogue (curve 2) versus the square of the protonated indicator concentration in the *tert*-butanol–trichloroacetic acid system

the conditions of reaction (5.12) are changed. Along portion I the dissociation of complex ZnChl proceeds all the way to acid H_2Chl, while over portion II protonated form H_4Chl^{2+} occurs.

The concentration of IH^+ ions could not be determined during dissociation of complex CdChl, therefore Fig. 5.6 illustrates the dependence of the effective reaction rate constant on the CCl_3COOH concentration, according to which the reaction is definitely of second order. This must be due to the fact that ionic complexes HgChl, CdChl, CaChl, and Na_2Chl may dissociate under the effect of not only solvated protons but also the acid molecules themselves. During dissociation of these complexes the concentration of H^+_{solv} is so small that the more concentrated but less active species becomes predominant.

Given in Table 5.6 are the dissociation rates of Mg, Zn, and Cd complexes with chlorophyllic acid, as well as the kinetic parameters of the dissociation reaction.[399] Note the wide difference in the kinetic stability of the complexes. ZnChl, which is one of the least stable metalloporphyrins, is much more stable kinetically than chlorophyll *a*. In *tert*-butanol with 0.350 M CCl_3COOH at 25 °C chlorophyll dissociates very rapidly at $k_{eff} = 15\ h^{-1}$, while ZnChl does not show any signs of dissociation. An approximate estimation by way of extrapolation suggests that the kinetic stability of complex ZnChl is at least three to four orders of magnitude greater than that of MgChl. Thus ZnChl is primarily a covalent complex. The difference in stability between ZnChl and MgChl in ethanol–acetic acid media is still wider (see Table 5.6), hence the strong dependence of the kinetics stability on the nature of the solvent, its structure, solvating

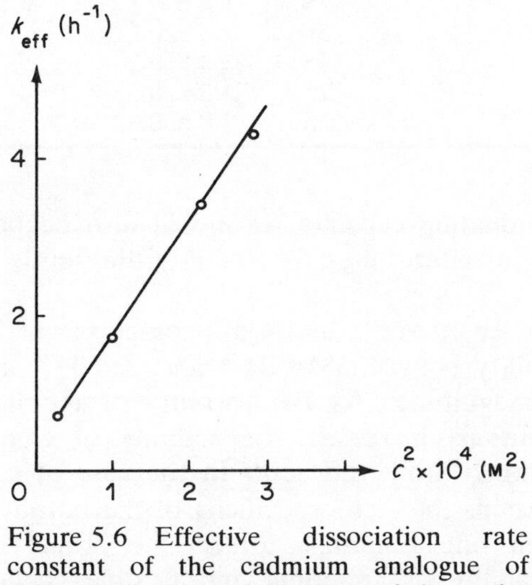

Figure 5.6 Effective dissociation rate constant of the cadmium analogue of chlorophyll versus the square of the molar concentration of trichloroacetic acid in *tert*-butanol

Table 5.6 Kinetic parameters of dissociation of chlorophyll and its zinc and cadmium analogues in the t-C_4H_9OH–CCl_3COOH system

Complex	$c^0_{CCl_3COOH}$ (M)	$c^2_{IH^+} \times 10^6$ (M)	t (°C)	k_{eff} (h^{-1})	E (kcal mol^{-1})	$\Delta S^{\ddagger}$ (e.u.)
ZnChl	1.10	0.220	40	0.212 ± 0.008		
			50	0.36 ± 0.03		
			60	0.72 ± 0.05		
			25	0.067	13.1 ± 0.5	−36 ± 2
	1.72	0.870	25	0.50 ± 0.03		
	2.06	1.85	25	1.13 ± 0.04	14.3 ± 0.7	−26 ± 3
			35	2.48 ± 0.09		
			45	5.17 ± 0.05		
	2.30	2.34	25	2.7 ± 0.1		
	2.50	3.06	15	1.73 ± 0.04		
			20	2.66 ± 0.09		
			25	3.67 ± 0.05	13.3 ± 0.7	−28 ± 2
MgChl	0.100		25	1.19 ± 0.07	14.5 ± 0.5	−26 ± 2
			35	2.7 ± 0.2		
			45	5.4 ± 0.4		
	0.150	0.0008	25	2.0 ± 0.1	13.3 ± 0.5	−28 ± 3
			35	4.2 ± 0.1		
			45	8.3 ± 0.6		
	0.200	0.0025	25	4.1 ± 0.1	12.6	−26
	0.270	0.0043	25	8.4 ± 0.6		
	0.350	0.010	25	14.6 ± 1.1		
CdChl	0.005		25	0.56 ± 0.03	7.8 ± 1.0	−45 ± 4
			35	0.79 ± 0.05		
			45	1.3 ± 0.2		
	0.010		25	1.7 ± 0.1	6.8 ± 1.0	−51 ± 4
			35	2.5 ± 0.1		
			45	3.4 ± 0.2		
	0.015		25	3.3 ± 0.3		
	0.018		25	3.8 ± 0.3		

capacity, and protonating capacity. It should also be borne in mind that solvents of different nature alter the stability differently depending on the metal species.

If one turns to Zn^{2+}, Mg^{2+}, and Cd^{2+} complexes with ammonia,[328] the difference in stability between $MgNH_3^{2+}$ and $ZnNH_3^{2+}$ amounts to two or three orders of magnitude. As the capacity of the ligand for covalent binding of the cations increases, the stability of complexes of various metals becomes much more different. In the case of coordination of the Mg^{2+} and Zn^{2+} cations the stereospecificity of the porphyrin molecule does not manifest itself but materially affects the properties of CdChl and HgChl. If the stability of ammonia and flexible chelate ligands of the ethylenediamine type changes as $Hg^{2+} \gg Cd^{2+}$, $Zn^{2+} \gg Mg^{2+}$, in the case of metalloporphyrins it changes as $Zn^{2+} \gg Mg^{2+} \gg Cd^{2+} \gg Hg^{2+}$.

The porphyrin ligand imposes the most stringent requirements on the covalent radius and electronic hybridization. Covalent radius $5s5p^35d^2$ of the Cd^{2+} ion would have greatly exceeded the dimensions of the inner coordination space, therefore no covalent complex with such a structure is formed. A predominantly ionic complex is formed instead, which can be seen from the dissociation constants. Extrapolation of the data presented in Tables 5.5 and 5.6 suggests that the kinetic stability of CdChl in *tert*-butanol is by at least three to four orders of magnitude lower than that of chlorophyll. In the ethanol–acetic acid medium the difference is as great as five to seven orders of magnitude. Thus CdChl differs from ZnChl in stability by a factor of eight to 10.

The kinetic stability of complex HgChl is still poorer. Its dissociation is almost instantaneous in cases where CdChl dissociates very slowly. In the ethanol–acetic acid medium the rate constant is approximately three orders of magnitude greater than that of CdChl. It may be assumed that porphyrin complexes with mercury and cadmium are sterically hindered and entail certain changes in the geometric configuration of the free porphyrin ligand. According to Hoard[206] a prerequisite for formation of a stable covalent complex is that the sum of the covalent radii of the nitrogen atoms should not greatly exceed 2.00 Å. In view of the fact that the radius of nitrogen is 0.65 Å,[400] that of the metal should not exceed 1.35–1.40 Å in the case of formation of a covalent complex. The radii of metals Mg, Zn, Cd, and Hg are 1.40, 1.31, 1.48, and 1.48 Å, respectively.

Possessing high (Cd^{2+}) and extremely high (Hg^{2+}) capacity to form complexes with simple and flexible nitrogen-containing ligands, the latter two ions are not capable of strong binding with the porphyrin because of their large size (the octahedral covalent radius of Cd is longer than the tetrahedral one by 1.48 Å), nor are they capable of d^2sp^3 or dsp^2 hybridization (Hg atom). In the case of complexes CdChl and HgChl only two linear covalent bonds N—M—N may emerge. Such a coordination system in porphyrins would have had an extremely poor kinetic stability even in weak proton-donor media and would have been closely similar to an ionic complex.

In the case of Zn^{2+}, Cu^{2+}, Fe^{3+}, Co^{2+}, Ni^{2+}, Ag^{2+}, Pd^{2+}, and other platinum metals both the steric and electronic factors favour formation of stable covalent porphyrin complexes. Presented in Table 5.7 are the conditions and kinetic parameters of dissociation of stable metal analogues of chlorophyll in glacial acetic acid.[71,376,379,398,401]

The dissociation of stable chlorophyllic acid complexes is slow in glacial acetic acid with 0.2–5 M H_2SO_4 monohydrate (Ni^{2+}, Cu^{2+}, Pd^{2+}, Fe^{3+}). The rate constants were measured in the temperature range 50–90 °C. The dissociation rates of Co^{2+}, Ni^{2+}, Cu^{2+}, Ag^{2+}, and Zn^{2+} complexes were measured in the same medium without H_2SO_4. The measurements were taken at temperatures as high as 80–110 °C (Ni^{2+}, Cu^{2+}, Co^{2+}) and 40–70 °C (Ag^{2+}). The dissociation of the Zn^{2+} complex can easily be

Table 5.7 Kinetic parameters of dissociation of chlorophyllic acid *a* complexes in glacial acetic acid

Complex	Addition	k_{eff}^{25} (h^{-1})	E (kcal mol^{-1})	$\Delta S^{\ddagger}$ (e.u.)
NiChl	0.5 M H_2SO_4	1.20×10^{-2}	16.5 ± 3.0	-28 ± 10
	1.0 M H_2SO_4	1.67×10^{-2}	17.0 ± 1.3	-26 ± 4
	—	1.3×10^{-6}	27.4 ± 1.4	-11 ± 6
CuChl	0.2 M H_2SO_4	1.23×10^{-2}	16.4 ± 2.8	-28 ± 9
	0.5 M H_2SO_4	1.92×10^{-2}	16.3 ± 1.5	-31 ± 5
	—	9.8×10^{-5}	19.0 ± 1.0	-29 ± 3
PdChl	1.0 M H_2SO_4	2.18×10^{-3}	17.1 ± 1.0	-29 ± 3
AcFeChl	0.5 M H_2SO_4	8.95×10^{-4}	17.3 ± 0.5	-29 ± 2
	2.0 M H_2SO_4	0.16 ± 0.004	14.1 ± 1.0	-35 ± 3
	5.0 M H_2SO_4	1.14 ± 0.10	12.7 ± 0.6	-31 ± 1
CoChl	—	1.61×10^{-3}	12.8 ± 0.5	-44 ± 3
ZnChl	—	0.976	10.1 ± 0.2	-41 ± 1
	10% H_2O	1.37	12.0 ± 0.5	-34 ± 2
	10% C_2H_5OH	0.425	12.7 ± 0.3	-33 ± 1
AgChl	50% C_2H_5OH	Very slow	—	—
	—	0.0225	19.6 ± 0.5	-16 ± 2
	5% H_2O	0.226	20.0 ± 1.3	-12 ± 4
	10% H_2O	0.430	20.5 ± 1.5	-8 ± 4

Note: The dash in the second column indicates that nothing was added to glacial acetic acid.

measured at 20–40 °C. Addition of small amounts of water (5–10%) promotes the dissociation, whereas addition of ethanol invariably slows down the reaction or stops it completely.

As can be seen from Table 5.7, the dissociation of stable chlorophyllic acid complexes in glacial acetic acid involves activation energies ranging from 10 to 20 kcal mol^{-1} and, as a rule, low entropies of activation (from -16 to -40 e.u.). Addition of water increases $\Delta S^{\ddagger}$. The low values of $\Delta S^{\ddagger}$ are due to the higher polarity of the transition state (see below) as compared to the initial state. The activation energy generally does not increase with the covalence of a complex. For example, in the case of ZnChl in glacial HAc it equals 10 kcal mol^{-1}, whereas the more ionic MgChl complex exhibits a value of 13.7 kcal mol^{-1} in ethanol–acetic acid. In glacial HAc with 0.5 M H_2SO_4 the activation energies of the Ni^{2+}, Cu^{2+}, and Fe^{2+} complexes are practically the same, while the kinetic stability of the latter is by one order of magnitude greater than that of the former two.

The kinetic stability, energy, and entropy of activation are strongly dependent on the nature of the solvent. For example, in glacial HAc with 0.5 M H_2SO_4 the stability of complexes NiChl and CuChl is almost the same, while the nickel complex is 100 times more stable than the copper one in glacial HAc (see Table 5.7). The strong dependence of the complex stability on the solvent is a feature common to all metalloporphyrins. In glacial acetic acid, its mixtures with ethanol and water, and with addition of small amounts of H_2SO_4 the kinetic stability of complexes of chlorophyllic acid *a* diminishes as follows: AcFeChl, PdChl > NiChl > CuChl > CoChl > AgChl > ZnChl ⪢ MgChl ⪢ CdChl ⪢ HgChl. In the case of phthalocyanine complexes[37] in aqueous concentrated H_2SO_4 the kinetic stability diminishes in a somewhat different sequence: PtPc, PdPc > CuPc > NiPc > CoPc > ZnPc ⪢ MgPc. Both sequences have been derived from the values of $\Delta G^{\ddagger}_{25}$, that is under standard conditions. The values of E obtained in the present author's laboratory[284] for all examined metal phthalocyanines and metalloporphyrins enable similar stability series to be derived at any temperature (up to 155 °C).

Comparison of the above kinetic stability series of metalloporphyrins and metal phthalocyanines, which can be accepted as a reference, suggests that the order in which the complexes are arranged is to a great extent dependent on the porphyrin species. For example, cations Cu^{2+} and Ni^{2+} have interchanged their positions in these series. As opposed to AcFeChl, complex AcFePc is highly unstable; it is closer to labile rather than stable phthalocyanines.[402] All iron porphyrins are among the most stable complexes. Unlike AgChl no phthalocyanine complex with Ag^{2+} could be produced. Complexes MgPc, Ag_2Pc, CdPc, and HgPc, just as the corresponding metalloporphyrins, are labile. Of particular interest is comparison of the kinetic stabilities of complexes of chlorophyllic acids *a* and *b* (Table 5.8).[248]

The transition from MChl*a* to MChl*b*, associated with the replacement of CH_3 by a CHO group at C3 in the second pyrrole ring, slows down the protolytic dissociation of the complex (300 times in the case of Zn^{2+} and 25 times in the case of Cu^{2+}), increases the activation energy by 1–3 kcal mol^{-1}, and changes the entropy of activation by 4–9 e.u. As was shown above (see Chapter 4), the complex formation rate in the case of H_2Chl*b* is also much slower than in the case of H_2Chl*a*. These interesting facts can only be explained by the +C effect of the formyl group or, to be more precise, the porphyrin molecule fragment including a semi-insulated double bond in the second pyrrole ring and a formyl group. As a result, the electron density at the nitrogen atoms increases, and the N—H and N—M bonds in porphyrin become stronger. In agreement with this

Table 5.8 Kinetic parameters of dissociation of copper and zinc complexes of chlorophyllic acids *a* and *b*

Complex	k^{25}_{eff} (h^{-1})	E (kcal mol^{-1})	$\Delta S^{\ddagger}$ (e.u.)
CuChl*b*	3.8×10^{-6}	23.4 ± 0.15	-25 ± 1
CuChl*a*	0.98×10^{-4}	19.0 ± 1.0	-29 ± 3
ZnChl*b*	3×10^{-3}	10.9 ± 0.5	-49 ± 2
ZnChl*a*	0.976	10.1 ± 0.2	-41 ± 1

hypothesis are the lower acid ionization constant of H_2Chlb and the stronger tendency of the porphyrin aromatic system to filling of the antibonding orbitals with electrons.[9,223–227]

The opposite point of view expressed by some investigators[23,403] attributes the properties of formyl derivatives to −C or I effects of the CHO group. Its proponents proceed exclusively from the basic properties of porphyrins, that is their protonating capacity.

However, the processes of protonation and complex formation in porphyrins differ in that the former do not involve any considerable stretching of the N—H bonds, while the latter are associated with their breakage. Falk[23] and Phillips[403] ascribe the lower complexing rates of porphyrins as a result of inclusion of a formyl group to the lower electron density at the nitrogen atoms. This assumption agrees well with the decreasing capacity for protonation of porphyrins with a CHO group. However, none of the investigators conducting similar studies have carried out polythermal experiments with protonation and measurements of changes in entropy. This is why the effect of solvation, which is an essential factor determining the direction of reactions in solutions, remains unknown as far as different porphin substituents are concerned. This problem has to be solved in the near future.

The available data suggest that the changes in the electron density at the N atoms of porphyrins under the effect of functional substituents influence the rate of complex formation only through the changing state of the N—H bonds, whereas the effect of the increasing (or decreasing) electron density at the pyrrolenine (=N—) nitrogen atoms on the coordination reaction rates is practically nil, otherwise it would have been strongly dependent on the electron-acceptor capacity of the metal of reagent salt $MX_2(solv)_{n-2}$.

The data presented in Chapter 4 indicate that the rate of reaction (4.50) is the same for acetates of Cu^{2+} and Cd^{2+}, Hg^{2+} and Pd^{2+}, Fe^{2+} and Ag^{+}, that is elements with widely different electron-acceptor capacity and the tendency to coordination to nitrogen-containing ligands. Therefore we attribute the slower coordination rate of metals in the case of H_2Chlb as opposed to H_2Chla to the increasing electron density at the nitrogen atoms and N—H bonds under the effect of the formyl group since the major portion of the activation energy of reaction (4.50) is involved in the stretching of the N—H bonds.

In the case of dissociation most of the energy will probably be expended in the breakage of the M—N bonds. One cannot explain, in terms of the electron-acceptor properties of the CHO group,[23,403] the greater stability of MgChl*b*, ZnChl*b*, and even CuChl*b*, the weaker ionization of H_2Chlb, and other behaviour.

It is of particular interest to compare the kinetic stabilities of complexes ClFeChl and ClFePP (haemin), as well as other complexes of two important naturally occurring porphyrins—protoporphyrin and chlorophyllic acid *a*. Haemin was found to be an extremely stable complex.

In pure glacial HAc it does not dissociate at a measurable rate even at elevated temperatures (116 °C maintained for 40 h). Metal-free protoporphyrin dissociates completely under the same conditions to less coloured products. The dissociation rate of haemin could be measured only in a highly protogenic medium, such as a mixture of acetic acid with 2–5 M H_2SO_4 at 60–90 °C (Figs 5.7 and 5.8).[76]

Addition of H_2SO_4 to the haemin solution in glacial HAc reduces the absorption intensity over the entire spectrum. A similar decrease in intensity was more clearly observed in stable H_2Chl complexes. In this case the long-wave (620–660 nm) band decreased in intensity (Fig. 5.9). The decrease is more pronounced as the complex becomes less covalent and as

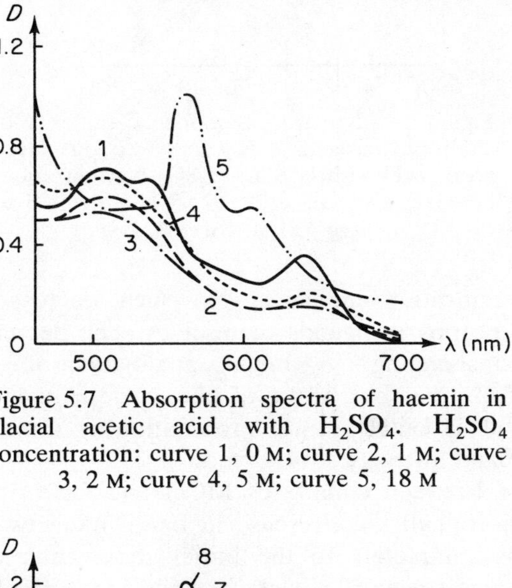

Figure 5.7 Absorption spectra of haemin in glacial acetic acid with H_2SO_4. H_2SO_4 concentration: curve 1, 0 M; curve 2, 1 M; curve 3, 2 M; curve 4, 5 M; curve 5, 18 M

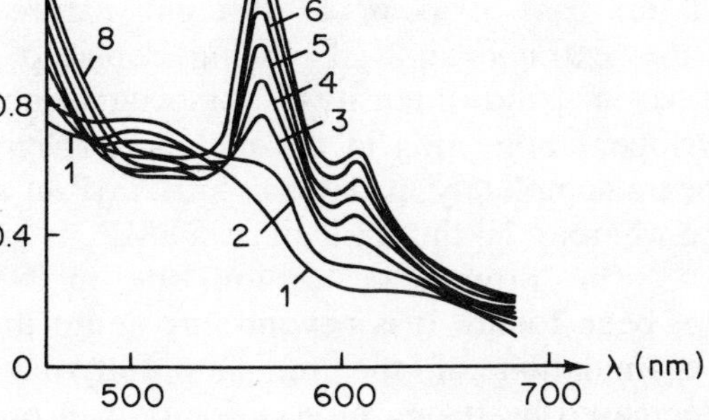

Figure 5.8 Dissociation of haemin in glacial acetic acid with 5 M H_2SO_4 at 70 °C. Time: curve 1, 0 h; curve 2, 0.25 h; curve 3, 0.70 h; curve 4, 1.50 h; curve 5, 2.50 h; curve 6, 4.0 h; curve 7, 6.0 h; curve 8, 8.0 h

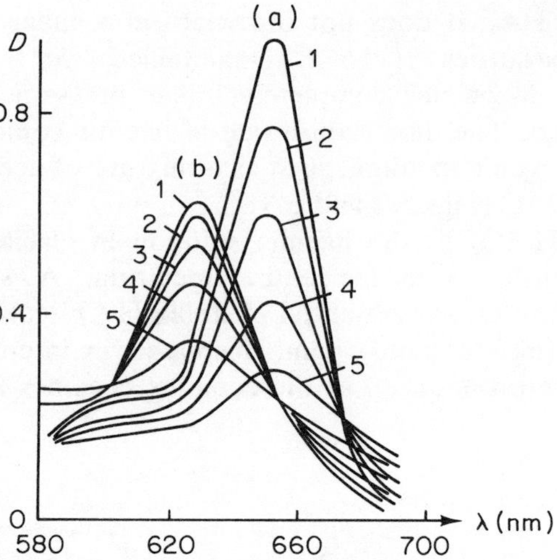

Figure 5.9 Spectra of copper complexes of chlorophyllic acids *a* (a) and *b* (b) in glacial acetic acid with H_2SO_4. H_2SO_4 concentration: curve 1, 0 M; curve 2, 0.2 M; curve 3, 0.5 M; curve 4, 1 M; curve 5, 2 M

the H_2SO_4 concentration increases. No such decrease in intensity is observed in the porphyrin ligands themselves. The decreasing intensity of the bands corresponding to metal analogues of chlorophyll was attributed[71,247,376,398] to protonation of the cyclopentanone ring. In free protonated ligand $H_2ChlH_2^{2+}$ such protonation is impossible, hence no decrease in the band intensity is observed.

The difference between complexes MChl and ClFePP is that in metal analogues of chlorophyll the decrease in band intensity is reversible (the spectrum returns completely to the initial state when the solutions are diluted with organic solvents or water), while in haemin it is irreversible. Hence, in the latter case it is of a different nature and is possibly determined by the extra ligand Cl^- being replaced by HSO_4^-. The reversible decrease in band intensity is associated not only with the presence of a cyclopentanone ring in the metalloporphyrin molecule. The first band disappears completely in the case of NiMP in glacial acetic acid with 5 M H_2SO_4, whereas in the case of AcFeMP its intensity does not decrease at all.[383] The proton localization site in NiMP and similar complexes has not been found. It is beyond any doubt that systematic and comprehensive examination of the metalloporphyrin absorption band intensity being decreased by strong proton donors will provide a wealth of valuable data on the origin of the bands and energy levels typical of porphyrins, their molecular structure, and chemical bonds in metalloporphyrins.

Figure 5.7 shows that the dissolution of haemin in a strong sulphuric acid solution leads to its rapid dissociation, yielding protonated protoporphyrin:

$$ClFePP + 4H^+_{solv} \rightarrow H_2PPH^{2+} + FeCl^{2+}_{solv}. \tag{5.16}$$

The protonation of protoporphyrin stabilizes its molecule. Thus heating of H_2PP in a mixture of acetic acid with 5 M H_2SO_4 at 70 °C, that is under extremely rigorous conditions, does not bring about any perceptible changes in the spectrum. Therefore the changes in the haemin spectra (see Fig. 5.8) must be due entirely to its dissociation in accordance with equation (5.16).

Chlorophyllhaemin or ClFeChl*a* (so-called iron phaeophytinate), which is a structural analogue of haemin, was found to be a less stable complex than haemin. Its dissociation in acetic acid with 2 M H_2SO_4 could be measured at 25–45 °C. Under such conditions the dissociation of iron phaeophytinate is accompanied by instantaneous decomposition of H_2Chl*a* immediately after the dissociation of the complex in accordance with equation (5.16). In the case the dissociation proceeds according to the following general equation:

$$MP + 2H^+_{solv} \rightarrow H_2P^* + M^{2+}_{solv}, \tag{5.17}$$

$$H_2P^* \rightarrow \text{decomposition products.} \tag{5.18}$$

The above path of the reaction involving decomposition of metalloporphyrins with disappearance of the porphyrin molecules in the course of decomposition and their conversion to weakly coloured products was established by the present author in studies into the kinetic stability of metal analogues of chlorophyll. It was found that at certain acidity and temperature the same complexes, such as ZnChl*b*, may dissociate, both yielding porphyrin in a protonated form (5.17) and causing its complete disappearance (5.18). All labile, or ionic, complexes dissociate under any conditions yielding H_2P and H_4P^{2+}.

Most reactions of such covalent chlorophyllic acid complexes as CuChl, NiChl, PdChl, and CoChl in glacial acetic acid with H_2SO_4 involve decomposition of chlorophyllic acid during the dissociation. The highly covalent complex XFeChl characterized by a geometric configuration different from that of other stable MChl complexes (the Fe atom does not lie strictly in the plane of the MN_4 group but is 0.25–0.5 Å above this plane[404–405]) dissociates under the same conditions, yielding chlorophyllic acid.

Complexes ZnChl*a* and AgChl*a*, which are less covalent than CuChl, always dissociate yielding invariable H_2Chl or $H_2ChlH_2^{2+}$. Their decomposition discontinues at step (5.17). The following findings are of particular interest for understanding the dissociation reaction mechanism.[76,247] Complex ZnChl*a* dissociates down to H_2Chl*a* in all proton-donor media. Its structural analogue ZnChl*b* dissociates in glacial acetic acid in accordance with equations (5.17) and (5.18) with decomposition of chlorophyllic acid; however, suffice it to add a small amount of H_2SO_4 (10^{-2} M) and the spectrum of H_2ChlH^+ is restored during dissociation. Chlorophyllhaemin XFeChl*a* dissociates at 15–45 °C in acetic acid with H_2SO_4 with the porphyrin ligand decomposing, and at a high temperature (above 80 °C) $H_2ChlH_2^{2+}$ is produced.

These facts indicate that the transformations according to (5.18) have nothing to do with the effect of the functional substituents in the porphyrin molecule but result from the drastic rearrangement of the conjugation in the macroring during activation of the complex. Depending on the porphyrin species, activity of solvated proton H^+_{solv}, covalence of the M—N bond, and temperature, the active molecule of porphyrin H_2P^* produced by dissociation may be stabilized owing to the transition $H_2P^* \rightarrow H_2P \rightarrow H_4P^{2+}$. However, this transition may not occur under less favourable conditions, in which case H_2P decomposes. The possibility of the reaction proceeding along two paths in the case of iron phaeophytinate is shown in Fig. 5.10. Due to the fact that haemin dissociates under more stringent conditions (5 M H_2SO_4), invariable ligand $H_2PPH_2^{2+}$ is produced in accordance with (5.16). It may be assumed that at lower acidities and temperatures paths (5.17) and (5.18) are involved. This, however, has not been corroborated because of the slow reaction rates. Table 5.9 presents the effective rate constants, energies, and entropies of activation of dissociation of haemin and chlorophyllhaemin, obtained under identical conditions.[76]

The dissociation constant of haemin is 100 times lower than that of its chlorophyll analogue, which is indicative of the higher bond covalence in the former. Since addition of 2 M H_2SO_4 makes the reactions of dissociation of both complexes isoentropic, the difference in the activation energies of these structural analogues suggests that the M—N bond in

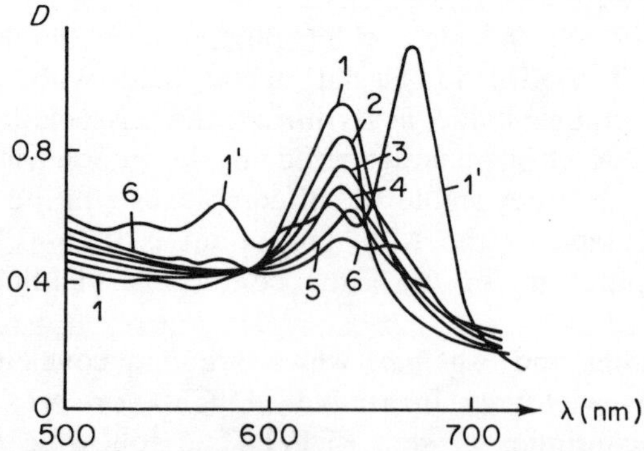

Figure 5.10 Dissociation of chlorophyllhaemin in glacial acetic acid with 2 M H_2SO_4 at 25 °C. Time: curve 1, 0 h; curve 1′, 0 h at 80 °C; curve 2, 2 h; curve 3, 5 h; curve 4, 7 h; curve 5, 10 h; curve 6, 20 h

Table 5.9 Kinetic parameters of dissociation of protohaemin and chlorophyllhaemin in glacial acetic acid

t (°C)	$c_{H_2SO_4}$ (M)	k_{eff} (h^{-1})	E (kcal mol^{-1})	$\Delta S^{\ddagger}$ (e.u.)
(Cl)FePP				
25	2	0.00177	15.9 ± 0.6	−34 ± 2
70	2	0.056 ± 0.005		
80	2	0.103 ± 0.017		
90	2	0.200 ± 0.028		
25	5	0.00845	18.0 ± 1.0	−24 ± 3
60	5	0.205 ± 0.020		
70	5	0.455 ± 0.046		
80	5	1.04 ± 0.090		
90	5	1.88 ± 0.080		
*(Cl)FeChl*a				
25	2	0.164 ± 0.004	14.1 ± 1.0	−35 ± 3
35	2	0.387 ± 0.013		
45	2	0.740 ± 0.053		
15	5	0.54 ± 0.06	12.7 ± 0.6	—
25	5	1.14 ± 0.10		
35	5	2.27 ± 0.25		

haemin is 0.5 kcal mol^{-1} stronger than that in chlorophyllhaemin. The stabilization of haemin is due to the greater strength of the donor–acceptor $M \leftarrow N$ bond. As a consequence of the non-coplanar position of the Fe^{3+} ion and its high-spin state, the formation of backward dative $Fe \rightarrow N$ π bonds in ferroporphyrins is impossible and their stability is determined entirely by the δ bond. This is the reason why one should expect, and the scant available data corroborate this, that the Fe^{2+} complexes with porphyrin will be less stable in an acidic medium than the Fe^{3+} complexes. Therefore ferroporphyrins must dissociate much more readily in the presence of reducing agents.

From the experimental point of view this problem remains practically unsolved. Just as in the case of Zn^{2+} complexes, porphyrin complexes with Fe^{2+} and Fe^{3+} involve 4*d* rather than 3*d* orbitals. Hence phthalocyanine complexes with iron ions belong to the labile group similarly to complexes with Mg^{2+} and Cd^{2+}. The steric hindrances at the MN_4 group of phthalocyanine do not favour coordination $(n)d(n)s(n)p^2$ if the ion is larger than Sn^{4+}.

As can be inferred from the above, the higher covalence of complex ClFePP as compared to ClFeChl*a* stems from the more pronounced electron-donor properties of protoporphyrin as compared to chlorophyllic acid, that is higher electron density at the nitrogen atoms of H_2PP. Here again we see that the $CH{=}CH_2$ group exhibits electron-donor properties with respect to the conjugated system of porphyrin.

Complex CoPP dissociates in glacial acetic acid at $k_{eff}^{25} = 6.1 \times 10^{-2}\ h^{-1}$,

$E = 7.7 \pm 0.1$ kcal mol^{-1}, and $\Delta S^{\ddagger} = -44 \pm 1$ e.u. The dissociation is accompanied by decomposition of the protoporphyrin molecule[398] in accordance with reactions (5.17) and (5.18). Complex CoPP is much inferior in stability to haemin but is kinetically much more stable than ZnPP.

The dissociation rates of protoporphyrin complexes with Zn^{2+}, Mg^{2+}, and Cd^{2+} were measured[336,339] in glacial acetic acid and its aqueous mixtures as well as in *tert*-butanol–trichloroacetic acid. The results were used in calculating the energies and entropies of activation summarized in Table 5.10.

As has been established by the above method, the reaction of dissociation of protoporphyrin complexes with Zn^{2+} and Mg^{2+} is of the second order with respect to H^{+}_{solv}. Earlier[375] it was established for chlorophyll *a* and its analogue with Zn^{2+}. Comparison of Tables 5.6 and 5.10 suggests that the kinetic stability of protoporphyrin complexes with Zn and Mg differs to a lesser extent than that of complexes of chlorophyllic acid *a*. This interesting fact is due to the stability of complexes in the transition MChl → MPP increasing in the case of Mg^{2+} and decreasing in the case of Zn^{2+} and Cd^{2+}. In *tert*-butanol with 0.35 M CCl_3COOH at 25 °C the difference in stability between MgPP and ZnPP is only 20-fold. The kinetic stabilities of Mg^{2+} and Cd^{2+} protoporphyrins differ to a greater extent than those of the corresponding MChl*a* complexes.

These data indicate that structural changes in the porphyrin molecule may alter the stability variation pattern of complexes of various metals (even Zn^{2+} and Mg^{2+} having many similar properties) in an opposite manner. It can also be inferred that the kinetic and thermodynamic stability of metalloporphyrins is determined not only by the electron-donor properties of the porphyrin but also by geometric factors, namely fine details in the structure of the coordination centre of its molecule, which are favourable for some cations and unfavourable for others. This is one of the reasons for the great variety of properties of metalloporphyrins and their biological importance. As will be shown later, another contributing factor is the unusually strong dependence of the stability of metalloporphyrins on the nature of the solvent.

In view of the above specific features of porphyrin molecules one should compare the stabilities of their complexes and explain the differences in stability in terms of the electronic influence of the substituents, taking porphyrins as structurally close as possible in strictly identical media. Now follows a summary of the kinetic stability of mesoporphyrin complexes with Zn^{2+}, Cd^{2+}, and Mg^{2+} in *tert*-butanol with CCl_3COOH,[382] Cu^{2+} and Zn^{2+} in glacial acetic acid with H_2O or H_2SO_4,[381] as well as Ni^{2+} and Fe^{3+} in HAc + H_2SO_4 [383] (see Table 5.11).

The most stable among the mesoporphyrin complexes examined is the Fe^{3+} complex (so-called mesohaemin); complex NiMP dissociates five times

Table 5.10 Kinetic parameters of dissociation of zinc, magnesium and cadmium complexes of protoporphyrin in the t-C_4H_9OH–CCl_3COOH system

Complex	$c^0_{CCl_3COOH}$ (M)	$c^2_{\Gamma H^+} \times 10^8$ (M)	t (°C)	k_{eff} (h^{-1})	E (kcal mol^{-1})	$\Delta S^‡$ (e.u.)
ZnPP	0.35		25	0.40[a]		
	0.78	9.6	25	0.98 ± 0.03	12.5 ± 0.5	−32 ± 2
			35	1.91 ± 0.04		
			45	3.7 ± 0.2		
	1.10	22	25	3.1 ± 0.1	12.0 ± 0.5	−32 ± 2
			35	5.9 ± 0.2		
			45	10.9 ± 0.3		
	1.26	33	25	4.3 ± 0.2	12.0 ± 0.5	−31 ± 2
			35	8.4 ± 0.4		
			45	15.4 ± 0.9		
	1.32	38	25	6.7 ± 0.2		
	1.38	48	25	10.7 ± 0.5		
	1.51	55	25	16.2 ± 1.2		
	1.58	65	25	21.4 ± 1.2		
MgPP	0.20	0.25	25	1.8 ± 0.1	12.1 ± 0.2	−32 ± 1
			35	3.6 ± 0.1		
			45	6.7 ± 0.3		
	0.22	0.28	25	2.5 ± 0.1		
	0.25	0.43	25	3.3 ± 0.2	12.4 ± 0.3	−31 ± 1
			35	6.4 ± 0.2		
			45	12.4 ± 0.5		
	0.30	0.72	25	6.2 ± 0.3		
	0.35	1.0	25	8.4 ± 0.5		
CdPP	5.0×10^{-4}		25	3.9 ± 0.2	0	−71
			35	3.7 ± 0.3		
			45	3.4 ± 0.4		
	1.0×10^{-4}		25	12.5 ± 1.1	0	−70
			35	12.2 ± 0.8		
			45	12.1 ± 0.9		
	1.2×10^{-3}		25	20.1 ± 1.4		
	1.9×10^{-3}		25	31.1 ± 1.4		
	2.4×10^{-3}		25	40.1 ± 2.0		
	5.0×10^{-3}		25	91 ± 4		

[a]Found by extrapolation.

faster, and CuMP is the least stable of all porphyrin complexes with copper. Under conditions in which other complexes dissociate slowly (2 M H_2SO_4) the dissociation of NiMP and CuMP is practically instantaneous. As can be seen from Table 5.11, complex CuMP also features the lowest activation energy, lower even than that of ZnMP.

Complex MgMP (a structural analogue of chlorophyll) is only 20–30 times less stable than ZnMP. The latter, in turn, is kinetically five times less stable than magnesium protoporphyrin and two times less stable than chlorophyll. Complex CdMP is twice as stable as CdPP. Zinc

Table 5.11 Kinetic parameters of dissociation of mesoporphyrin complexes in glacial acetic acid with proton-donor additives

Complex	Additive concentration	k_{eff}^{25} (h^{-1})	E (kcal mol^{-1})	$\Delta S^{\ddagger}$ (e.u.)
CuMP	0.5 M H_2SO_4	2.61 ± 0.11	11.1 ± 0.5	-35 ± 2
	1.0 M H_2SO_4	25.1 ± 0.7	11.2 ± 0.2	-31 ± 2
NiMP	2.0 M H_2SO_4	0.09 ± 0.001	12.9 ± 0.3	-36 ± 1
(Ac)FeMP	2.0 M H_2SO_4	0.0174 ± 0.0012	15.4 ± 0.1	-30 ± 1
	5.0 M H_2SO_4	0.43 ± 0.01	15.4 ± 1.3	-24 ± 4
ZnMP	No additive	25.1 ± 1.6	12.3 ± 0.5	-27 ± 2
	1% H_2O	57.2 ± 2.1	—	—
	2% H_2O	94.2 ± 2.8	12.4 ± 0.1	-24 ± 1
	5% H_2O	214 ± 17	11.2 ± 0.5	-26 ± 2
ZnMP	0.40 M CCl_3COOH	0.76 ± 0.03	11.5 ± 0.5	-36 ± 2
	0.51 M CCl_3COOH	1.16 ± 0.07	11.6 ± 0.5	-35 ± 2
	0.62 M CCl_3COOH	2.3 ± 0.1	11.3 ± 0.3	-35 ± 1
	0.73 M CCl_3COOH	4.2 ± 0.2	—	—
	1.10 M CCl_3COOH	27.8 ± 0.9	—	—
MgMP	0.05 M CCl_3COOH	0.622 ± 0.031	9.6 ± 0.4	-42 ± 1
	0.10 M CCl_3COOH	2.20 ± 0.10	10.3 ± 0.4	-38 ± 1
	0.20 M CCl_3COOH	7.7 ± 0.4	—	—
CdMP	10^{-4} M CCl_3COOH	2.6 ± 0.3	0	−72
	2×10^{-4} M CCl_3COOH	7.0 ± 0.2	0	−70
	5×10^{-4} M CCl_3COOH	38.8 ± 0.3	—	—

mesoporphyrin is 10 times less stable than zinc protoporphyrin. Thus all covalent (Fe, Ni^{2+}, Cu^{2+}, Zn^{2+}) or highly covalent (Mg^{2+}) mesoporphyrin complexes are much less stable than the corresponding complexes of protoporphyrin.

This important fact corroborates the less pronounced electron-donor properties of the mesoporphyrin ligand as opposed to protoporphyrin. It also accounts for the difference in total acid ionization constants of H_2PP and H_2MP. In the latter it is five times greater (see Table 4.11). The pronounced electron-donor properties of H_2PP can only be a result of the +C effect of the vinyl group with respect to the aromatic system of the porphyrin macroring. Unlike porphyrin complexes, all of the examined mesoporphyrin complexes dissociate as follows:

$$\text{MMP} + 2\text{H}^+_{\text{solv}} \rightarrow \text{M}^{2+}_{\text{solv}} + \text{H}_2\text{MP} \xrightarrow{2\text{H}^+_{\text{solv}}} \text{H}_2\text{MPH}_2^{2+}, \quad (5.19)$$

without decomposition of the porphyrin molecule. This property is determined by the lower covalence of mesoporphyrin complexes.

The results of systematic studies into the kinetic stability of tetraphenylporphin complexes were reported earlier[252,384] and are summarized in Table 5.12.

Table 5.12 Kinetic parameters of dissociation of tetraphenylporphin complexes in glacial acetic acid

Complex	$c_{H_2SO_4}$ (M)	k_{eff}^{25} (h^{-1})	E (kcal mol^{-1})	$\Delta S^{\ddagger}$ (e.u.)
(Ac)FeTPP	2.0	0.0084	16.2 ± 0.6	−30 ± 2
	5.0	0.35 ± 0.01	19.7 ± 0.7	−10 ± 2
NiTPP	2.0	0.0370	18.1 ± 0.2	−20 ± 1
CuTPP	2.0	1.85 ± 0.05	17 ± 1.0	−17 ± 4
	1.0	0.0760	17 ± 1.0	−23 ± 4
	0.5	0.00545	18.1 ± 0.6	−24 ± 2
	2.0[a]	0.823 ± 0.034	16.9 ± 0.6	−18 ± 2
ZnTPP	—	0.93 ± 0.08	11.0 ± 0.5	−37 ± 2

[a] With 5 wt% H_2O.

The stability series of tetraphenylporphin complexes $AcFe^{2+} > Ni^{2+} > Cu^{2+} > Zn^{2+}$ is similar to those of chlorophyllic acid, protoporphyrin, and mesoporphyrin complexes. Tables 5.11 and 5.12 show that the dissociation of all tetraphenylporphin complexes is slower than that of metal mesoporphyrins and is characterized by slightly higher energies and entropies of activation. The only exception is ZnTPP which, being 25 times more stable than ZnMP, has lower values of E and $\Delta S^{\ddagger}$. This is another manifestation of the high specificity of the complexing properties of all porphyrins and the delicate mechanisms of dissociation of their complexes.

The isoentropism of dissociation of AcFeTPP and AcFeMP in HAc + 2 M H_2SO_4 permits one to establish from the difference in the activation energies that the Fe ← N bond in tetraphenylporphin is stronger (by 0.2 kcal) than in mesoporphyrin, which can be ascribed to its higher covalence. Hence, the phenyl residues are probably electron donors with respect to porphyrins, and the total electronic charge of the M—N bond increases. The factors responsible for the higher kinetic stability of the other MTPPs seem to be the same.

The fact that the C_6H_5 groups are electron donors is corroborated by H_2TPP having the lowest acid ionization constants of all examined porphyrins (see Table 4.11). This finding is a bit surprising since it is well known that the benzene rings do not lie in the plane of the macroring. The same behaviour is displayed by the phenyl nuclei in polyphenyls in which the C_6H_5 groups are also electron donors with respect to one another, although their arrangement is not coplanar. The N—H bond vibration frequencies in infrared spectra, presented in Table 4.12, do not reveal any differences in the state of these bonds in mesoporphyrin, protoporphyrin, and tetraphenylporphin, but indicate that the phenyl groups act as electron donors (see Chapter 4).

In the literature no systematic data on the kinetics and mechanism of dissociation of complexes of other porphyrins, other than those discussed above, are available. Only scant data can be found on copper etioporphyrin,[406] porphin,[380] zinc etioporphyrin, zinc *N*-methyletioporphyrin, zinc deuteroporphyrin and its disubstituents,[366] zinc *meso*-(tetra-4-*N*-methylpyridyl)porphin,[361] and magnesium deuteroporphyrin.[392]

The studies summarized in the above references are limited to the complexes mentioned; nevertheless, some of the findings may be useful in understanding different aspects of the kinetic behaviour of porphyrins. If one considers the work by Shears *et al.*,[366] for example, the following findings are of interest. In a methanol solution of HCl taken in high concentrations the dissociation rate of complex ZnEP obeys an equation of second order with respect to the acid, so that the effective rate constants are expressed as

$$k'_{\text{eff}} = k'_v(c^0_{\text{HCl}})^3/K' + c^0_{\text{HCl}}, \tag{5.20}$$

where $k'_v = (2.0 + 0.1) \times 10^5\ \text{l}^2\ \text{mol}^{-2}\ \text{s}^{-2}$ and $K' = (6.6 \pm 0.2) \times 10^{-3}\ \text{mol l}^{-1}$.

In the case of zinc *N*-methyletioporphyrin at the same HCl concentrations the following first-order equation is involved:

$$k''_{\text{eff}} = k''_v(c^0_{\text{HCl}})^2/K'' + c^0_{\text{HCl}} \tag{5.21}$$

with constants $k''_v = (2.5 \pm 0.2) \times 10^2\ \text{l mol}^{-1}\ \text{s}^{-1}$ and $K'' = (2.5 \pm 0.2) \times 10^{-1}\ \text{mol l}^{-1}$.

The unusual kinetic relations (5.20) and (5.21) represent the complex protolytic equilibria in the methanol–HCl solution which was observed in an ethanol–HAc solution during dissociation of MgChl*a* CdChl*a*. In this connection HCl in methanol cannot be considered as a suitable solvent for determining the order of the reaction with respect to the proton-donor particle. The accelerating effect of the Cl^- ion on dissociation, as opposed to the ClO_4^- anion, established by Shears *et al.*[366] has not yet found any explanation. One may assume that it is due to the effect of extra coordination on the state of the Zn—N bonds or, to be more precise, on the increase in the ionic component in these bonds. It is not believed that HCl is the attacking particle in the act of dissociation. Of course the $CH_3OH_2^+$ particle retains its advantages in the dissociation of sufficiently covalent zinc porphyrins. The stoicheiometric mechanism proposed by Shears *et al.*[366] is not likely to be involved (see below). Nor is it reasonable to ascribe to a purely empirical quantity ρ the physical meaning of the relationship between the dissociation rate constants of intermediate ZnP · *n*HCl and the rate of dissociation of the intermediate to free porphyrin. In this case the value of ρ would have been highly positive rather than negative.

Other important results following from the work by Shears *et al.*[366] include the substantiated conclusion that the order of dissociation of the *N*-methylporphyrin complex with respect to the proton is $n - 1$ as compared to the porphyrin itself and is determined by the anion charge. Hence, it may be expected that the dissociation of an *N,N′*-dialkylporphyrin complex will not depend on the acidity of the medium.

The dissociation rate constant ratios between *N*-alkylated and non-alkylated zinc complexes of porphyrins are somewhat unusual. The strangest thing is that k'_v is much greater than k''_v. A reverse relationship would be more likely. Maybe the above-cited work[336] contains errata in the numerical data. The authors' assertion that Zn^{2+} in zinc *N*-alkylporphyrin is tricoordinated lacks proof. It is beyond any doubt that the coordination number is much greater. Presented in the same work are data on the effective rate constants of protolytic dissociation of zinc complexes with porphyrins of different basicity in a methanol–HCl solution (Table 5.13).

A serious omission in Table 5.13 is the lack of temperature data, which together with inaccuracies in the presented numerical values and activation parameters make the tabulated data semi-quantitative. Another striking feature is the high kinetic stability of the protoporphyrin complex as opposed to the mesoporphyrin one, which is an indication of the above-mentioned effect of stabilization of the vinyl group. In contrast to mesoporphyrin **(XXV)** the two ethyl groups of deuteroporphyrin in positions 2 and 4 are substituted by H atoms. This substitution provides for a substantial kinetic stabilization which cannot be explained merely by the absence of a +I effect of the ethyl groups.

The stabilizing effect of the bromine atoms is particularly pronounced in positions 2 and 4. Shears *et al.*[366] attribute this factor to the lower basicity of porphyrin. No theoretical interpretation of these facts is provided. As can be inferred from Table 5.13, the basicity of porphyrins does not always correlate with the kinetic stabilization of a complex. For example, the introduction of *N*-methylpyridinium residues into the meso positions considerably reduces $pK_{PH_3^+}$, but the kinetic stability increases insignificantly. This problem will be discussed in greater detail in conjunction with the mechanism of dissociation of porphyrin complexes.

Shellgrove and Plane[392] investigated the dissociation of the complex

Table 5.13 Rates of dissociation of zinc complexes of various porphyrins in methanol–7.5 × 10^{-3} M HCl solution

Complex	k_{eff} (s^{-1})	p*K* of reaction (4.14)
Zinc mesoporphyrin	1.59	5.85
Zinc deuteroporphyrin	0.573	5.50
Zinc dibromodeuteroporphyrin	0.00130	3.0
Zinc protoporphyrin	0.130	4.80
Zinc *meso*-(tetra-4-*N*-methyl-pyridyl)porphin	0.067[a]	2.2[b]

[a]Determined in an aqueous HCl solution.
[b]The value of $pK_{PH_4^{2+}}$.

MgDP DME in aqueous–alcoholic solutions in the presence of 10^{-6}–10^{-3} M $HClO_4$. The rate of detachment of the Mg^{2+} cation is in accordance with equation (5.22). Rate constant k'_v increases in the following solvent series: ethanol < methanol < water. Pyridine in methanol was found to act catalytically on the dissociation of MgDP. In the presence of pyridine and water the kinetic equation takes the form

$$-\mathrm{d(MgP)}/\mathrm{d}t = [\mathrm{MgP}]\left\{\frac{k_1[\mathrm{H}^+]^3}{\rho_1 + [\mathrm{H}^+]} + \frac{k_2[\mathrm{Py}][\mathrm{H}^+]^3}{\rho_2 + [\mathrm{H}^+]} + k_3[\mathrm{Py}][\mathrm{H_2O}][\mathrm{H}^+]^2\right\}. \tag{5.22}$$

The kinetic parameters have been determined at a non-standard temperature of 42 °C. The value of k_1 is $1.2 \times 10^3\ \mathrm{s}^{-1}$ in ethanol, $3.1 \times 10^5\ \mathrm{s}^{-1}$ in methanol, and *c.* $10^7\ \mathrm{s}^{-1}$ in water. The values of other constants are as follows: $k_2 = 7.2 \times 10^8\ \mathrm{s}^{-1}$, $k_3 = 2.7 \times 10^9\ \mathrm{M}^{-4}\ \mathrm{s}^{-1}$, $\rho_1 = 10^{-5}$ M, and $\rho_2 = 5 \times 10^{-6}$ M. The enthalpy–entropy constants of activation are $\Delta H^\ddagger = 2.2$ kcal and $\Delta S^\ddagger = -27$ e.u. for k_1 and $\Delta H^\ddagger = 5$ kcal and $\Delta S^\ddagger = -1$ e.u. for k_2, respectively.

Shellgrove and Plane[392] have established many more valuable facts relating to the mechanism of dissociation of metalloporphyrins, which were used by the author of this monograph in a number of other works.[37,282,284,385,388]

Investigators are also intrigued by water-soluble porphyrins. In an extremely interesting work,[361] *meso*-(tetra-4-*N*-methylpyridyl)porphin (**LV**)

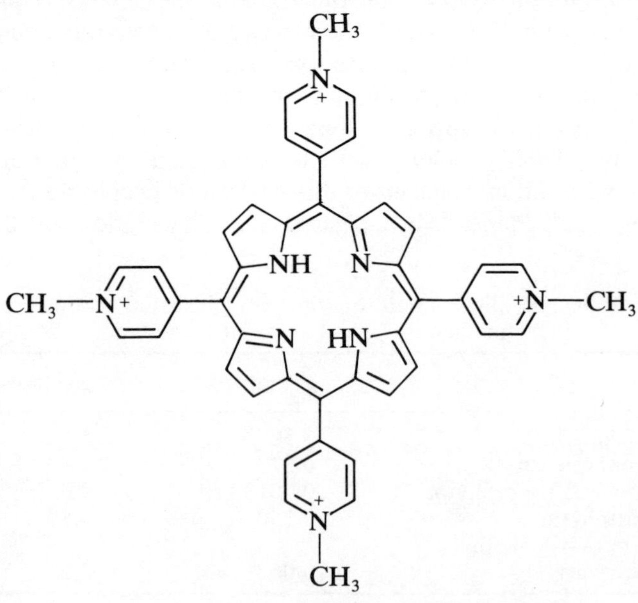

LV

was synthesized, which is water-soluble in highly acidic ($pH < 0$) and highly alkaline ($pH > 14$) media. The total constant of protonation from H_2P to H_4P^{2+} is equal to 2.2, that is porphyrin is not readily protonated (see Chapter 4), which may result from the positive charge being transferred from the nitrogen atoms of pyridine to those of the macroring.

This porphyrin dissociates as an acid much more readily than other porphyrins examined (see Table 4.11) and is similar in this respect to phthalocyanine. Spectrophotometric titration with NaOH has enabled the first-stage dissociation constant to be derived from equation (4.1) ($pK_{S1} = 12.9 \pm 0.2$).

The kinetics of dissociation of the Zn complex of this porphyrin was studied in an aqueous solution of HCl.[361] The dissociation was found to be strictly of second order with respect to the H_3O^+ ion and to be catalysed by chlorine ions so that the rate of dissociation in hydrochloric acid can be written as

$$v = k_v[\text{ZnP}][\text{H}^+]^2[\text{Cl}^-]^2. \tag{5.23}$$

In the same work,[361] the value of k_v is given as equal to $(6.75 \pm 0.32) \times 10^{-2}\ \text{s}^{-1}$ without indicating the temperature (I assume that all experiments reported in this work were conducted at room temperature). Complex ZnTMPP seems to be the most stable of all zinc complexes of porphyrin. For example, ZnTPP (see Table 5.12) in glacial HAc has $k_{\text{eff}}^{25} = 0.93\ \text{h}^{-1}$. In an aqueous 1 M HCl solution its dissociation would be instantaneous, while ZnTMPP has $k_{\text{eff}} = 2.42 \times 10^2\ \text{h}^{-1}$ under the same conditions.

However, rigorous comparison of these as well as other data obtained in different solvents is practically impossible in view of the extremely strong influence of the nature of the solvent on the rates and other parameters of dissociation of metalloporphyrins. Let us consider this influence at greater length.

5.3 Influence of the Solvent on Dissociation

The basic methods that take into account the effect of solvation, that is the nature of the solvent, on the rates of chemical reactions have been summarized in the well-known monograph by Entelis and Tiger.[359] The authors have vividly demonstrated that the effect of the solvent on the reaction kinetics boils down mainly to the varying degree and character of solvation of the starting reagents and transition state, thereby altering the mechanisms of activation of the reagents.

From the theoretical standpoint this problem remains unsolved. Its solution requires, first of all, accumulation of experimental data. One of the reliable ways to quantify the effect of the solvent on kinetics is by using a combination of thermodynamic parameters, including the heats of solution and mixing of the reagents in a number of solvents, and kinetic data, including the entropies of activation under the same conditions. The selection of one of the most inert, that is weakly solvating, solvents to be used as a reference permits determination of the relative thermodynamic and kinetic parameters of solvation from the following relations:

$$\Delta\Delta H_{solv} = \Delta H_{dissol(i)} - \Delta H_{dissol(0)}, \quad (5.24)$$

$$\Delta\Delta H^{\ddagger}_{solv} = \Delta\Delta H^{\ddagger} + \sum \Delta\Delta H_{solv}, \quad (5.25)$$

where $\Delta\Delta H_{solv}$ is the relative enthalpy of solvation of a starting reagent, $\sum\Delta\Delta H_{solv}$ is the relative enthalpy of solvation of all starting reagents, $\Delta H_{dissol(0)}$ and $\Delta H_{dissol(i)}$ are the enthalpies of dissolution of a starting reagent in the reference (0) or any other (i) solvent, $\Delta\Delta H^{\ddagger}$ is the relative enthalpy of activation of the reaction in both above-mentioned solvents, and $\Delta\Delta H^{\ddagger}_{solv}$ is the relative enthalpy of solvation of the transition state. Relations similar to (5.25) can be written for the free energies and entropies of activation of the solvation process.

Attempts to study the solvation of some dyes in solution, using equations similar to (5.24), have been made in a number of works[407–409] with the result that differences in solvation depending on the molecular structure of the solvent have been established. No such data have been obtained for porphyrins and metalloporphyrins as yet because of the unavailability of sizable amounts of these compounds. Therefore the effect of the solvent on the kinetics of reactions involving porphyrins is studied only qualitatively. Different aspects of the influence of the nature of the solvent on the kinetics and mechanism of metalloporphyrin-yielding reactions were discussed in Chapter 4 and are quite clear from the qualitative point of view.

In metalloporphyrin dissociation reactions the nature of the solvent determines the bonding of the proton, that is the activity of H^{+}_{solv} and its concentration, as well as the character of solvation of the MN_4 group due to extra coordination of the solvent at the central metal atom and the competitive interaction of the proton-donor part of the solvent molecule at nitrogen atoms. These factors determine the solvation (sign and value of $\Delta S^{\ddagger}$) of the transition state. In addition, a change in the state of the porphyrin molecule from MP to H_2P or H_4P^{2+} may lead to significant changes in the solvation of its functional groups and individual segments.

As far as phthalocyanine complexes are concerned, the only solvent in which their dissociation was studied is aqueous H_2SO_4. Recent studies of 4,4′,4″,4‴-tetracarboxylic acid of copper phthalocyanine have shown[410] that in a solution of 57–86% H_2SO_4 (which includes a mixture of unprotonated and monoprotonated tetracarboxylic acid) the mechanism of protolytic dissociation, established for MPc in stronger H_2SO_4, is involved. It has been found that in a solution of 43–53% H_2SO_4 phthalocyanine is not protonated and precipitates completely from the solution in the course of time. The effective rate constant decreases from 1.09×10^{-2} h^{-1} in 57.6% H_2SO_4 to 6×10^{-1} h^{-1} in 86.6% H_2SO_4 (at 25 °C).

The dissociation of metalloporphyrins was studied in glacial acetic acid with H_2SO_4 or ethanol, as well as in *tert*-butanol in the presence of CCl_3COOH. The comparative data are presented in Tables 5.14–5.16. Addition of water sharply accelerates the dissociation of ZnPP in glacial HAc. The dissociation rate increases threefold in the presence of 1% water and more than 100-fold when 20% water is added. This is due to the growing concentration of the solvated proton, in this case H_3O^+, which is probably more active than $CH_3COOH_2^+$ as a result of the more pronounced donor properties of H_2O in comparison with CH_3COOH.

However, other factors also make a considerable contribution, particularly the changing solvation of the porphyrin.

Introduction of water into glacial HAc reduces the activation energy of dissociation of ZnPP. The entropy of activation remains practically the same. The variation pattern of the activation parameters depends on the metal species. Addition of H_2O to the solution of the silver analogue of chlorophyll in glacial HAc accelerates the reaction considerably at first, then the effect of water on the dissociation rate diminishes (Table 5.15). Interestingly, the acceleration is in this case associated with an increase in $\Delta S^{\ddagger}$ because the activation energy increases somewhat.

Table 5.14 Kinetic parameters of dissociation of zinc protoporphyrin complex in glacial acetic acid[a]

H_2O (%)	k_{eff}^{25} (h^{-1})	E (kcal mol^{-1})	$\Delta S^{\ddagger}$ (e.u.)
0	0.936 ± 0.024	—	—
1	3.40 ± 0.30	12.0 ± 0.5	−32 ± 1
2	7.97 ± 0.30	12.4 ± 0.4	−29 ± 1
2.5	11.1 ± 0.50	—	—
5	24.7 ± 0.60	11.5 ± 0.5	−29 ± 2
8	46.9 ± 1.1	—	—
10	60.9 ± 1.4	10.7 ± 0.5	−30 ± 2
15	92.9 ± 1.6	—	—
20	96.2 ± 3.0	—	—

[a] $c^0_{ZnPP} = 5.6 \times 10^{-5}$ M.

Table 5.15 Kinetic parameters of dissociation of a zinc analogue of chlorophyll in glacial acetic acid with proton-donor additives

Additive	t (°C)	k_{eff} (h^{-1})	E (kcal mol^{-1})	$\Delta S^{\ddagger}$ (e.u.)
No additive	40	0.110 ± 0.014	19.6 ± 0.5	−16 ± 2
5% water	40	1.14 ± 0.20	20.0 ± 1.3	−11 ± 4
10% water	40	2.05 ± 0.20	20.5 ± 1.5	−7 ± 5
80% ethanol	60	0.011 ± 0.001	—	—
60% ethanol	60	0.19 ± 0.01	—	—
50% ethanol	60	0.40 ± 0.03	—	—

It should be emphasized that, unlike metalloporphyrin-yielding reactions in which the entropy of activation is true as it is calculated from a known chemical kinetics relation using true rate constants k_v, in reactions dissociating metalloporphyrins in organic solvents $\Delta S^{\ddagger}$ is an arbitrary value since it is calculated from effective rate constants. Metal phthalocyanines make an exception, their dissociation rate constants being determined in a solution of concentrated aqueous H_2SO_4 at known H_3O^+ concentrations. For the systems glacial HAc–H_2O, HAc–H_2SO_4, alcohols–CCl_3COOH, and HAc–C_2H_5OH no data are available to calculate the concentrations or activities of the solvated proton. Therefore, the effective constant is calculated from the kinetic equation $k_{eff} = k_v[H^+_{solv}]^n$.

The entropies of activation ($\Delta S^{\ddagger}_{arb}$) calculated using the constant k_{eff} instead of k_v differ from so-called true entropies by a value determined by the concentration of ions H^+_{solv}:

$$\Delta S^{\ddagger} = \Delta S^{\ddagger}_{arb} + R \ln[H^+_{solv}]^n. \tag{5.26}$$

In the case of stable phthalocyanine and porphyrin complexes, $n = 2$.

In most of the dissociation reactions considered above, particularly in the systems HAc–H_2SO_4 and t-C_4H_9OH–CCl_3COOH, $\Delta S^{\ddagger}$ and $\Delta S^{\ddagger}_{arb}$ differ but slightly (by about 1–5 e.u.). $\Delta S^{\ddagger}_{arb}$ can be used to gain an insight into the changes in solvation during dissociation, the polarity of the transition state, and differences in the solvation of MP with the same M and different porphyrins or with different M and the same porphyrin. A quantitative study into the proton-donor properties of mixed solvents, the necessity of which is now obvious, will permit true entropies of activation to be found from the value of $\Delta S^{\ddagger}_{arb}$ in equation (5.26). In spite of the fact that the values of $\Delta S^{\ddagger}$ are extremely useful in determining the reaction mechanisms, in some cases they must be used with caution. The objective reason is that in most cases one does not know all particles participating in the strong interactions of the reacting system in the transition state. This is why $\Delta S^{\ddagger}$ in solutions are more often than not effective arbitrary values. Nevertheless, their use in similar chemical processes is quite justified and necessary if the latter occur in similar media.

There are practically no data available on the effect of water addition on the kinetics of dissociation of metalloporphyrins in organic solvents in the presence of H_2SO_4. In one work,[384] additions of 5% H_2O were found to slow down the dissociation of complex CuTPP in glacial HAc with 2–3 M H_2SO_4. Possibly water is first incorporated into the general structure of the solvent H_2O–HAc–H_2SO_4 and only later leads to an increase in the concentration of H_3O^+. Unfortunately the properties of such ternary systems have not been studied quantitatively.

Another work[336] deals with the effect of H_2O on the dissociation of zinc complexes of various porphyrins in HAc (Table 5.16).

Note that addition of water produces substantially different effects on the dissociation rate of ZnP, depending on the porphyrin species. Addition

Table 5.16 Kinetic parameters of dissociation of zinc complexes of various porphyrins in glacial acetic acid with water addition

Complex	k_{eff} (h^{-1})	E (kcal mol^{-1})	$\Delta S^{\ddagger}$ (e.u.)
0.02% H_2O			
ZnPP	0.936 ± 0.02	—	—
ZnTPP	0.930 ± 0.08	11.0 ± 0.5	−37 ± 2
ZnChl*a*	0.976	10.1 ± 0.2	−40 ± 1
ZnChl*b*	3.0×10^{-3}	10.9 ± 0.5	−49 ± 2
ZnMP	25.1 ± 1.6	12.3 ± 0.5	−27 ± 2
5% H_2O			
ZnPP	24.7 ± 0.6	11.5 ± 0.5	−29 ± 2
ZnTPP	15.1 ± 0.2	10.3 ± 0.3	−34 ± 1
ZnChl*a*	1.2	12.0 ± 0.5	−34 ± 2
ZnMP	214 ± 17	11.2 ± 0.5	−26 ± 2

of 5% H_2O to glacial HAc speeds up the dissociation of zinc protoporphyrin 26.5-fold, that of zinc mesoporphyrin 8.5-fold, and that of zinc tetraphenylporphin 16-fold, while the dissociation rate of the zinc analogue of chlorophyll *a* increases only 1.2-fold. This example demonstrates that not only the activity of the H^+_{solv} ion is affected in going from one solvent to another but also the solvation of the porphyrin molecule undergoes considerable changes. The latter depends on the porphyrin structure and appreciably influences the activation process. Another important conclusion is that the kinetic stability series of metalloporphyrins in various proton-donor media may be widely different.

At present no predictions can be made to this effect because the role of solvation in various processes involving porphyrin molecules has not yet been quantified. The solvation of metalloporphyrins, whose important role is determined by the kinetic and thermodynamic compensation effects (see Chapter 4), includes specific and universal solvation to the same extent. Both types of solvation seem to affect the biological functions of metalloporphyrins significantly.

Some works[73,251,378,379] are concerned with the kinetics of dissociation of metalloporphyrins in glacial HAc with ethanol. In all cases ethanol slows down the dissociation reaction gradually and then sharply (Fig. 5.11).

In a 10% ethanol solution of HAc ZnTPP dissociates at the same rate as in glacial HAc. However, in a 30% ethanol solution of HAc the dissociation becomes twice as slow and in a 50% solution 16 times as slow. In this case, too, the effect of ethanol is to a great extent dependent on the porphyrin species. For example, the dissociation of zinc chlorophyll in a 10% ethanol solution of HAc is 2.4 times as slow, in a 20% solution nine times as slow, and in a 50% solution the dissociation is practically nil. The effect of ethanol is associated with the changing solvation of the MP molecules in the initial and transition states rather than with the decreasing

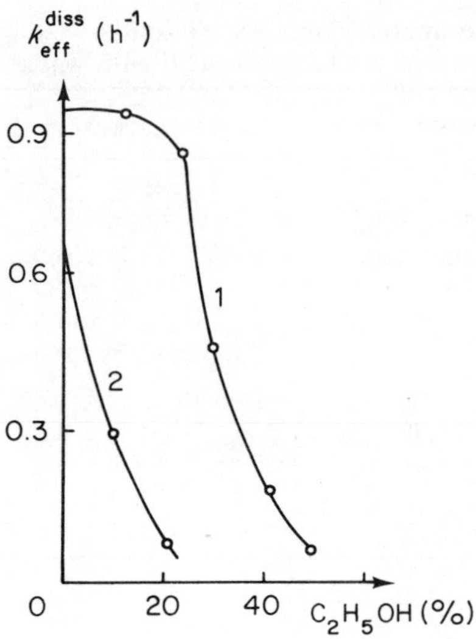

Figure 5.11 Dissociation rates of zinc complexes of tetraphenylporphin (curve 1) and chlorophyllic acid (curve 2) versus HAc–ethanol solvent composition

activity of H^+_{solv} and its concentration. The lack of systematic studies in this area does not permit elucidation of the above phenomena. Nevertheless, it cań be easily seen that the dissociation rate as a function of the solvent is dependent on the fine structural details of the MN_4 group and peripheral substituents of metalloporphyrins.

A rather unexpected relationship between the effective MP dissociation rate constant and H_2SO_4 concentration was established in an earlier work[384] (Fig. 5.12). $k_{eff} = f[H_2SO_4]$ was found to differ widely for various metal complexes of the same porphyrin. It is more pronounced in the case of CuTPP than (Ac)FeTPP. The different $k^{25}_{eff} = f[H_2SO_4]$ for complexes of the same metal with different porphyrins indicates that it is not the different order of the reaction with respect to the H^+_{solv} ion that matters. Of all the Fe^{3+} complexes the least sensible to changes in the H_2SO_4 concentration is Fe-protoporphyrin (haemin). The tetraphenylporphin complex is more sensitive and Fe^{3+}-chlorophyll is the most sensitive to addition of H_2SO_4. The series of sensitivity to H_2SO_4 is opposite to that of kinetic stability, (Ac)FePP > (Ac)FeTPP > (Ac)FeChl, which is another manifestation of the dependence of solvation of the initial and transition states of MP on their structure and on the structure of the solvent.

Characteristically, the $k_{eff} = f[H_2SO_4]$ lines converge in the region of low H_2SO_4 concentrations (see Fig. 5.12), where the structure of HAc as a solvent is practically invariable. It may be assumed that one of the reasons for constant k^{25}_{eff} being dependent on the porphyrin species is the

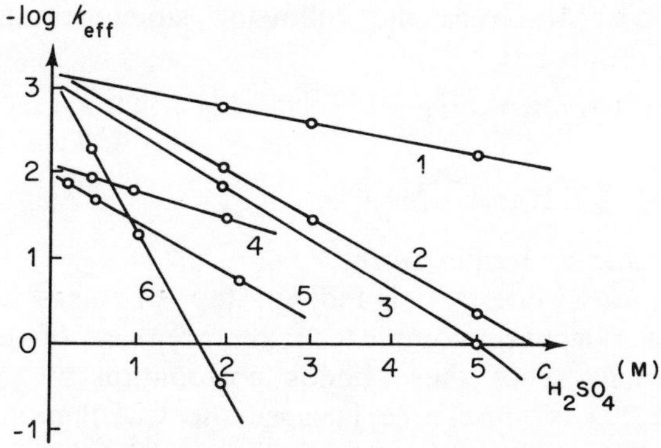

Figure 5.12 Effective dissociation rate constants of various metalloporphyrins versus H_2SO_4 concentration in HAc. MP: curve 1, AcFePP; curve 2, AcFeTPP; curve 3, AcFeChl; curve 4, NiChl; curve 5, CuChl; curve 6 CuTPP

protonation of the porphyrin molecule at its periphery. Such protonation, which leads to destruction of the MP complex, is more likely to occur at the C_6H_5 group of tetraphenylporphin and in the isocycle of chlorophyllic acid than at the $CH{=}CH_2$ groups of protoporphyrin. The protonation of the $CH_2CH_2COOCH_3$ groups destabilizes the M—N bond to a much lesser degree because of the presence of bridging CH_2 groups. The complex species, that is the nature of the metal, also affects considerably the depth of protonation of the porphyrin molecules, which was vividly demonstrated in earlier works.[71,247,376,383,398]

It is rather typical that in the case of stable metal phthalocyanines in aqueous H_2SO_4 all $k_{eff} = f[H_2SO_4]$ lines are strictly parallel[284] and do not depend on the metal. This is due to the similarity of the dissociating forms of metal phthalocyanines $MPcH^+$.

5.4 Metalloporphyrin Dissociation Mechanism

The concept of a metalloporphyrin dissociation mechanism implies the number of elementary steps involved in the transformation of the complex molecule to free porphyrin, as well as the structure–energy characteristic of the transition states. Depending on the nature of MP and the medium this characteristic determines the ratio between the rates of the elementary steps and the limiting step which is of particular practical interest.

The first attempt to describe the stoicheiometric mechanism (i.e. to define the number of elementary steps) was made in 1955.[406] In studying the kinetics of decomposition of the stable covalent copper etioporphyrin complex in glacial HAc–H_2SO_4 monohydrate, Caughey and Corwin established that the dissociation reaction is of second order with respect to the metalloporphyrin and fourth order with respect to the acidity of the solution (h_0). On this basis the following stoicheiometric dissociation mechanism was proposed:

$$CuEP + 2H^+ \overset{fast}{\rightleftarrows} CuEP \cdot 2H^+, \tag{5.27}$$

$$2CuEP \cdot 2H^+ \overset{slow}{\rightleftarrows} 2Cu^{2+} + 2H_2EP; \tag{5.28}$$

this was criticized in an earlier work.[385]

Indeed, the most energy-demanding step of dissociation of stable strongly covalent porphyrin complexes is the breakage of the M—N bonds. Without the breakage of these bonds protonation of the reactive site according to (5.27) is impossible because the CuEP molecule lacks free proton-acceptor sites. Therefore step (5.27) must be the slow limiting step of the whole process. Intermediate complex $CuEP \cdot 2H^+$ cannot accumulate in the system under conditions of a strong proton-donor medium, almost instantaneously detaching the cation, while porphyrin H_2EP is immediately protonated to $H_2EPH_2^{2+}$ (in concentrated H_2SO_4) or decomposes (in HAc with a small added amount of H_2SO_4).

In later kinetic studies using model molecules of metal phthalocyanines it was established that the dissociation of stable complexes in concentrated H_2SO_4 is always of first order with respect to the complex and second order with respect to the H_3O^+ ion.[385–388] The second-order dissociation with respect to the porphyrin complex was not observed in any of the later experiments with metalloporphyrins. This indicates that, having properly selected the proton-donor solvent (HAc + H_2SO_4), Caughey and Corwin[406] derived wrong kinetic values because of the complicated procedure of measuring the reaction rate and arrived at a wrong conclusion as to the reaction mechanism.

The concepts of the metalloporphyrin dissociation mechanism have been developed by the author of this monograph on the basis of studies into the kinetics of dissociation of numerous metal phthalocyanines as the most suitable model molecules. The results were unambiguous owing to the use of the most appropriate of all proton-donor media, namely concentrated H_2SO_4, in whose solutions the content of all main proton-donor particles is known.

The third order of the reaction of dissociation of stable phthalocyanine complexes, the values of the activation parameters, and the extremely high covalence of the M—N bonds have led to a conclusion[386] that the dissociation reaction involves a trimolecular mechanism, which was corroborated later.[37,284,388] In these studies it was pointed out that the dissociation mechanism is common to stable phthalocyanine and porphyrin complexes.

This is how the trimolecular S_N3 mechanism has been arrived at. In a sulphuric acid solution all stable phthalocyanines are in the form of monoprotonated $MPcH^+$ particles (**XLII**).

The proton is located at a peripheral nitrogen atom far from the MN_4

group and acts indirectly on the latter. This action may manifest itself in the four coordinate M—N bonds becoming non-equivalent. The M⇄N bonds seem to be somewhat weakened with respect to those in the unprotonated particle for the presence of a positive charge in the macroring considerably weakens the σ interaction and only slightly improves the conditions for π interaction in them. It is quite obvious that the mechanism of dissociation of $MPcH^+$ and MPc must be the same, the only difference being in the activation parameters.

H_3O^+ attacking a M—N bond causes deformation of the latter as a result of polarization of the electron cloud of the σ bond towards the hydroxonium proton. Because no π orbitals are available at the H atoms of hydroxonium, their interaction with the bonding π orbitals is negligible and cannot lead to a breakage of the M—N bond. The geometric configuration of the H_3O^+ particle[411] is favourable for the occurrence of two kinds of its interaction with the atoms of the M—N bond at a time. As the σ-electron pair of this bond is being freed and transferred to the antibonding orbital of H_3O^+, one of the H—O bonds in hydroxonium is weakened, with the result that the bonding σ-electron pair of this bond is freed and transferred to the bonding orbital of the metal ion. A charge separation at the $\overset{\delta-}{N}—\overset{\delta+}{M}$ and $\overset{\delta+}{H}—\overset{\delta-}{O}$ bonds takes place. As a consequence, the hydrogen atom enters into an acid–base interaction with the nitrogen atom, while the oxygen atom enters into a donor–acceptor interaction with the metal ion, in which another coordinate place becomes vacant in the course of activation and breakage of the M—N bond.

The occurrence of two interactions at a time during the activation of dissociation of metal phthalocyanine (metalloporphyrins) almost fully compensates for the energy expended in the bonds being broken due to solvation and is corroborated by facts. If the activation had been caused only by protonation of the σ-electron pair of the M—N bond, the dissociation rate would have grown with the H_2SO_4 concentration. In fact this does not take place since H_2SO_4 does not carry such a strongly solvating molecule as H_2O together with the proton. The presence of anion HSO_4^- after H^+ has been given up by the H_2SO_4 molecule or, in the case of $H_3SO_4^+$, the presence of H_2SO_4 near a vacant coordinate place of the cation is not sufficient to compensate for considerable energy expenditures in the breakage of the M—N and O—H bonds. The foregoing is also corroborated by interesting facts presented by other authors.[288,412]

It has been found that even labile and weakly covalent magnesium phthalocyanine does not show any signs of dissociation with protic acids in the gaseous phase[288] or in their solutions in non-coordinating solvents.[412] Thus it appears that the breakage of an M—N bond in covalent metalloporphyrins requires a sufficient concentration of solvated proton H^+_{solv} in a coordinating solvent such as water, alcohols, acetone, glacial HAc, DMSO, amines, DMFA, and sulpholane.

The breakage of the first M—N bond is an energy-demanding and slow reversible step of dissociation. The breakage of only one out of the four covalent M—N bonds still leaves the entire system of bonds rigid and strong. This is supported by the high stability of *N*-alkylporphyrin complexes with Cu, Zn, and other metals, in which only three M—N bonds may be assumed. Bonding with the fourth nitrogen atom is sterically hindered by the CH_3 group. Without breakage of the second M—N bond the cation cannot leave the plane of four heteroatoms. The second rather than first order of dissociation with respect to H_3O^+ or H^+_{solv} indicates that the breakage of the second bond consumes as much energy as that of the first. The coplanarity of three M—N bonds and the higher charge at the metal atom provide for stability of the coordination system. Therefore the first and second steps may be equivalent in the case of stable MPc and MP complexes, as far as their energy demand and rates are concerned.

Under the conditions in which the dissociation of MPc was studied it was only natural to assume that two H_3O^+ ions attack MPc from either side of the plane at a time, thereby providing the transition state of dissociation (see diagram):

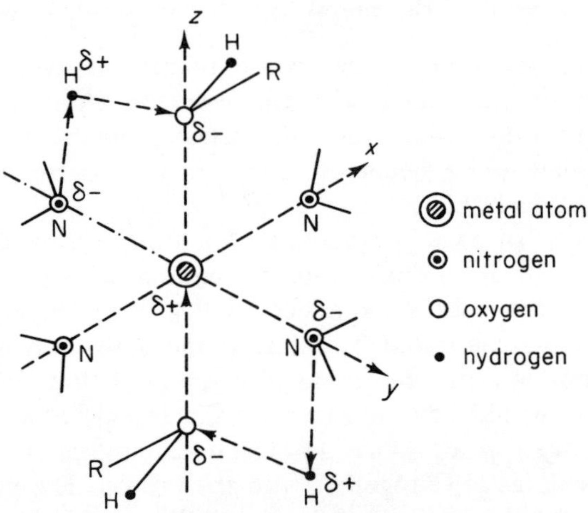

The point is that 75–90% H_2SO_4 is essentially a melt of hydroxonium bisulphate $H_3O^+\cdots HSO_4^-$ with an extremely high concentration of hydroxonium (up to 13 M H_3O^+). The postulation of a trimolecular mechanism under such conditions is quite justified, as is shown elsewhere.[37,388] In the diagram the arrows indicate the electron density circulation between the bonding and antibonding orbitals of atoms, occurring during activation of the system.

Naturally, in media with a low H^+_{solv} concentration (e.g. HAc, alcohols + HAc, HAc + H_2SO_4) it is not always appropriate to assume a simultaneous attack by $2H^+_{solv}$. The steps of breakage of the first and second M—N bonds may not coincide in time and follow each other at slow but commensurate rates:

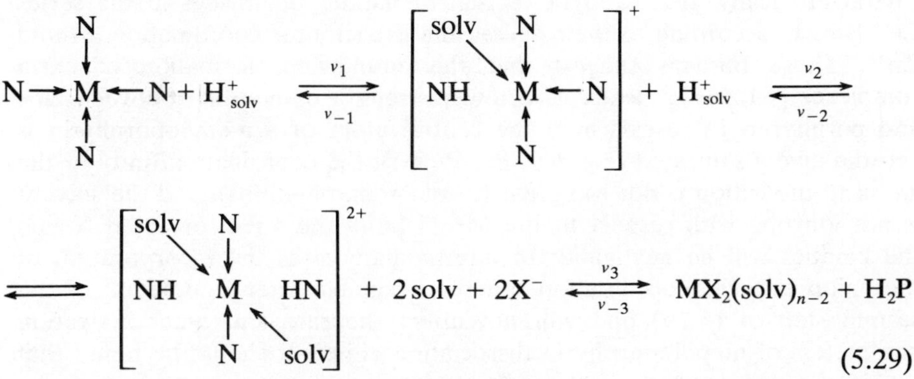

(5.29)

Each step will have transition states similar to those in the schematic representation. Therein, $\nu_1 \approx \nu_2, \nu_3 \gg \nu_1, \nu_1 \ll \nu_{-1}$, and $\nu_3 \approx \nu_3$. The transition states of the first two limiting steps of dissociation repeat those of the final two limiting steps of the complexing reaction. Therefore the equilibrium constants of MPc and MP formation, derived from kinetic data, agree well with the normally observed stability series of complexes with nitrogen-containing ligands (see Chapter 6) and are corroborated by the spectral criterion of stability (see Chapter 7).

The final steps of dissociation, included in the third term of expression (5.29), are in all probability dissimilar to the initial steps of MP formation because the conditions for formation and dissociation of MP are not the same for known reasons. MP complexes are formed under such mild conditions that all the reverse steps in expression (5.29) practically combine into a single elementary step, which was shown earlier (Chapter 4) for most cases. Only rare cases where metalloporphyrin intermediates are formed are at variance with the common consistent mechanism of their formation.

After breakage of two out of the four M—N bonds the coordination system MPH_2^{2+} becomes unstable. The metal ion readily leaves the MN_4 group under the effect of the solvent molecules. The more polar the proton-donor solvent, the faster the final third step of dissociation. Chapter 4 contains sufficient evidence to support this conclusion. Of particular interest are the questions: at which activation stage are anions X^- incorporated into the coordination sphere of the cation? and, which electronic mechanism of protonation is involved? The answer to the first question will depend on the nature of the metal and solvent. In cases where the central metal atom in a porphyrin complex is coordinatively unsaturated (e.g. Mg^{2+}, Cd^{2+}, Hg^{2+}, Zn^{2+}), the anion is already included in the coordination sphere before the act of dissociation if its coordinate affinity to the cation is great enough. In this case the anion concentration is included in the kinetic equation.

Similar facts were observed in the case of zinc porphyrins.[361,366] Characteristically, the 'catalytic' action of halides diminishes in the series Cl^-, Br^-, I^- according to their decreasing capacity for coordination around Zn^{2+}. These findings suggest that the equilibrium formation of extra complexes $[Cl_2ZnP]^{2-}$ leads to a lower strength of bonding between zinc and porphyrin. In cases where the central atom of a metalloporphyrin is coordinatively saturated (e.g. Cu, Pd, Pt, Ni), the coordinate affinity of the anion to the cation is not too great (most ordinary anions), and the solvent is not ionizing with respect to the M—N bond the effect of anion X^- on the kinetics will be negligible. In intermediate cases the incorporation of the anion into the coordination sphere is possible after completion of the second step of (5.29) and will not affect the rate and other activation parameters of metalloporphyrin dissociation either. It should be noted that the intriguing problem of the effect of the anionic composition of the medium on dissociation of porphyrin complexes has hardly been studied.

The electronic mechanism of protonation is one of the most important components of the overall delicate mechanism of the limiting step of MP dissociation. As far as the proton transfer is concerned, everything is absolutely clear from the above.

The most essential thing to establish is which orbital of the nitrogen or any other atom of MP supplies electrons to the 1*s* orbital of H^+. This orbital (first antibonding, then vacant bonding orbital) probably overlaps with the filled *s*, p_z, or hybrid *sp* orbital of the nitrogen atom. At present there is no consensus to this effect. According to Albert[413] the weak basicity of *n* electrons or absolute lack of it is typical of *s* electrons. To support this point an example with N_2 and $(CN)_2$ is given. In so doing the author forgets the more convincing examples with NH_3 and saturated amines (*s*-electron pair), aniline (p_z-electron pair), pyridine (*s*-electron pair), and pyrrole (p_z-electron pair). Comparison of the acid–base properties of the above compounds immediately refutes the assumption that the basicity of the *s*-electron pair is weak. The reason for *n* electrons losing their basicity is their delocalization over the π-electron system or vacant *d* orbitals. Analysis of conjugated and aromatic systems indicates that the p_z-electron pair (pyrrole, aniline) participates in occupation of the lowest antibonding orbital with the result that the heteroatom loses its basic properties partially or completely.

As to the *s* electrons of heteroatoms, first they convert to the *p* state, then they are distributed among the atoms of the conjugated system. The presence of a proton near such an atom minimizes the possibility of *sp* states being mixed, and the *s*-electron pair is adequately protonated (pyridine, imidazole, etc.). The presence of vacant lower *d* orbitals (atoms S, P) causes *sd* mixing of the orbitals of *n* electrons with the result that such *n*-electron pairs lose their capacity for protonation (thiophene, dialkyl sulphides). The *ns* electrons of diazines and triazines losing their capacity for protonation is most probably due to the accumulation of the positive charge at the π skeleton, which promotes the $s \rightarrow p_z$ conversion of n electrons.

In porphyrins and phthalocyanines n electrons occupy s orbitals strongly mixed with p_z orbitals because of the possible participation of the latter in a strong conjugation (particularly in phthalocyanines), which is facilitated by the low-lying vacant ψ_π^* orbitals. In a phthalocyanine ligand all intracyclic N atoms are equivalent. The p_z-electron pairs are not protonated because they are involved in the conjugation (see Chapter 3). Then, it is absolutely out of question that they can be protonated in the molecules of stable phthalocyanines in which the n electrons are distributed among the orbitals of the metal. Protonation of only one of the extracyclic N atoms is possible (see Chapter 4). Therefore the dissociation process begins with the competition of the attacking proton with the metal atom for the bonding s-electron pair. The removal of this pair from the bonding orbital of M—N, demanding great amounts of energy, and its transfer to the H^+ orbital determine the slow reaction rates.

A somewhat different picture is observed in the case of dissociation of stable metalloporphyrins. In a porphyrin ligand the nitrogen atoms are not equivalent. The pyrrole N—H groups have a nitrogen atom in the sp^2-hybridization state and n electrons in the p_z state. These n electrons are not capable of protonation. The pyrrolenine N atoms are in the p^2-hybridization state and have an s-electron pair which is in a weak $\sigma \rightleftarrows \pi$ conversion with the p_z orbital. Therefore the protonation of these n electrons is relatively easy. For this reason porphyrins H_2P are diprotonated in the centre of the molecule. An additional attachment of $2H^+$ is practically impossible.

In stable metalloporphyrins all of the four nitrogen atoms are levelled off,[8,179] and a unique conjugated system of the phthalocyanine type is formed. The hybrid dsp^2 orbitals of the metal are filled with the sp-electron pairs of the porphyrin dianion. The heteroatoms are left only with p_z electrons heavily involved in the conjugation, that is in the filling of the ψ_π^* orbitals. In this case the attack by the proton leads to localization of p_z orbitals at the nitrogen atoms, their interaction with the proton, deformation of the M—N bonds as a result of the hybridization of the nitrogen atoms changing from sp^2 to sp^3, as well as rearrangement of all conjugated bonds in the aromatic system of porphyrin. The breakage of the M—N bonds is followed by distortion of the p^3 state of the nitrogen atoms, which then changes to the sp^2 state with the protons being arranged in the molecular plane.

The rearrangement and strong deformation of the bonds lead (see the beginning of this chapter) in some cases to fragmentation of porphyrin systems. This is promoted by the increasing covalence of the M—N bonds and insufficient acidity of the solution. Of particular interest are cases where dissociation either involves fragmentation of the porphyrin or proceeds without it, depending on temperature and acidity. These cases are indicative of instability of porphyrin molecules with nitrogen atoms which are in the sp^3 state. The latter, in turn, results in deformation of the coplanar aromatic system of the porphyrin and further loss of stability. The limiting step proceeding faster with higher acidity, rising temperature, and decreasing covalence of the M—N bond provides for proton exchange and rules out fragmentation of the porphyrin macroring.

The foregoing does not suggest a deformation of all pyrrole nuclei as strong as that assumed by Shears *et al.*[366] since it is not justified either geometrically or energetically. Naturally, in the case of predominantly ionic complexes (Mg^{2+}, Cd^{2+}, Hg^{2+}), the sp^3 hybridization state of nitrogen atoms does not occur. Therefore labile complexes always dissociate without decomposition of the porphyrin ligand. The ligands themselves in an acid medium, that is H_4P^{2+}, are extremely stable.

The distinguishing features of the metalloporphyrin dissociation mechanism, namely trimolecularity and simultaneous protonation and solvation, presented in this chapter were demonstrated and reported by the present author more than 10 years ago. Similar interpretations of the reaction mechanism appeared in a number of papers published outside the USSR[366,392] several years later with proper references.

The MPc and MP dissociation mechanism proposed by the present author belongs to the self-consistent processes of ligand substitution with associative–dissociative activation. During activation of a reacting system an equally important role is played both by the complex and by the solvated proton. Hence, in accordance with earlier classifications,[321,352] the mechanism may be referred to as $I_{a,d}$. It should be remembered that the mechanism of metalloporphyrin formation is the same, which is quite natural in view of the reversibility of the dissociation reaction. In my early works[385] I used the inaccurate term 'hydrolysis of metal phthalocyanines'.† Unfortunately this term has taken root in non-Soviet literature with a slight modification, namely 'acid-catalytic solvolysis of complexes'. It should be noted that this term is alien to the chemistry of coordination compounds which include metalloporphyrins and metal phthalocyanines. Protons do not act as catalysts of MP dissociation at all for they disappear in the course of protonation of the complexes. Nor is there any solvolysis involved, for the solvent molecules are not split into the proton and residue but act as a solvating agent. It is therefore more appropriate to refer to the reactions in which the cation in MP is exchanged for protons as protolytic dissociation of complexes.

In conclusion I should like to emphasize that all kinetic parameters—rate constants, energies, and entropies of activation of MP and MPc dissociation—are in good agreement with the above concepts. The moderate activation energies (10–20 kcal mol^{-1}) at moderately low

†This term is not appropriate because in sulphuric acid solutions the dissociation of MPc is followed by hydrolytic splitting of the H_2Pc ligand.

entropies of activation (−30 to −40 e.u.) are indicative of high polarity of the transition state as compared to the initial one, absence of any significant non-coplanarity of the macroring system during dissociation, the breakage of the M—N bond as being the most energy-demanding step of the dissociation process, and the important role of extra coordination and solvation in this process.

5.5 Kinetic Criterion of Stability of Porphyrin and Phthalocyanine Complexes

In the case of some extremely stable complexes, including those with rigid cyclic ligands such as porphyrins, phthalocyanine, and some of its analogues, estimating the stability in solution with the aid of stability constants is practically impossible. The unidirectionality of the reactions of their formation and dissociation,[389] the poor solubility or insolubility of the ligands and complexes in most solvents, the decomposition of the ligand in proton-donor media after most complexes have passed through the limiting step of dissociation—all these factors lead to what seem to be unsurmountable obstacles in measuring the equilibrium constants with a view to determining the thermodynamic stability of these complexes in solutions.

In this connection attempts have been made, starting with the first works by Linstead *et al.*[414] on metal phthalocyanines, to correlate the stability of MPc complexes with some other physicochemical properties of the latter. At present these include the electronic absorption spectra and complex dissociation rates in proton-donor media.[23,191,403] However, as has been shown previously,[88,289] methods based on these parameters suffer from serious limitations.

The most promising method for the study of MP and MPc complexes is the kinetic technique providing two kinds of information about them. In all cases it permits study of the kinetic stability of complexes of the same type, which is characterized by free activation energy of complex dissociation under standard conditions

$$\Delta G^{\ddagger}_{25} = -RT \ln k^{25}_{\text{diss}} - RT \ln (KT/h) \tag{5.30}$$

or by dissociation rate constant k^{25}_{diss}. In some cases it is possible to measure the rates of the forward and backward MP dissociation reactions and to determine, on their basis, the thermodynamic stability of the complexes. So far it has not been possible to obtain the latter parameter for metal phthalocyanines because of the insuperable difficulties in measuring the rates of their formation in solutions. This has prompted investigators to look for a correlation between the variations in the kinetic and thermodynamic stability of phthalocyanine and porphyrin complexes in solutions. This problem was formulated and discussed for the first time in conjunction with MP and MPc complexes in an earlier work.[387]

The question as to whether there is a direct relationship between the kinetic and thermodynamic characteristics of chemical reactions has long intrigued investigators. In one of his works[415] Hammet writes that the idea of a simple relationship existing between rate and equilibrium is one of the most staunchly defended and at the same time most hotly disputed ones in theoretical chemistry. This idea has already been put to successful use by many organic chemists in studying similar reactions despite the fact that the relationship between rate and equilibrium has no theoretical foundation and the only way in which they are related is by the well-known principle of detailed balance.[416]

Nevertheless simple empirical relations have been derived for similar reactions. A case in point[415] is the broad group of second-order reactions between reagent S and acid catalyst HA

$$HA + S \overset{k_v}{\rightleftarrows} SH^+ + A^- \overset{\text{fast}}{\rightarrow} HA + \text{products.} \tag{5.31}$$

The rate constant is related to the acid dissociation constant (K_S), which varies directly with the equilibrium constant of reaction (5.31) simply as

$$k_v = AK_S \tag{5.32}$$

Relation (5.32), applicable to alkyl transfer reactions $R\text{-}COOCH_3 + N(CH_3)_3 \rightleftarrows (R\text{-}COO^- + N(CH_3)_4^+$, has been firmly established. The simple relationship between rate and equilibrium is particularly essential in the case of inert complex compounds. In the dissociation of such complexes and substitution of the ligand by solvent molecules the limiting step is, as was shown in Section 5.4, the dissociation of the covalent donor–acceptor bond.

The dissociation rate of the metal–ligand bond and the equilibrium of the process must depend on the nature of the metal (with the same ligand), the ligand (with the same metal), and the solvent to the same extent in all cases where the metals, ligands, and solvents of the series differ widely in electronic structure and steric effects. Generally, as one goes from water to organic solvents, the dissociation rate and equilibrium constant are reduced, that is both the kinetic and thermodynamic stabilities of the complexes increase. This is due to the fact that the strength of the chemical donor–acceptor bond depends on the solvent species and in solutions it may differ considerably from that under standard conditions (in the gaseous phase).

A polar solvent produces a differentiating effect on the bond. For example, water breaks ionic bonds almost completely (typically ionic compounds dissociate completely in an aqueous solution) and perceptibly weakens polar covalent bonds, thereby providing for a more or less pronounced equilibrium of the dissociation process, while weakly polar and purely covalent bonds remain practically unaffected so that the dissociation of such compounds is virutally nil. Similarly, the covalence of a bond influences the kinetics of its dissociation. An increase in the covalence of breaking bonds invariably result in slower dissociation rates.[417,418]

Such parallelism between the rate and equilibrium of dissociation of complexes permits the free energy of the dissociation reaction

$$\Delta G_{\text{diss}} = -RT \ln K_{\text{H}} \tag{5.33}$$

or the free activation energy $\Delta G^{\ddagger}$ (see equation (5.30)) of the reaction to be taken as a quantitative measure of reactivity.

As was established in a number of works concerning the *trans* and *cis* effect of the ligands in complexes of the platinum group,[419,420] both the substitution rate and the equilibrium constants vary correspondingly. This indicates that the covalent bonding of the *trans* partner is weakened under the effect of the *trans* ligand. As far as metalloporphyrins are concerned, a qualitative correlation between the dissociation rate and the stability was postulated in earlier works.[403,406] However, their stability was estimated using a spectral criterion without taking into account the complicated way in which it is related to the type of the metal–porphyrin bond. Therefore a kinetic stability series $Mg^{2+} < Zn^{2+} < Cu^{2+} < Co^{2+} < Ni^{2+} < Pd^{2+}$ was proposed in which the mutual positions of Cu, Co, and Ni are doubtful.

Reliable quantitative data on dissociation rates were obtained for the first time for phthalocyanine complexes[37,386,388] and complexes of chlorophyllic acid.[88,401,421] From the values of k_v^{25} for metal phthalocyanines (see Table 5.3) and k_{eff}^{25} for metal analogues of chlorophyll (see Table 5.7) one can write the following kinetic stability series:

$$\begin{gathered}\text{AcFeChl, PdChl} > \text{NiChl} > \text{CuChl} > \text{CoChl} > \text{AgChl} > \\ > \text{ZnChl} \gg \text{MgChl} \gg \text{CdChl} \gg \text{HgChl};\end{gathered} \tag{5.34}$$

$$\begin{gathered}\text{PtPc, PdPc} > \text{CuPc} > \text{NiPc} > \text{CoPc} > \text{ZnPc} \ggg \text{FePc} > \\ \text{MgPc, Ag}_2\text{Pc, CdPc, HgPc;} \\ (\text{HSO}_4)\text{RuPc} > (\text{HSO}_4)\text{RhPc} > (\text{HSO}_4)\text{AlPc} > \\ (\text{HSO}_4)\text{IrPc} > (\text{HSO}_4)\text{GaPc} \ggg (\text{HSO}_4)\text{FePc}; \\ \text{O=OsPc} > \text{O=SiPc} > \text{O=GePc} > \text{O=VPc} > \text{O=SnPc, O=TiPc.}\end{gathered} \tag{5.35}$$

In series (5.34) the position of all complexes corresponds strictly to the dissociation rate constants. In series (5.35) a similarly strict correspondence applies only to stable complexes. The kinetic stability of labile phthalocyanine complexes with Mg^{2+}, Ag^{+}, Cd^{2+}, Hg^{2+}, Fe^{2+}, and Fe^{3+} was determined under different conditions, namely in the solid phase.[130] The position of complexes in the kinetic series indicates either a greater or lesser covalence of the metal–ligand bond if the difference in rate constants is sufficiently great (by at least one order of magnitude).

Since the same mechanism is involved in the dissociation of phthalocyanine and porphyrin complexes, it is possible to quantify the difference in covalence of similar complexes depending on the ligand. Thus the enormous difference in kinetic stability between stable phthalocyanine complexes (Pd^{2+}, Ni^{2+}, Co^{2+}, Cu^{2+}, Zn^{2+}) and respective porphyrins is the best proof of the much higher strength of the covalent M—N bonds in MPc. The kinetic data are unambiguously indicative of the ionic nature of labile MP and MPc complexes, the considerable covalence of the Mg–porphyrin bonds, the high covalence of bonds in ferroporphyrins, and its predominantly ionic character in ferrophthalocyanines.

At the same time one cannot disregard the effect of solvation, particularly that which is due to extra coordination of the solvent at the central atom, on the kinetic stability series of complexes in various proton-donor media. For example, the dissociation rates of CuChl and NiChl in glacial HAc with H_2SO_4 monohydrate differ but slightly, whereas in pure HAc the dissociation rate of NiChl is almost 100 times as low as that of CuChl.[398] Hence the kinetic stability series have no absolute value. For different solvents these series may be substantially different, which is why they must be associated with a specific solvent and a particular temperature.

The kinetic stability series alone are not sufficient to characterize complexes fully in the reactions of their dissociation because the rate does not represent the intensity of the process. Therefore the value of $\Delta G^{\ddagger}$ is representative only of kinetic reactivity and provides the answer to the question 'how fast?'. To answer the question 'how intensive?' necessitates thermodynamic data.

The kinetic criterion of stability of complexes becomes meaningful if, apart from the rate constants, the energies and entropies of activation are also known for similar complexes. They permit establishment of the temperature dependence of the stability series and the role of solvation in these processes. At the same time the activation energy alone tells nothing about the kinetic stability. It is well known that if the dissociation reaction proceeds in the gaseous phase, the activation energy approaches the dissociation energy in magnitude for it is determined only by the breakage of a particular bond. In solvents the picture becomes much more complicated due to the effect of the solvent.

In this case the value of the activation energy may characterize the bond strength in reactions proceeding according to a dissociative mechanism (S_N1 or D) but only qualitatively, and the energy will be expended primarily to detach the ligand. In S_N2 and S_N3 ($I_{a,d}$) dissociation reactions the activation energy is associated mainly with deformation of the reacting particles,[387] therefore it is only indirectly dependent on the strength of the breaking bonds. For example, in MPc dissociation reactions (see Table 5.3) there is no correlation between the activation energy and kinetic stability variations. Particularly stable complexes are formed by the platinum group elements and still the activation energy in the case of PtPc and PdPc is lower than in the case of CuPc, and if one takes the complex O═OsPc, its activation energy is lower than that of O═VPc and equals that of the least stable complex O═SnPc. This is due to the fact that the activation energy is essentially a sum,

$$E = E_{es} + E_{M—N} + E_{O—H} - E_{M—O} - E_{N—H} - E_{solv}, \quad (5.36)$$

which includes the energies of electrostatic repulsion of cations H_3O^+ and $MPcH^+$, deformation of the M—N and O—H bonds, coordinative interaction M—O, acid–base interaction N—H, and additional solvation of the reagents in the transition state. The values of E_{es}, $E_{O—H}$, $E_{M—O}$, $E_{N—H}$, and E_{solv} do not remain constant in the phthalocyanine series since they depend on the effective charge at the central ion. Therefore the activation energy is higher in cases where the coordinating capacity of the metal ion has been used up (Cu, Pt, Pd) and decreases as the effective charge grows (Ga, Al, Zn, Co, Os, etc.). In the MPc as well as the MP series the activation energy does not characterize stability even qualitatively.

The entropy of activation is related in a known manner to two energy values, one of which ($\Delta G^{\ddagger}$) is determined by the complex stability and the other (E) varies at random in the stability series; therefore the absence of any regular changes in $\Delta S^{\ddagger}$ with the stability of MPc complexes is quite natural.

Chapter 6

Thermodynamics of porphyrin and phthalocyanine complex formation

This chapter is a brief summary of the results of direct measurements of the equilibria of metalloporphyrin-yielding reactions and subsequent coordinations of extra ligands, the results of calculations of equilibrium constants from kinetic data, and other aspects of thermodynamic stability of these complexes in solutions.

6.1 Results of Direct Equilibrium Measurements

As has already been mentioned, direct measurements of metalloporphyrin dissociation (formation) equilibria involve a number of practically insurmountable difficulties, which is why the field of physical chemistry that can be called thermodynamics of complexing in porphyrin molecules is still in its infancy. No data on this subject can be found in the fundamental reviews.[23,403] Nor have such data become available in later years. The main reason for this is the extremely high stability of most metalloporphyins, as a result of which practically all of the porphyrin molecule is bound into a complex even if the solution contains an equivalent amount of metal. Many metalloporphyins dissociate into ions only under the effect of molecules with pronounced proton-donor properties. These molecules cause protonation of the porphyrin ligands with the result that the equilibrium shifts to one of its extreme positions. Because of the unavailability of most metalloporphyrins or their insufficiently high solubility in proton-donor media, the equilibrium concentrations of one of the dissociation products are so small that they cannot be measured by the existing techniques. The situation is further aggravated in cases where the ligand is destroyed during dissociation (see reactions (5.17) and (5.18)). Such cases were described in Chapter 5 in conjunction with stable metalloporphyrins. They rule out completely the possibility of direct measurement of the dissociation equilibria of these complexes.

In the case of stable phthalocyanine complexes attempts to measure equilibria directly are absolutely futile because of the decomposition of the ligand (see reactions (5.2) and (5.3)). The only possible way to determine their thermodynamic stability in such cases is by the kinetic approach considered in the following section.

As far as labile porphyrin and phthalocyanine complexes are concerned, the situation is not as hopeless as with stable complexes. One publication[422] presents the instability constant of zinc mesoporphyrin in aqueous solution with detergents ($K^{20} = 10^{-29}$). In spite of this value being highly approximate because of the aggregation of porphyrins and their complexes in the above media and the lack of an exact value of the acid ionization constant of the mesoporphyrin in question, it demonstrates the extremely high stability of porphyrin complexes. For comparison consider the instability constants of the following complexes: 10^{-11}, zinc ethylenediamine; 3×10^{-10}, zinc alaninate; and 2.6×10^{-17}, zinc ethylenediaminetetraacetate.[328]

In view of the above one should evaluate the so-called extraction method for determining the instability constants of metalloporphyrins, proposed by Kukhtevich.[240] In his work Kukhtevich examined the equilibrium in a two-layer system: an aqueous metal acetate solution with pH = 0.6–6.2 and a chlorophyllic acid solution in *n*-butanol were poured into a vessel, shaken at 100 °C, cooled to room temperature, and analysed. The MChl complex remained in the organic phase together with H_2Chl, was extracted, its concentration determined from the optical density, and from the concentration found those of the particles involved in the equilibrium were calculated.

The mixed extraction chemical equilibrium constant

$$K_e = [\mathrm{MChl}]_{\mathrm{org}}[\mathrm{H}^+]^2_{\mathrm{aq}}/[\mathrm{H_2Chl}]_{\mathrm{org}}[\mathrm{M}^{2+}]_{\mathrm{aq}} \tag{6.1}$$

was used in calculating the extraction constants of instability of the MChl complex from the relation

$$K_{(\mathrm{org/aq})} = K_{\mathrm{S1,2}}(\mathrm{H_2Chl})/K_e \tag{6.2}$$

As was mentioned in an earlier paper,[244] in determining equilibrium constant (6.1) Kukhtevich did not define the constants of distribution of the H^+ and M^{2+} ions between the aqueous and organic phases, with the result that the exact concentrations of these ions were not known. In addition, complete ionization of the metal acetates was assumed. Because of this, and as a result of using an incorrect value of the total acid ionization constant $K_{\mathrm{S1,2}}$ of chlorophyllic acid (see Chapter 4), the equilibrium constants obtained were too low:

Complex	k	Complex	k
CuChl	2×10^{-23}	ZnChl	5×10^{-20}
CoChl	5.2×10^{-20}	CdChl	1.7×10^{-11}

These values, taken from Kukhtevich's work,[240] are not, as will be shown later, adequately representative of the differences in stability between the complexes in question. Therefore, even if a more reliable value of $K_{S1,2}$ for H_2Chl (see Chapter 4) is used in calculations, exact values of instability constants still cannot be derived.

At the same time, careful analysis of the extraction method indicates that with a more rigorous approach it can be used for exact definition of the equilibrium constants of porphyrin complexes.

6.2 Calculation of Equilibrium Constants From Kinetic Data

To obviate the difficulties encountered in direct equilibrium measurements involving porphyrin complexes two versions of the kinetic method were proposed and successfully applied by the author to determine their thermodynamic characteristics.[88,389]

In the first version the equilibrium constants are calculated for metalloporphyrins from known values of the rate constants of dissociation and formation of these complexes, obtained under identical conditions (temperature, solvent, anionic composition of the medium). Such a calculation is possible owing to the microreversibility principle (the principle of detailed balance) being applicable to metalloporphyrins. Their formation and dissociation under selected conditions involve one and the same transition state (see Chapters 4 and 5).

Therefore, despite the different forms that can be taken by total equations of metalloporphyin formation

$$MX_2 + H_2P \rightleftarrows MP + 2HX \tag{6.3}$$

and dissociation

$$MP + 2H^+_{solv} \rightleftarrows M^{2+}_{solv} + H_2P, \tag{6.4}$$

the ratio between the dissociation and formation rate constants gives the equilibrium constant of the dissociation reaction:

$$K_e^{diss} = k^{25}_{diss}/k^{25}_{form}. \tag{6.5}$$

The equilibrium constants calculated from equation (6.5) are arbitrary in view of the fact that the true values of K^{25}_{diss} are unknown and are replaced by effective constants in the calculation. For various metalloporphyrins the latter constants have been determined under the same conditions (temperature, solvent, proton-donor reagent concentration), hence the dissociation equilibrium constants are truly representative of the thermodynamic stability series of porphyrin complexes and can be used for evaluating the effective metalloporphyrin instability constants from the equation

$$K^{MP}_{inst} = K_e^{diss} K_{S1,2}. \tag{6.6}$$

In this way equilibrium and instability constants have been calculated for some porphyrin complexes in glacial HAc (Table 6.1) and binary solvents (Table 6.2).

The formation rate constants of complexes PdChl, AcFeChl, and AgChl could not be obtained under the conditions in which dissociation was studied. Therefore their equilibrium and instability constants have been determined using the second version of the kinetic method. In this case use is made of the weak dependence of the complex formation rate on the metal. For Zn^{2+}, Cu^{2+}, Fe^{2+}, Ag^{2+}, Cd^{2+}, and Pd^{2+} k_v^{form} may be assumed to be practically invariable on the basis of the data of Chapter 4. Then the following expression can be written for the complex of interest (M′P) and the reference complex (MP):

$$K^0_{\text{MP}}/K^i_{\text{M'P}} \approx k^0_{\text{diss}}/k^i_{\text{diss}}, \tag{6.7}$$

correlating the dissociation equilibrium constants (K) and dissociation rate constants (k_{diss}) measured under identical conditions.

If the equilibrium constant for the reference complex is assumed equal to unity, one can find the relative equilibrium constant

$$K^i_{\text{M'P}} = k^i_{\text{diss}}/k^0_{\text{diss}}. \tag{6.8}$$

This is how the relative dissociation equilibrium constants have been determined for phthalocyanine complexes (Table 6.3) whose formation rate constants are still unknown, the reason being the negligibly low solubility of the ligand of phthalocyanine and its complexes in the currently used solvents as well as the similarity of their absorption spectra.

Knowledge of the reference complex formation rate constant will enable the equilibrium constants to be calculated from the relative constants. Thus, from the known value of the dissociation equilibrium constant of CuChl taken as a reference complex, absolute equilibrium constant values were derived for AgChl, PdChl, and AcFeChl (see Table 6.1).

Table 6.1 Dissociation equilibrium constants (K_e^{diss}) and instability constants ($K_{\text{inst}}^{\text{MP}}$) of porphyrin complexes in glacial acetic acid[a] at 25 °C

Complex	K_e^{diss}	$K_{\text{inst}}^{\text{MP}}$	Complex	K_e^{diss}	$K_{\text{inst}}^{\text{MP}}$
ZnChl*a*	7.0×10^{-5}	3.5×10^{-34}	ZnChl*b*	2.0×10^{-5}	3.0×10^{-35}
CoChl*a*	1.0×10^{-5}	5.0×10^{-35}	ZnMP	1.3×10^{-5}	6.0×10^{-36}
AgChl*a*	2.0×10^{-7}	1.0×10^{-36}	ZnPP	1.3×10^{-6}	1.0×10^{-37}
NiChl*a*	3.0×10^{-8}	1.5×10^{-37}	ZnTPP	1.0×10^{-5}	2.0×10^{-38}
CuChl*a*	5.0×10^{-9}	2.5×10^{-38}	ZnChl*a*[b]	5×10^{-4}	—
PdChl*a*	8.0×10^{-11}	4.0×10^{-40}	ZnMP[b]	1.0×10^{-2}	—
(Ac)FeChl*a*	5.0×10^{-11}	2.5×10^{-40}	ZnPP[b]	1.7×10^{-3}	—
CoPP	5.0×10^{-6}	4.5×10^{-37}	ZnTPP[b]	6.0×10^{-3}	—
CuChl*b*	1.6×10^{-9}	2.5×10^{-29}			

[a]Contains 0.02% H_2O.
[b]Contains 5% H_2O.

Table 6.2 Dissociation equilibrium constants (K_e^{diss}) and instability constants (K_{inst}^{MP}) of porphyrin complexes in binary solvents at 25 °C

Complex	Added CCl_3COOH (M)	K_e^{diss}	K_{inst}^{MP}
t-C_4H_9OH			
ZnP	1.1	4.4×10^{-5}	—
ZnChl*a*	1.1	2.7×10^{-4}	1.35×10^{-33}
ZnMP	1.1	2.2×10^{-2}	1.1×10^{-32}
ZnPP	1.1	1.3×10^{-3}	1.2×10^{-34}
ZnPP	0.78	3.0×10^{-4}	2.7×10^{-35}
ZnPP	1.38	7.0×10^{-3}	6.3×10^{-34}
CdChl*a*	0.005	5.0×10^{-3}	2.5×10^{-32}
CdMP	0.0005	3.0×10^{-3}	1.5×10^{-33}
CdPP	0.005	1.1×10^{-2}	1.0×10^{-32}
CdPP	0.0019	2.5×10^{-3}	2.25×10^{-34}
CdPP	0.0024	5.7×10^{-3}	5.1×10^{-34}
Glacial CH_3COOH			
ZnPP	0	1.3×10^{-6}	1.2×10^{-37}
ZnPP	1	1.5×10^{-5}	1.35×10^{-36}
ZnPP	2	7.0×10^{-5}	6.3×10^{-36}
ZnPP	5	1.7×10^{-3}	1.5×10^{-34}
ZnMP	0	1.3×10^{-5}	6.5×10^{-36}
ZnMP	1	1.9×10^{-5}	9.5×10^{-36}
ZnMP	2	3.3×10^{-3}	1.6×10^{-33}
ZnMP	5	1.1×10^{-2}	5.5×10^{-33}

Table 6.3 Relative dissociation equilibrium constants of phthalocyanine complexes in 80–96% H_2SO_4 solutions at 25 °C

Complex	K_{rel}	Complex	K_{rel}	Complex	K_{rel}
$ZnPcH^+$	1	$PtPcH^+$	15200	$(HSO_4)_2RuPcH^+$	7500
$CoPcH^+$	17.9	$(HSO_4)GaPcH^+$	1	$(HSO_4)_2SnPcH^+$	1
$NiPcH^+$	21.7	$(HSO_4)IrPcH^+$	5.8	$(HSO_4)_2VPcH^+$	130
$CuPcH^+$	2900	$(HSO_4)AlPcH^+$	40.5	$(HSO_4)_2OsPcH^+$	16400
$PdPcH^+$	17000	$(HSO_4)RhPcH^+$	1800		

The instability constants obtained and presented here for a sufficiently large number of porphyin complexes adequately represent the relationship between their thermodynamic stability and the nature of the metal. This relationship is similar to that observed in the case of ordinary monodentate and polydentate ligands.[328] It may be assumed that these constants differ slightly from the true values which must be four to six orders of magnitude lower in glacial acetic acid than the values listed in Tables 6.1 and 6.2. The arbitrary nature of the instability constants, that is their being related to a strictly predetermined concentration of H^+_{solv} ions, is responsible for their perceptible variations depending on the solvent nature and composition (particularly acid concentration). The arbitrary instability constants in glacial acetic acid are two to three orders of magnitude lower than in *tert*-butanol–trichloracetic acid mixtures. At the same time porphyrin molecules are highly specific both thermodynamically and kinetically. Their complexes with the same metal respond differently to changes in the composition of the medium. In the above example the least significant changes occur in ZnChl*a* complexes and the most significant ones in ZnMP.

The increase in the values of the instability constants with the concentration of CCl_3COOH in an alcohol or with the amounts of H_2O added to glacial acetic acid is due primarily to the growing concentration of H^+_{solv} and the complex dissociation rate. Naturally one cannot disregard the different tendency of complexes to solvation as well as the dependence of the acid dissociation constant $K_{S1,2}$ of the porphyrin on the solvent composition. Since this relationship could not be taken into account in calculating K^{MH}_{inst} (see Tables 6.1 and 6.2), these constants are approximate values. This, however, does not present an obstacle in analysis of the thermodynamic stability of the various metalloporphyrins studied in identical media.

The evaluation of the constants will become much more accurate as more data appear on the quantitative ionic composition of organic acids and their mixtures with other solvents. At present practically no such data are available.[423]

6.3 Stabilitity Series. Factors Responsible for High Stability of Macrocyclic Complexes

Attempts to arrange porphyrins in stability series have a long history. The results are summarized in a review by Phillips,[268] in which the following stability series is presented on the basis of spectral and kinetic criteria: $Pt^{2+} > Pd^{2+} > Ni^{2+} > Co^{2+} > Ag^{2+} > Cu^{2+} > Zn^{2+} > Mg^{2+} > Cd^{2+} > Sn^{2+} > Li^{1+} > Na^{1+} > Ba^{2+} > K^{1+} > Ag^{1+}$.

Phillips does not draw a clearly defined distinction between the kinetic and thermodynamic stabilities of complexes and does not analyse the underlying factors. It is hardly appropriate to include in the same series ions of different charges for they are characterized by different dissociation mechanisms and coordination sphere structures. The present author disagrees with Phillips' implication that one cannot construct valid stability series irrespective of the factors affecting the complexes. In Chapter 5 kinetic stability series were presented for porphyrin and phthalocyanine complexes. The kinetic stability varies directly with the time during which the coordination sphere of a complex remains unchanged and the complex is characterized quantitatively by the rate constant.

The data of Tables 6.1 and 6.3 can be used in deriving a thermodynamic stability series for porphyrin and phthalocyanine complexes. The position of metals from Zn to Fe in this series is determined by strictly quantitative factors. The series may, in addition, include complexes with Cd^{2+}, Mg^{2+}, and Hg^{2+} if the following is taken into account. The widely different kinetic stability of porphyrin complexes with Zn^{2+} and Cd^{2+} rules out the possibility of determining their constants in the same proton-donor medium. For practical purposes the dissociation constants of CdChl were measured in ethanol with tiny amounts of HAc added, while the constants of ZnChl were determined in HAc with small amounts of ethanol. Extrapolation of these constants to 50% ethanol HAc gives approximate but comparable values of equilibrium constants $K_e \approx 0.5$ for CdChl and $K_e \approx 5 \times 10^{-7}$ for ZnChl. Hence, in an ethanol–acetic acid medium and in glacial HAc complexes, CdChl and ZnChl differ in thermodynamic stability by six orders of magnitude.

The findings of the present author and the data available in the literature permit the dissociation equilibrium constant of chlorophyll to be roughly estimated in ethanol–acetic acid. From the dissociation rate constant of chlorophyll (see Table 5.5), measured in ethanol with 2 M HAc, and from the MgMP formation rate constant, converted with the aid of the data presented in Tables 4.23 and 4.31 to a value typical of an ethanol medium and equal to $k_v^{form} = 3.6 \times 10^{-5}\, l\, mol^{-1}\, h^{-1}$, one can obtain a dissociation equilibrium constant $K_e^{diss} \approx 360$ for chlorophyll in an ethanol–acetic acid medium.

The corresponding value for chlorophyll in 10 M ethanol–HAc is $K_e^{diss} = 10^6$, which exceeds that for ZnChl by 12 orders of magnitude and for CdChl by more than six orders.

Thus one arrives at the following thermodynamic stability series: AcFeChl, PdChl > CuChl > NiChl > AgChl > CoChl > ZnChl ≫ CdChl ≫ MgChl, HgChl. This thermodynamic metalloporphyrin stability series was derived for the first time for chlorophyllic acid in glacial HAc at 25 °C.

It may be assumed that a similar series can in principle be derived for other porphyrins and in other solvents. However, bearing in mind the strong dependence of the metalloporphyrin dissociation and formation rates (see Chapters 4 and 5) on the medium, it should be admitted that the mutual position of individual adjacent members of the series may be different. Therefore it is impossible to provide a thermodynamic stability series applicable to all metalloporphyrins (absolute stability series). In fact one is dealing with a plurality of such series which can form a generalized one. How close the similarity between this and the above series will be is hard to tell at this moment.

The position of complexes in the thermodynamic stability series characterizes the ability of a complex to maintain its coordination sphere unchanged in the equilibrium state, that is independently of time.

Of particular interest is comparison of the kinetic and thermodynamic stabilities of chlorophyllic acid complexes. First, the position of MChl in these series is different when the complex formation rates deviate from the normal pattern. For example, the chlorophyll formation rate is seven to eight orders of magnitude lower than that of its metal analogues. Therefore, although having a greater kinetic stability than CdChl, chlorophyll is much inferior to the latter in thermodynamic stability. Some irregularities in the NiChl formation rates make it less thermodynamically stable than CuChl, but the kinetic stability of NiChl is greater than that of CuChl. The same applies to complexes CoChl and AgChl. The former is kinetically more stable than the latter but, because of the abnormally slow rate of formation of CoChl, AgChl has a greater thermodynamic stability.

In view of the above it should be emphasized that evaluation of the thermodynamic stability from the dissociation rates of similar complexes is a very rough approximation.

Second, the position of metalloporphyrin in the thermodynamic stability series is in many cases determined by the specific pattern of its formation in solution. As can be inferred from studies into the mechanism and kinetic parameters of metalloporphyrin formation, the stability (covalence) of the forming complex does not make any significant contribution to the rate of its formation. This is why the thermodynamic stability series of metalloporphyrins (unlike the kinetic stability series) tells nothing about the covalence of the metal–porphyrin bond.

Third, note the specific position of chlorophyll in the kinetic and thermodynamic stability series. As has been shown, the kinetic stability is determined predominantly by the covalence of the coordinate metal–porphyrin bond. MgChl is much more stable kinetically than CdChl and, of course, CaChl, therefore the Mg–porphyrin bond is much more covalent than the Cd– and Ca–porphyrin bonds. At the same time chlorophyll is much less covalent than, for example, ZnChl, which is one of the least stable metalloporphyrins of the stable group. The intermediate position of MgChl in the stability series provides for a sufficient effective charge on Mg, which favours coordination at the central atom of various molecules forming part of the photosynthesis system.[6]

For a better understanding of the factors responsible for the extremely high kinetic and thermodynamic stability of porphyrin and phthalocyanine complexes one should consider the factors determining stability in general. A classification of complexing agents and central atoms into distinct groups is offered by Yatsimirsky, along with careful analysis of the factors determining the stability of complexes within each group.[24,328,428,429] However, the complexity of porphyrin and phthalocyanine ligands makes it impossible to characterize them using simple generalized parameters, such as charge, radius, and ionization potential, describing the particle as a whole. This is due to the fact that the charge of anions P^{2-} and Pc^{2-} is delocalized among all atoms of the conjugated system of the molecule, the ligand itself is concentrically distributed around the central atom, and the ionization potential is in a complex relationship with the properties of the coordinating N atoms because of the presence of conjugated π orbitals.

Among the factors determining the stability of porphyrin and phthalocyanine complexes the most important are the spatial ones, that is the congruity of the covalent radius of the cation with the inner space confined between the four intracyclic nitrogen atoms.

This problem was discussed for the first time by Linstead[94] using phthalocyanine complexes as an example. As was correctly pointed out, stable metal phthalocyanines are formed only by those cations whose covalent radius is close to the size of the inner coordination space of phthalocyanine with a radius of 1.35 Å (the distance between a coordinating nitrogen atom and the centre of the inner space). These include Cu^{2+}, Zn^{2+}, Co^{2+}, Ni^{2+}, Pt^{2+}, Pd^{2+}, Al^{3+}, Ga^{3+}, and VO^{2+}. Later it was shown[108–110] that such a relationship favours formation of strong dative σ and π bonds of the covalent type. The high covalence of the dative M—N bonds is one of the reasons for the extremely high kinetic and thermodynamic stability of MPc. It was corroborated by studies into not only the kinetics of protolytic dissociation (see Chapter 5) but also proton-acceptor properties of MPc (see Chapter 4).

These studies have shown that the high stability of Zn^{2+}, Al^{3+}, Ga^{3+}, VO^{2+}, and Sn^{4+} complexes with phthalocyanine is determined by the formation of only dsp^2 or d^2sp^3 hybrid bonds with a planar MN_4 group. The high polarizability of the common σπ-electron cloud of the phthalocyanine molecule ensures a high degree of occupation of the vacant σ orbitals of the metal by the ligand electrons. Low effective charges on the metal in MPc may be assumed[109] on the basis of the kinetic inertness of the complexes and their electronic absorption spectra (see Chapter 7). These properties suggest that the charge is much greater on the metal atoms in HSO_4AlPc and $(HSO_4)_2SnPc$.[37,385,424]

The presence of extremely low-energy antibonding π* orbitals (see Chapters 3 and 7) in the macroring, including the coordinating nitrogen atoms, creates highly favourable conditions for formation of MPc of backward dative π bonds M → N of an exceptionally high strength in cases where the metal atoms have at least one of the atomic d_{xy}, d_{xz}, d_{yz} orbitals filled with an electron pair. Such bonds are formed by metal atoms in complexes XRuPc, XRhPc, PtPc, PdPc, X_2OsPc, CuPc, NiPc, CoPc, and XIrPc (listed here in the assumed order of weakening π bond strength). The high covalence of M—N bonds results in extremely slow coordination rates providing for a very high kinetic and, as follows from equations (6.5) and (6.6), thermodynamic stability of all of the above-listed complexes. Also responsible for the high stability of these complexes are such factors as the chelate effect,[425,426] that is formation of four conjugated cycles; strong screening of the MN_4 group by a heavy π-electron cloud of the conjugated system, which provides for an unsurpassed rigidity of the phthalocyanine ligand; and screening of the MN_4 group atoms against attack by reagents (solvent, H^+_{solv}).

As was shown recently,[427] tetradentate nitrogen-containing macrocylic porphyrin ligands without double bonds in the macroring react at rates 10^3 times faster than porphyrins. However, as soon as double bonds are introduced into the ring near the coordinating nitrogen atoms, the reaction slows down sharply. Opening of such a cycle, on the contrary, increases the coordination rate by four orders of magnitude. This example serves as a vivid demonstration of the high screening capacity of the atomic skeleton of the cyclic ligand and the extreme rigidity of such a coordination sphere with respect to various deformations of the ligand.

The main factor responsible for the high stability of phthalocyanine complexes is the strong covalent metal–ligand bond. All of the above factors are also involved in numerous porphyrin complexes. However, the dimensions of the coordination space of porphyrin, less favourable for the formation of strong σ bonds (see Chapter 3), bring about a perceptible deformation of the porphyrin ligand in the course of coordination to metals forming stable complexes (Zn^{2+}, Al^{3+}, Sn^{4+}, Cu^{2+}). The coordination may result in significant changes in the valence angles of the C—N bonds, which is energetically unproductive. It should be noted that the planarity of the ligand remains the same, as suggested by X-ray analysis data. The decreased aromaticity of porphyrins as opposed to phthalocyanines, mentioned in Chapter 3, leads to an increase in the ψ_π^* energy and reduces the capacity of porphyrins to form dative π bonds. This is why both σ and $\sigma\pi$ complexes of porphyrins instantaneously dissociated under the rigorous conditions (concentrated H_2SO_4) in which respective metal phthalocyanines stay practically unchanged for many months.

The determining role of bond covalence in the kinetic and thermodynamic stability of MPc and MP manifests itself in the lability of ionic complexes of alkali and alkali earth metals, including Be^{2+} and Mg^{2+}, as well as Ag^+, Sn^{2+}, Cd^{2+}, Hg^{2+}, Cu^+, Cr^{2+}, Ga^{2+}, La^{3+}, and lanthanides.[110] These metal ions are not capable of providing for planar arrangement of the coordinate M—N bonds because of the *d* orbitals being inaccessible to the ligand electrons. In this case predominantly ionic complexes are formed whose central atom is easily ejected from the plane of the MN_4 group under the effect of H_2O molecules, H_3O^+ ions, and other acids.

Of particular interest are the differences in kinetic and thermodynamic stability of porphyrin and phthalocyanine complexes with iron. Porphyrin complexes with Fe^{3+} (protohaemin, chlorophyllhaemin, mesohaemin, and other haemins) exhibit the highest kinetic stability comparable to that of platinum and palladium porphyrins. In contrast, Fe^{3+} phthalocyanine is unstable and occupies an intermediate position between the labile and stable groups of metal phthalocyanines (see Chapter 5). The same applies to Mn^{2+} complexes.

Thus the electronic $3d^5$ configuration is not favourable to strong coordination to such a rigid cyclic ligand as phthalocyanine. The stability of this configuration leads to formation of outer-orbit high-spin complexes $4s4p^34d^2$ or $4s4p^24d$. The greater covalent radius of the central atom in such a complex seems to exceed the radius of the coordination space of phthalocyanine, which renders the coordination sphere strained and the complex unstable. In porphyrins the coordination space radius (see Chapter 3) must be much greater, which will enhance the stability of the coordination sphere of complexes FeP, (X)FeP, MnP, and (X)MnP. These complexes fall in the category of stable high-spin metalloporphyrins,[28,430,431] provided the fifth and sixth coordinate places remain practically free. Extra coordination to strong ligands brings them to a low-spin state.

Unfortunately the stability of Fe and Mn porphyrins of all degrees of oxidation has not yet been studied despite the enormous interest in this group of complexes.

6.4 Extra Coordination in Porphyrin Complexes

Extra coordination, that is attachment of the fifth and sixth ligands along the z axis perpendicular to the plane of the metalloporphyrin molecule, is probably the main function of metalloporphyrins if one is speaking of such biologically active species as chlorophyll, protohaem, and enzymes of the tetrapyrrole type:

$$\mathrm{MP} + \mathrm{L} \overset{K_1}{\rightleftarrows} \mathrm{MP{-}L} + \mathrm{L} \overset{K_2}{\rightleftarrows} \mathrm{L{-}MP{-}L} \tag{6.9}$$

Chlorophyll is capable of photosynthesis only in a complex with protein which is assumed to act as an extra ligand.[2] Protohaem as part of the blood haemoglobin is essentially a complex which enters, in the course of saturation of blood with oxygen, into extra coordination to the latter and, at the same time, to the imidazole residue in accordance with (6.9). In view of the importance of this process it becomes clear why so many works have been concerned with extra coordination in metalloporphyrins.[23,28,268,432,433]

Analysis of the large body of literature on the problem which deserves treatment in a separate monograph is beyond the scope of this work, and I shall restrict myself to a brief review of its main aspects and current state. Particular interest is aroused by the way in which the structure of porphyrin and extra ligand molecules, electronic structure of the metal atom, and the nature of the solvent are correlated with the stability of extra complexes, as well as by the nature of the chemical bonding in them. It should be remembered that all of these aspects of the major and highly important problem are treated here to a most limited extent.

Depending on the intensity of the coordinative metal–porphyrin interaction, determined by the covalence of the dative M—N bonds and by the formal charge on the free metal cation, metals are divided into coordinatively saturated and coordinately unsaturated ones. The former have a formal charge of 2+ and form extremely stable covalent complexes of the σπ type (Cu^{2+}, Ni^{2+}, Pt^{2+}, Pd^{2+}). The porphyrin complexes with these metals lack low-energy antibonding orbitals at the metal, and the formal charge on them is so small that they display no tendency to extra ligand attachment irrespective of the nature of the porphyrin molecule and solvent. In the gaseous phase, though, the attachment of extra ligands does take place.[434] The interaction with extra ligands in solution is so insignificant that even in pyridine complexes CuTPP and NiTPP attach only one molecule of the latter. An extra complex is formed with stability constant $K_1 = 5 \times 10^{-2}$. Attachment of another pyridine molecule has never been experimentally observed.

Coordinatively unsaturated metals in porphyrin complexes are those which have no tendency to dsp^2 hybridization at a formal charge of 2+ (Zn^{2+}, Fe^{2+}, Cr^{2+}, Mn^{2+}), feature unstable oxidation states (Co^{2+}, Re^{2+}, Cu^0), or yield predominantly ionic complexes with a slightly neutralized charge (Be^{2+}, Mg^{2+}, Cd^{2+}, Hg^{2+}, Ca^{2+}, Ga^{2+}). They also include metals with a formal charge exceeding 2+ irrespective of the covalence of the complexes they form (Al^{3+}, Ga^{3+}, Fe^{3+}, Mn^{3+}, Co^{3+}, Ru^{3+}, Os^{4+}, Sn^{4+}, Ti^{4+}, VO^{2+}, UO_2^{2+}), as well as non-metals As(III), Si(IV), and Ge(IV).

It is difficult to agree with the central atoms in metalloporphyrins being divided into two groups with coordination numbers 5 and 6, as was done in some works.[23,268] The point is that normally, as a result of heavy saturation of the chemical affinity of the metal with a porphyrin molecule in the case of M^{2+} and also with counterion (X^-, X^{2-}) in the case of M^{3+} and M^{4+}, the energy of extra coordination of additional ligands is so small that complete extra interaction cannot always be detected and quantitatively measured.

For example, in the case of the Mg^{2+} ion exhibiting a universal capacity for extra coordination to various molecules the attachment of the first Py molecule in MgTPP takes place with stability constant $K_1 = 2 \times 10^3$, while that of another Py molcule involves $K_2 = 0.58$.[268] In MgTPChH these constants differ to a greater extent ($K_1 = 4 \times 10^3$ and $K_2 = 0.83$). The first constants of attachment of pyridine to Zn, Cd, and Hg in their tetraphenylporphins are, respectively, 3.7×10^3, 2.7×10^3, and 12, whereas the second constants have not been determined because of the weak interaction.

Extra complexes of metalloporphyrins represent one of the specific groups of mixed complex compounds. The regularities of formation of such complexes with a mixed coordination sphere and indications of ligand compatability have been summarized in a well-known review.[435] According to the authors the inner coordination sphere more readily incorporates ligands close in electronic nature (e.g. σ-donor, π-acceptor, and electrostatic ligands). Hence it may be expected that the coordination sphere of porphyrin complexes will more easily incorporate such π ligands as Py, CN^-, DMSO, CO, O_2, and the like. They exhibit both σ-donor and π-acceptor properties, and in the case of metals with a vacant d shell they may enter into a competitive interaction with d_{xz} and d_{yz} orbitals. As a result, the d_π–p_π interaction between the metal and porphyrin decreases while the σ interaction increases, which manifests itself in the form of a hypsochromic shift of bands in the electronic absorption spectra in the course of reaction (6.9).[432] The changes in electronic spectra indicate that instead of simple electrostatic MP–ligand interactions a donor–acceptor interaction between the ligand and d^2sp^3 hybrid orbitals of the metal takes place along with a π–π interaction.

In cases where the central metal atom in a porphyrin complex has closed stable electronic orbitals (Mg^{2+}, Zn^{2+}, Cd^{2+}, Hg^{2+}) the formation of dative π bonds becomes practically impossible.[436] Here the extra interaction is limited to σ bonding involving d^2sp^3 orbitals or an ion–dipole interaction. In both cases the extra ligand increases the electron density at the metal atom and reduces the degree of its σ interaction with the porphyrin, resulting in a bathochromic shift of the first band in the electronic absorption spectrum of the complex. Such cases were observed[438,439] in solutions of zinc complexes of various porphyrins in CCl_4 with DMSO (Fig. 6.1).

Unfortunately no systematic data are available in the literature on the effect of the properties of cations and extra ligands on the kinetic and thermodynamic parameters of formation or dissociation of extra complexes. This also applies to the spectral and other physicochemical properties. More detailed information is available on extra complexes of porphyrins with Mg^{2+}, Fe^{3+}, and Fe^{2+} thanks to their greater biological importance. However, without data on other metals one cannot understand the physics and regularities of extra coordination.

The Mg atom in porphyrin complexes[23] coordinates to oxygen-containing ligands as readily as to nitrogen-containing ones. This follows from roughly approximate stability constants of attachment of the first extra ligand by chlorophyll *a* in benzene (Table 6.4). The constants have been determined

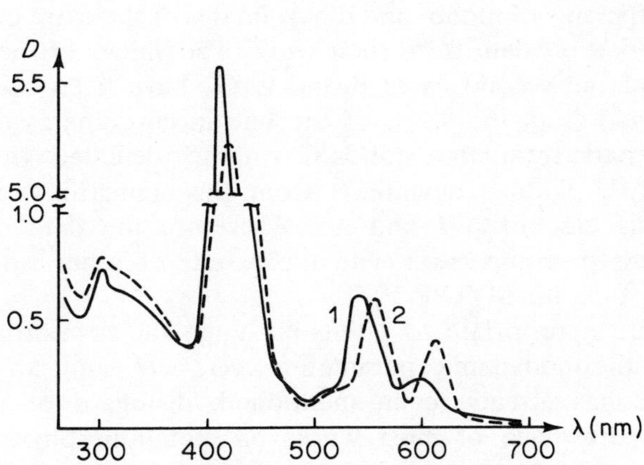

Figure 6.1 Electronic absorption spectra of ZnTMPP (curve 1) and its extra complex (DMSO)ZnTMPP (curve 2) in carbon tetrachloride at 25 °C

Table 6.4 Stability constants of the first extra complex of chlorophyll *a*

Ligand	log K_1	Ligand	log K_1
Dimethylaniline	+1.0	Quinoline	+4.1
Phenol	+1.2	Benzyl amine	+4.4
Aniline	+1.7	Water	+4.5
Benzyl alcohol	+3.5	*n*-Heptyl amine	+5.2
n-Octyl alcohol	+3.7		

from the fluorescence intensity, which increases as a greater amount of the extra ligand is added to the solution of MgChl in absolute benzene. It should be borne in mind that the data of Table 6.4 do not necessarily represent true values of stability constants because fluorescence is not intensified only by disaggregation of the chlorophyll associates due to extra coordination. The complexity of the phenomenon may stem from several mechanisms of formation and decomposition of MgChl associates being involved (see Chapter 7). However, it may be assumed that the values listed in Table 6.4 are in general adequately representative of the tendency of the above-mentioned molecules to extra coordination to Mg.

Emission spectroscopy techniques similar to the one just described may be successfully used in various modifications along with electronic absorption spectroscopy in investigating extra complexes of porphyrin. An attempt has been made[438] to determine the stability of extra pyridinate complexes of Mg-porphyrins from the temperature dependence of absorption spectra as a function of the porphyrin structure. Storm *et al.*[438] have defined constant K_2 of attachment of the second ligand to Py·MgP and other thermodynamic parameters (Table 6.5) using the molar extinction coefficients of mono- and dipyridinates of the extra complex. Here I present rounded off data from their work. The values of the free energy of formation of the second extra ligand (ΔG) have been calculated by a graphical method from the slope of the line in coordinates log K_2 versus $1/T$. They depart from the standard values calculated from equation $\Delta G = -RT \ln K_2$. Similar deviations from the standard values are also observed in the case of ΔH and ΔS. Therefore the data of Table 6.5 cannot be used for comparison with the results of other works (see, for example, $K_2 = 0.58$ for MgTPP[268]).

It is therefore appropriate to point out again the necessity of reducing the values of thermodynamic parameters ΔG, ΔH, and ΔS to standard conditions. As has already been mentioned, disregard of this essential condition is a drawback of most works on metalloporphyrins. The only conclusion that can be drawn from Table 6.5 is the weakness of the coordinative interaction between L·MgP and the second ligand. The additional solvation occurring in the series of Mg porphyrins results in lower values of ΔH.

Table 6.5 Comparative stability characteristics of the second pyridine extra complex of Mg-porphyrins

Complex	K_2	ΔG (kcal mol^{-1})	ΔH (kcal mol^{-1})	ΔS (e.u.)
$MgMP(Py)_2$	0.070	1.60	−4.0	−18
$MgEP(Py)_2$	0.068	1.60	−3.0	−15
$MgPP(Py)_2$	0.242	0.84	−4.3	−17
$(Py)_2$Mg-deoxophylloerythrine ME	0.074	1.60	−2.6	−14
$MgTPP(Py)_2$	0.161	1.10	−2.9	−11
$MgDP(Py)_2$ DME	0.228	0.85	−3.9	−16
$(Py)_2$Mg-2,4-dichlorodeuteroporphyrin DME	0.282	0.76	−3.9	−15
$(Py)_2$Mg-2,4-dinitrodeuteroporphyrin DME	0.326	0.68	−4.0	—

The values of ΔH, which are almost the same for various porphyrins, tell nothing about the effect of the porphyrin structure on extra coordination except that this effect is probably weak. As can be easily seen in the case of dueteroporphyrin, even the introduction of two $\sigma\pi$-electron-acceptor substituents (NO_2) into positions 2 and 4 does not alter the values of K_2 and ΔH (see Table 6.5).

The iron atom in porphyrin complexes features a particularly close affinity for anions of the CN^- type. The degree of oxidation of the initial iron atom plays a significant role in this case.[23] Thus, in the case of extra complex $[(CN)_2FePP]^-$, the total ligand attachment constant is $K_{1,2} = 1.4 \times 10^{12}$, while in the case of $[(CN)_2FePP]^{2-}$ it is $K_{1,2} = 8 \times 10^9$. Characteristically, the respective constants for the dipyridine complex are $K_1 = 0.8$ and $K_2 = 5 \times 10^7$. These facts are indicative of the important role of both σ and π bonding of extra ligands in the case of iron porphyrins. Also, the accumulation of the negative charge in the porphyrin macroring destabilizes the system. Therefore anionic complexes are more stable in the case of Fe^{3+} porphyrins and less stable in the case of Fe^{2+} porphyrins.

Since the *d* electrons become less mobile as the positive charge on the central atom grows (the stability of the *d* orbital sharply increases[432]), ferriporphyrins as opposed to ferroporphyrins lose their capacity for extra coordination of molecules of the pyridine or oxygen type, apt to form dative π bonds. This capacity is restored if one of the extra ligands is a strong electron pair donor. This is why mixed cyanopyridinates of Fe^{3+} porphyrins are more stable than those of Fe^{2+} porphyrins ($K_{1,2} = 3.6 \times 10^{12}$ for (CN)(Py)FePP and $K_{1,2} = 8 \times 10^{10}$ for $[(CN)(Py)FePP]^{-1}$). In this case mixed π ligands show a substantial gain in ΔG as compared to homogeneous ligands (2Py, $2CN^-$).

The iron atoms in ferri- and ferroporphyrins exhibit close affinity for other nitrogen-containing molecules (particularly for imidazole) as well as other oxygen-containing ligands (H_2O, OH^-, H_2O_2, CO, O_2). The investigation of extra complexes with H_2O and OH^- in an alkaline medium is considerably hampered by dimerization and a higher degree of aggregation of metalloporphyrins, which makes it extremely difficult to measure the thermodynamic parameters of extra coordination.

Ways to prevent aggregation of porphyrins are described in the well-known monograph by Blumenfeld.[26] The data presented in many works, for example by Falk,[23] do not lend themselves easily to analysis because more often than not they have been obtained in different media, which makes it difficult to establish the identity of the initial state of a reaction. Therefore I shall restrict myself to a brief qualitative summary of some more reliable results.

Interestingly, in spite of the high strength of the CN^-–iron bond in protohaem, one cyanide is easily displaced under the effect of CO. As a result, mixed extra complex (CO)FePP(CN) is formed. This is indicative of a *trans* influence similar to the well known *trans* effect discovered and examined by Chernyaev[419] on platinum compounds. The electronic nature of this phenomenon, considered in the light of recent findings,[437] may serve as a basis for examining extra coordination in iron porphyrins. This has already been done in detail by Yan[28] on a ferroporphyrin complex with imidazole and oxygen (see the following scheme):

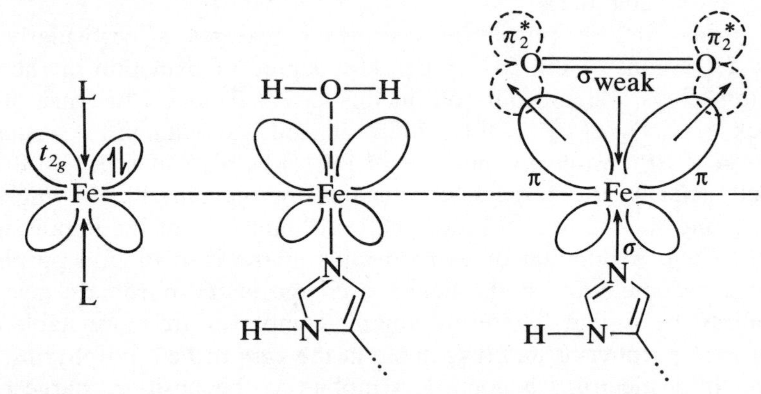

a **b** **c**

If the fifth and sixth coordinate places are occupied by similar ligands, the d_{xy} and d_{yz} orbitals of iron remain symmetrical (**a**). But if both ligands are dissimilar and one of them is a stronger electron donor, that is creates a stronger field, than the other, the t_{2g} orbitals shift as a result of electrostatic factors towards the second ligand (**b**). The latter cannot approach to form a strong bond if it is only a σ donor (H_2O, NH_3). This explains the weak linkage of water in haemochromes (H_2O)-haem(imidazole) and (H_2O)haem(pyridine). When a π acceptor is in a *trans* position with respect to the strong σ donor, its attachment is considerably facilitated. Favourable conditions are created for attachment of even such weak σ donors as oxygen, whose molecule accepts t_{2g}-electron pairs at the antibonding π_z^* orbitals. The resulting dative π bond provides for adequate binding of oxygen. Imidazole is a stronger σ donor than pyridine, which is why haemochrome (H_2O)haem(Py) could be isolated, while (H_2O)haem(imidazole) is unknown.[28]

If imidazole lies in the *yz* plane, the energy of the d_{yz} orbital increases and the latter is deformed toward O_2 (**c**). At the same time the energy of the d_{xz} orbital decreases and its deformation is in the opposite direction, thereby providing for a π bond with the imidazole molecule. Comparison of the properties of imidazole and other nitrogen-containing protein residues suggests that none of them creates more favourable conditions for attachment of O_2 than imidazole.

In the present author's laboratory the equilibria of extra coordination of DMSO to some zinc complexes or porphyrin have been studied.[440] The formula of haematoporphyrin is shown (**LVI**). The complexes were

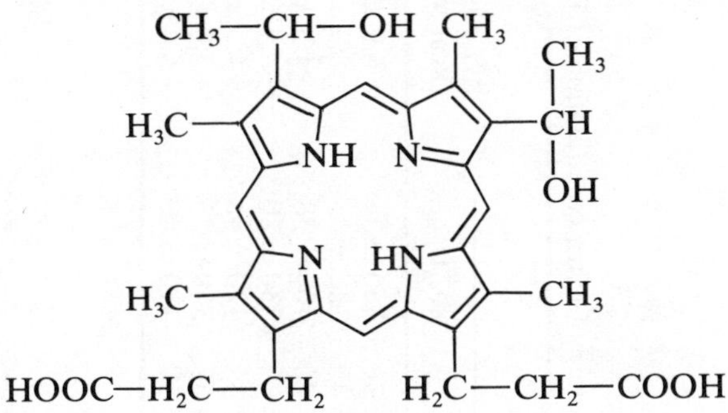

LVI

synthesized from free porphyrins and acetates of metals in ethanol. To remove the extra ligands their solutions in CCl_4 were passed through a column with Al_2O_3. The extra coordination of zinc complexes of porphyrins to DMSO is accompanied by a bathochromic shift of the absorption spectra by 70–100 Å (see Fig. 6.1), which permits using spectrophotometry to calculate the extra complex formation equilibrium constants and to determine the number of attached extra ligands. The calculation method and results are given elsewhere.[440] The zinc complexes of porphyrin examined attach, under predetermined experimental conditions (the DMSO concentration was as high as $0.02\ \mathrm{mol\ l^{-1}}$, 15–45 °C), not more than a single DMSO molecule. The second coordinate place is occupied by weakly bound CCl_4. The experimental values of equilibrium constants K_1 and the values of ΔH, ΔG, and ΔS calculated from them are listed in Table 6.6.

As can be inferred from the tabulated data, the equilibrium constant of attachment of the first DMSO molecule is relatively high; however, it is slightly lower than the first constant of attachment of pyridine to Mg

Table 6.6 Thermodynamic parameters of extra coordination of DMSO to zinc porphyrins

Complex	K_1				ΔH (kcal mol^{-1})	ΔG (kcal mol^{-1})	ΔS (e.u.)
	15 °C	25 °C	35 °C	45 °C			
ZnTPP	1150 ± 50	760 ± 30	530 ± 30	360 ± 35	−6.9 ± 0.4	−3.93 ± 0.02	−10
ZnTMPP	780 ± 76	550 ± 60	390 ± 40	280 ± 25	−5.9 ± 0.3	−3.72 ± 0.05	−9
ZnChl*a*	900 ± 80	650 ± 60	490 ± 40	400 ± 38	−4.8 ± 0.5	−3.83 ± 0.02	−3
ZnMP	370 ± 30	280 ± 25	230 ± 20	230 ± 20	−3.9 ± 0.3	−3.33 ± 0.01	−2
ZnPP	950 ± 90	710 ± 70	590 ± 45	420 ± 45	−4.5 ± 0.2	−3.88 ± 0.02	−2
ZnHP	1140 ± 100	770 ± 70	490 ± 35	350 ± 23	−7.0 ± 0.3	−3.93 ± 0.09	−4
ZnDP	1000 ± 90	750 ± 72	520 ± 50	410 ± 40	−5.4 ± 0.3	−3.90 ± 0.02	−2

porphyrins. The stability of complexes (DMSO)ZnP is almost independent of the porphyrin structure. Only in the case of ZnMP is the stability 2–2.5 times lower as compared to other porphyrins.

The substitution of DMSO for CCl_4 involves an almost zero entropy (the mean value of ΔS is -4 ± 3 e.u.) with decreasing free (ΔG) and total (ΔH) energies of the system. The deviations of ΔH depending on the molecular structure of porphyrin are insignificant (-5.5 ± 0.3 kcal mol^{-1} on the average). Analysis of the thermodynamic values obtained spectrophotometrically does not permit any conclusions to be drawn concerning the relationship between the extra coordination of DMSO in zinc porphyrin complexes and their molecular structure.

At present a large number of extra complexes with metal phthalocyanines are known (see Chapter 2). However, absolutely no quantitative data on their thermodynamic stability are available. It may be assumed that in phthalocyanine complexes the extra ligand bonding strength will be virtually the same as in metalloporphyrins; the extra complexes are basically of the same type.

Interestingly, oxygen carriers can be found among metal phthalocyanines. For example, MnPc in a pyridine solution absorbs O_2, in the cold.[105] The olive-green solution becomes dark blue ($\lambda = 712 \rightarrow 620$ nm). Heating results in reversible detachment of oxygen. Elvidge and Lever[105] propose the following scheme of its binding:

$$\text{MnPc} + \text{Py} + \text{O}_2 \underset{t^\circ}{\overset{20\ ^\circ\text{C}}{\rightleftarrows}} \text{Py–MnPc–O}_2\text{–MnPc–Py} \rightleftarrows \text{Py–MnPc=O}$$

$$\underset{\text{Py}}{\overset{180\ ^\circ\text{C}}{\rightleftarrows}} \cdots \text{O–MnPc–O–MnPc–O–MnPc–O} \cdots. \quad (6.10)$$

The end-product of the reaction—tetravalent manganese phthalocyanine polymer—in which the polymer chain develops owing to extra coordination to the bridging oxygen $(\text{OMnPc})_n$ again yields MnPc capable of absorbing oxygen when heated in pyridine.

A study of extra complexes of phthalocyanines with tri- and tetravalent metals and a hydroxy group has revealed cases of facile acid ionization of the coordinated OH groups.[439] This relevation has led one to assume the possibility of conjugation of a new type, which was termed perpendicular conjugation. In this case the filled π-electron orbitals of the oxygen atom may overlap at a right angle with the antibonding π orbitals of the nitrogen atoms included in the conjugated system of the phthalocyanine macroring. At the same time they may overlap with the t_{2g} orbitals of Cr, V, Ti, and other atoms.

Valuable data on the state of extra ligands and the character of their bonding with the metal atom can be obtained by the kinetic method. However, only a few works concerning the substitution of ligands in extra complexes have appeared so far.

From the rate of substitution of one protein by another in a haem–protein complex saturated with O_2 or CO Blumenfeld[26] established that the rate of detachment of globin from oxygenated protohaem $(O_2)FePP$ is limited by the rate of detachment of O_2, that is by the rate of the reaction $(O_2)FePP \rightleftarrows O_2 + FePP$ with constant $k_v = (5 \pm 2) \times 10^5\ s^{-1}$. The high rates of detachment of O_2 indicate that the covalent Fe—O_2 bond is weak. The activation energy of reaction (globin)-$FePP + O_2 \rightleftarrows$ (globin)$FePP(O_2)$ approaches zero, while in the case of the O_2 detachment reaction it equals 21 ± 1.5 kcal mol^{-1}, which is due to the simultaneous detachment of O_2 and the protein imidazole residue from the Fe atom. Blumenfeld also convincingly demonstrates the broad possibilities of the kinetic method as applied to phenomena based on extra coordination.

Akopov *et al.*[823] have studied the kinetics of substitution of chlorine by an OH^- group in silicon dichlorophthalocyanine:

$$Cl_2SiPc_{(solid)} + 2OH^- \rightarrow (HO)_2SiPc_{(solid)}. \tag{6.11}$$

The substitution rate constant in 0.1 M NaOH is $2.1 \times 10^{-2}\ h^{-1}$, $E = 15.6$ kcal mol^{-1}, and $\Delta S^{\ddagger} = 19$ e.u.

The low chlorine substitution rates in the extra complex, which were found to be by two to six orders of magnitude lower than the rates of hydrolysis of organochlorosilanes, are due to the sharp decrease in the charge (positive) at the silicon atom as a result of an extremely strong coordinate σ bond Si $\leftarrow$ N and a dative π bond of the same direction. The latter is formed by the vacant d_π orbital of silicon being filled with the p_z electrons of chlorine. All this minimizes the tendency of the Si atom to nucleophilic substitution reactions.

The formation of extra complexes of porphyrin and phthalocyanine not only underlies the vitally important processes of breathing, photosynthesis, and enzyme catalysis but also determines the possibility of association of many porphyrin and phthalocyanine complexes as well as their polymerization due to bridging extra ligands of the Cl^-, O^{2-}, and SO_4^{2-} types. This is why the studies into their kinetic and thermodynamic stability and chemical bonding are among the most important tasks of porphyrin chemistry. However, successful accomplishment of these tasks will to a great extent depend on the availability of reliable methods of synthesizing extra complexes, including unstable ones, and radically new techniques for measuring the equilibria and rates of the reactions in which they participate.

Chapter 7

Electron-optical properties and their correlation with other properties of porphyrin and phthalocyanine complexes

The electronic absorption (and emission) spectra of porphyrins are among the most important characteristic features of the latter. They reliably characterize the structure of porphyrins and adequately represent the state of their intramolecular energies. In this sense they correlate with many properties of porphyrins and metalloporphyrins. Different spectroscopic aspects, including the theory of origin of absorption spectra, have been exhaustively covered in a monograph by Gurinovich, Sevchenko, and Solovyev.[8] Therefore this chapter deals primarily with certain aspects that could not be discussed at length in the above monograph, namely the correlation between the electronic absorption spectra and coordination properties of porphyrins, as well as the limits of applicability of the spectral criterion of complex stability. The chapter also includes some general considerations pertaining to the spectra of porphyrins, their associates, and polymeric phthalocyanines.

7.1 On the Origin of Bands in Electronic Absorption Spectra of Porphyrins and Their Complexes

The fundamental critical review of the problem in the above-mentioned monograph[8] reduces the present task to drawing some conclusions.

The electronic absorption spectra of porphyrins proper are characterized by complexity, variability of the band position and intensity as a function of structure, low molar extinction coefficients (ε) of the visible bands, high intensity of the Soret band (*c.* 400 nm), low sensitivity of the bands to the nature of inert solvents, and extremely high sensitivity to protonating and complexing agents as well as strong bases.

Porphyrin itself, and its substituents, feature electronic absorption spectra classified by Stern into three types (for more detail see reference 8): the etio-type (Fig. 7.1) of symmetrically substituted porphyrin with the visible

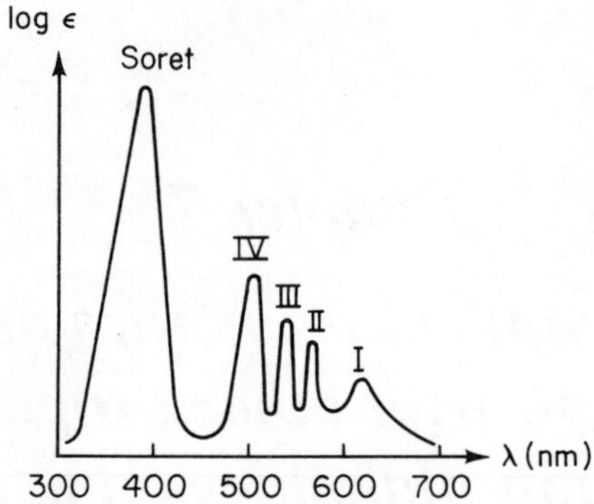

Figure 7.1 Etio-type of porphyrin spectrum (characteristic bands)

band intensity diminishing as IV > III > II > I; the phyllo-type characteristic of porphin and its monoalkyl substituents in one of its meso positions (i.e. α, β, γ, δ) with the band intensity diminishing as IV > II > III > I; and the rhodo-type of porphyrins substituted by a formyl or carboxyl group participating in direct conjugation with the π-electron system of the molecule, with the band intensity diminishing as III > IV > II > I.

Most of the widely occurring porphyrins belong to the etio-type (protoporphyrin, mesoporphyrin, haematoporphyrin, deuteroporphyrin, etioporphyrin I and II, coproporphyrin, uroporphyrin, tetraphenylporphin, pyrroporphyrin, etc.). Characteristically, the intensity ratio of individual bands and their position may vary significantly from one porphyrin to another.

The phyllo-type includes, apart from porphin, phylloporphyrin (**L**), 1,4-diethyl-2,3-dimethylporphin, porphin-1,4-dipropionic acid, and similar compounds.

The currently most acceptable physical interpretation of electronic absorption spectra is that based on profound theoretical analysis of the experimental and quantum-chemical data of the previously mentioned monograph.[8] According to this interpretation the absorption spectra of porphyrins do not exhibit bands of $n \rightarrow \pi^*$ transitions because of the symmetry of the n orbitals and antisymmetry of the π orbitals with respect to the plane of the porphyrin molecule. All bands are of $\pi \rightarrow \pi^*$ origin. The Soret band is due to an electronic ${}^1A_{1g} \rightarrow {}^1E'_u$ transition to the highest-energy vacant π^* orbital.

In substituted porphyrins E'_u is split into two states B'_{2u} and B'_{3u} which are close in energy, as a result of the reduced symmetry of the π-electron cloud. The electronic transitions to state E'_u (or B'_{2u} and B'_{3u}) are allowed, therefore the intensity of the Soret band is always very high ($\varepsilon \approx 10^5$).

Bands I and III of the visible region belong to quasi-forbidden electronic transitions. The reasons for transitions ${}^1A_{1g} \rightarrow B_{3u}$ and ${}^1A_{1g} \rightarrow B_{2u}$ being allowed, which is responsible for the appearance of bands I and III, have not yet been established. In the case of MP complex formation, protonation to $H_2PH_2^{2+}$, or complete ionization to P^{2-}, energy levels B_{2u} and B_{3u} combine into a single doubly degenerate level E_u as a consequence of the symmetry of the π cloud being enhanced from D_{2h} to D_{4h}. As a result bands I and III merge into one. In this case the effect of the functional substituents in positions 1–8, as well as meso positions, is not sufficient for dividing the E_u state into two independent states.[267] Bands II and IV are of vibrational origin, that is they are vibrational satellites of bands I and III, respectively. In metalloporphyrins bands II and IV disappear as a result of degeneration of the B_{3u} and B_{2u} states, and a new vibrational satellite emerges, corresponding to energy state E_u. The foregoing is illustrated in Fig. 7.2, which is reproduced from reference 8.

As can be seen from the illustration, band I (denoted as the α band in non-Soviet literature) corresponds to electronic transition ${}^1A_{1g} \rightarrow E_u$, while band II (β band) is its vibrational satellite. As far as dihydroporphyrins (chlorins) are concerned, including chlorophyll (II), the same level pattern can in general be applied since it is most probable that no new levels appear during hydrogenation.[8] The number of bands in the visible region of the spectra of chlorins and their complexes is the same as in the case of porphyrins (four and two). However, as a result of hydrogenation and particularly appearance of a cyclopentanone ring (in chlorophyll and its derivatives), new features appear in the spectra.[8] In this case the incorporation of a metal into the porphyrin molecules forming the basis of chlorophylls *a* and *b* does not lead to symmetry D_{4h}, and the number of bands in the visible region of the spectra of their complexes exceeds two

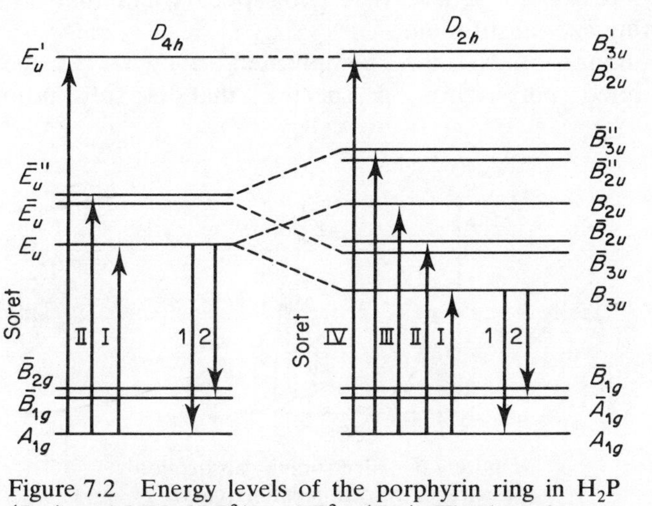

Figure 7.2 Energy levels of the porphyrin ring in H_2P (D_{2h}) and MH, $H_2P_2^{2+}$ and P^{2-} (D_{4h}). The bar above a symbol indicates a vibrational satellite

(incomplete degeneration). The weakest first band in porphyrins becomes the most intensive in chlorins (in etioporphyrin $\varepsilon_1 \approx 5 \times 10^3$; in etiochlorin $\varepsilon \approx 5 \times 10^4$).

Hydroporphyrins feature spectra of their own types in addition to Stern's classification for porphyrins. The theoretical problems involved in the interpretation of chlorin spectra, particularly those of chlorins containing a cyclopentanone ring, have not been solved. The main reason is the lack of experimentally derived spectra of these systems.

Aza- and benzosubstituted porphyrins have electronic spectra substantially different from those of porphyrins proper. One can trace the generic relation between the spectra of porphyrins proper and their aza-substituents.[8,267] As more aza bridges are being introduced, the quasi-forbiddenness is lifted from electronic transitions $A_{1g} \rightarrow B_{3u}$ and $A_{1g} \rightarrow B_{2u}$; the intensity of the corresponding bands I and III increases considerably and reaches a maximum in the case of tetrazaporphyrins. At the same time both bands undergo a significant bathochromic shift. The vibrational satellites of these bands practically vanish because of the higher aromaticity and rigidity of the π skeleton. The electronic vibrators corresponding to both transitions are polarized at right angles to each other and lie in the molecular plane.

Incorporation of a metal atom into tetrazaporphin enhances the symmetry of the π system from D_{2h} to D_{4h} and causes degeneration of the lower vacant electronic states B_{3u} and B_{2u}, resulting in an E_u state corresponding to which is the single intensive band in the visible region of the MTAP spectrum (Fig. 7.3). The Soret band generically associated with the presence of pyrrole nuclei in the porphyrin molecule shifts hypsochromically into the region of 330 nm in the case of aza substitution. There is no reason to believe that it disappears and that another band appears during aza substitution.[8]

The situation is much more complicated in so far as the effect of benzosubstituted porphyrins on spectra, that is substitution of the

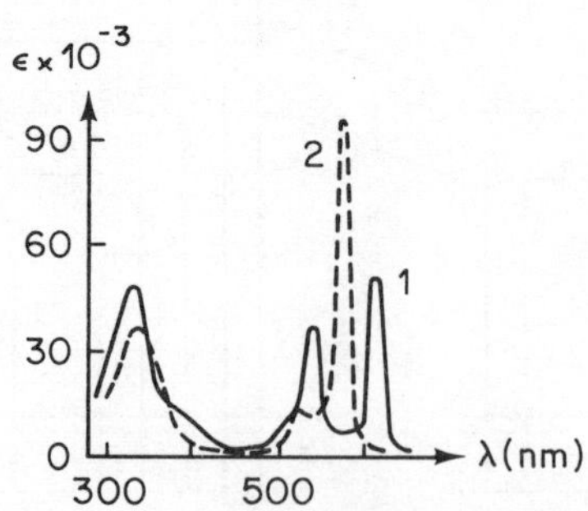

Figure 7.3 Electronic absorption spectra of tetrazaporphin (curve 1) and its copper complex (curve 2) in chlorobenzene

β,β′-pyrrole carbon atoms by benzene nuclei (**V**), is concerned. Here, the band position will depend on the molecule of porphin or tetrazaporphin in which the benzo substitution occurs. In the course of porphyrin-tetrabenzoporphin substitution (**XLIV**) the quasi-forbiddenness is lifted from the electronic transitions in the visible region, and the absorption bands become more intensive. A clearly defined bathochromic shift is observed, including that of the Soret band, which is split into two components. Corresponding to the $A_{1g} \rightarrow B_{3u}$ transition is a single symmetrical band, while the band corresponding to the $A_{1g} \rightarrow B_{2u}$ transition is split into three components. This is conclusively explained by Gurinovich *et al.*[8]

The incorporation of a metal into tetrabenzoporphin results in fusion of both visible bands into a single more intensive one. The Soret band is shown to degenerate (Fig. 7.4). The benzo substitution in tetrazaporphin

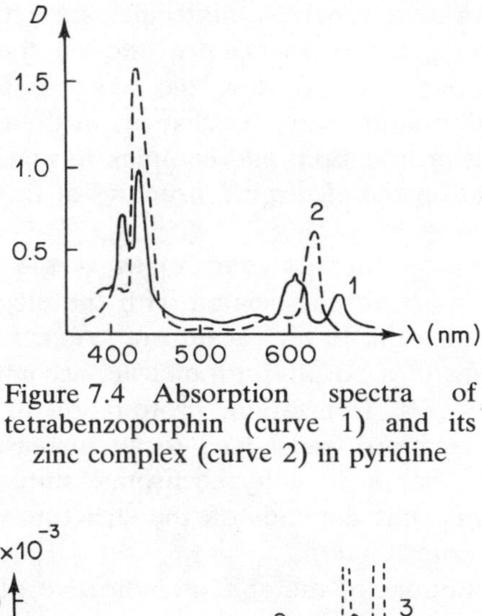

Figure 7.4 Absorption spectra of tetrabenzoporphin (curve 1) and its zinc complex (curve 2) in pyridine

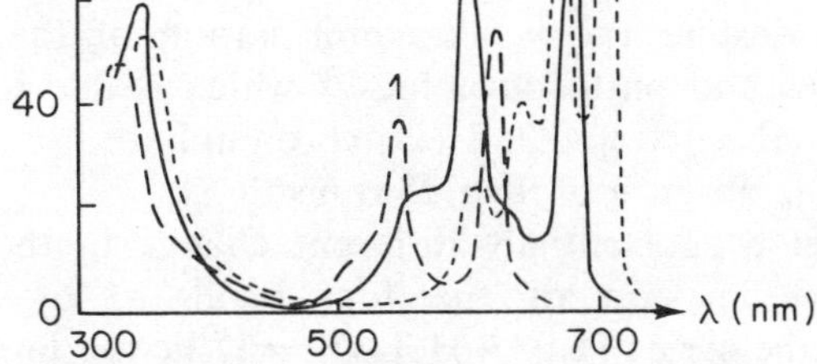

Figure 7.5 Electronic absorption spectra of tetrazaporphin (curve 1), tribenzo-tetrazaporphin (curve 2 in chloro-benzene, and phthalocyanine (curve 3) in α-chloronaphthalene,[267]

leads to a pronounced bathochromic shift of both visible bands, which is as great as 100 nm in the case of phthalocyanine.

The two visible absorption bands of phthalocyanine come close together and their intensity increases (Fig. 7.5). The bands approaching each other is indicative of a smaller difference in energies of the $A_{1g} \rightarrow B_{3u}$ and $A_{1g} \rightarrow B_{2u}$ transitions polarized in the molecular plane along axes x and y and may be due to the specific structure of the centre of the phthalocyanine molecule (see Chapter 3), particularly to the smaller difference in the states of the inner nitrogen atoms in H_2Pc. The formation of phthalocyanine complexes with metals also brings about fusion of two long-wave bands into a single intensive one (see Fig. 4.4).

7.2 Manifestation of Ionization and Coordination Properties of Porphyrins in Absorption Spectra

The absorption spectra of tetrapyrrole molecules are extremely sensitive to the interactions taking place in the centre and on the periphery of the molecule. Of particular interest for the coordination chemistry of porphyrins and phthalocyanines are the changes in their spectra, occurring during acid ionization, protonation, and complex formation. These changes are strongly dependent on the electronic structure of the molecules and the reagent.

Particularly sensitive to the above changes is the first band in the electronic absorption spectrum, associated with the electron transfer from the upper filled orbital to the lower vacant one. As can be seen from Fig. 7.2, the transformations of a porphyrin molecule, accompanying changes in symmetry from D_{2h} to D_{4h} (ionization, protonation in the centre of the molecule, complex formation), must lead to an increase in energy in the first absorption band, that is to a hypsochromic shift. Experience shows that the amount of this shift depends on the structure of H_2P and on the metal involved in the coordination.

In most cases the amount of the shift is indicative of the nature of the forming chemical bonds and of the structure of the reaction products. It is natural to assume that the data used for comparison must be obtained in the same solvent despite the weak manifestation of the solvatochromic effect in porphyrins and phthalocyanines,[22] which is due to the rigidity of the skeleton of the conjugated π chromophore of the examined molecules[441] and the absence of $n \rightarrow \pi^*$ transitions.

One may expect a substantially different change in the spectra (I am now referring to the shift of the first band λ_I) of porphyrins and azaporphyrins. In the series $H_2P \rightarrow H_2TAP \rightarrow H_2Pc$ the intracyclic nitrogen atoms, that is of $N_2(NH)_2$, become more equivalent owing to the decreasing difference in the electronegativities of the C and N atoms of the macroring. The link N⩦C⩦C⩦C⩦N in porphyrins is different from the link N⩦C⩦N⩦C⩦N in phthalocyanine and tetrazaporphin in which the carbon atom is associated with two more electronegative nitrogen atoms, which results in its greater electronegativity. This is probably what makes the conjugated system of phthalocyanine unique and leads to an intramolecule ionization of the N—H bonds as well as eliminating the differences in the states of the nitrogen atoms.

The mutual approach of the B_{3u} and B_{2u} levels in the H_2Pc molecule and the slight hypsochromic shift during transition $H_2Pc \rightarrow MPc$ (as opposed to porphyrins) is convincing proof of the above considerations.

The satisfactory results of quantum-chemical calculations of the energy levels in the phthalocyanine molecule and its other parameters, which are obtained assuming that the electronegativities of the macrocyclic C and N atoms are equivalent,[179,220,442] also corroborate the foregoing.

Consider now the spectral manifestations of acid ionization of $H_2P \rightarrow P^{2-}$. Proceeding from the data of Chapter 3 on the specific structural features of porphyrin molecules, it may be assumed that the stabilization of anions will be determined by the energy of the antibonding ψ_{π}^{*} orbitals, which can be derived from the energy in the first absorption band. It is to be expected that the greater the value of λ_I, the more likely the distribution of the anionic charge among the antibonding orbitals and the greater the value of K_S at the same degree of solvation of the anion. Compare the values of $pK_{S_{1,2}}$ and λ_I of the porphyrins listed in Table 7.1.

In the case of porphyrins proper (H_2MP, H_2PP and H_2EP, H_2TPP, and H_2TMPP) no correlation can be found between these values although it

Table 7.1 Ionization constants of protonated forms and wavelengths of the first absorption band of various porphyrins

	λ_I (nm)					
Porphyrin	H_2P^a	H_3P^{+b}	H_4P^{2+b}	$pK_{S_{1,2}}$	$pK_{PH_3^+}$	Ref.
H_2Chl*a*	668	—	—	29.3	1.75	8, 68
H_2Chl*b*	659.5	—	—	29.8	—	68
H_2MP	619	601.5	592	30.3	5.94	443
H_2PP	630	609.5	602	31.05	4.89	443
H_2EP	621	—	—	31.6	—	23
H_2TPP	646	—	—	32.7	—	23
H_2TMOPP	—	—	—	33.0	—	
Deuteroporphyrin	619	598.5	590	—	5.63	443
Deuteroporphyrin derivatives	641	—	—	—	3.90	23
2,4-$(CHO)_2$	650	—	—	—	2.90	23
2,4-$(CH_3CO)_2$	640	616	610	—	3.50	443
Uroporphyrin[c]	627	—	—	—	—	23

[a]Protonation in $CHCl_3$.
[b]Protonation in aqueous dodecyl sulphate.
[c]Porphin-1,3,5,7-tetraacetic-2,4,6,8-tetrapropionic acid.

seems to exist in chlorophyllic acids *a* and *b*. Hence the effect of entropy factors prevails over that of the energy ones in all cases where the difference in energies is not great. The same conclusion follows from the data of Chapter 4.

Besides, if the differences in λ_I of porphyrins are not too great (10–20 nm), the position of the energy levels of the optical electron transition may be determined by the deviations of individual atoms or their groups from coplanarity rather than by the magnitude of the π-electron charge of the chromophore and its distribution among individual atoms and bonds (see Chapter 3). The magnitude of the π-electron charge of the chromophore is determined not only by the number of its own π electrons but also by the participation of the *n* and π electrons of the conjugated functional substituents.

In view of this it may be expected that comparison of the values of λ_I and pK_S will be productive only in the case of unsubstituted porphyrin molecules such as H_2P ($\lambda_I = 616.5$ nm), H_2TAP ($\lambda_I = 624$ nm), H_2TBP ($\lambda_1 = 666$ nm), H_2Pc ($\lambda_1 = 698$ nm), and H_2ChH ($\lambda_1 = 638$ nm). These figures are taken from the work by Weiss *et al.*[443] Unfortunately no values of K_S for these compounds have been obtained. It is only known that the value of K_S for phthalocyanine is many orders of magnitude greater than for porphyrins. Attributing this fact to the more pronounced aromaticity of the phthalocyanine macroring, which is responsible for the lowest π^* orbitals being accessible to the *n* electrons of anion Pc^{2-}, one may arrive at the following acidity series on the basis of γ_I: $H_2P \approx H_2TAP < H_2TBP \ll H_2Pc$.

The absorption spectra of various porphyrins and azaporphyrins are related in a particular way to the structure of their cationic forms: formation of H_2P^{2+} changes the four-band visible spectrum of H_2P to a two-band one[23] and in some cases (e.g. porphin, tetrabenzoporphin, tetraphenylporphin) the second band of H_4P^{2+} has such a low intensity that the spectrum may be considered to contain a single band. In the literature there is practically no indication as to the position of the absorption bands of acidic forms of porphyrins under comparable conditions, except for some scant data (see Table 7.1) which are far from being sufficient to discuss this problem. This calls for extensive systematic studies into the spectral and other properties of protonated porphyrins. The specific behaviour of tetrazaporphyrins and phthalocyanines during protonation was discussed in Chapter 4.

Nor are there any systematic data available on the relationship between the electronic absorption spectra of porphyrins and their coordinating capacity. Whatever data are available suggest that there is no simple relationship between the coordination rate of metal salts and the spectral type of porphyrin as well as the position of the first band in the absorption spectra. It may be assumed that the coordination rate is practically independent of the energy of the bonding and antibonding π orbitals of porphyrins. Even the significant spectral, hence electronic and energetic, differences between porphin and chlorin and their numerous derivatives do not affect the formation rates of complexes of the same type (see Table 4.31). Unfortunately no data have been obtained so far on the rates of coordination of metal salts to H_2TBP, H_2TAP, and H_2Pc which could provide the final answer to the question under consideration.

It is certain, though, that there must be a relationship between the electronic absorption spectra and stability of porphyrin complexes for complex formation brings about a rearrangement of porphyrin electrons among the ligand and cation orbitals. As was shown in the case of phthalocyanine, the electron distribution pattern depends on the strength of the σ and π bonds forming between the metal and ligand, that is on the electronic coordination effect (see Chapter 4). It has been established that the σ effect of coordination increases free energy ΔG_S of the dissolution reaction. Similar manifestations have the effect of the central atom charge, the forward dative π effect, and the steric effect of disturbed coplanarity of the π ligand. According to Fig. 4.6, the backward dative π effect increases the electron density at the antibonding *n* orbitals of the nitrogen atoms and reduces the value of ΔG_S.

Since quantity ΔG_S is related to the free energies of the lattice (ΔG_{lat}), solvation (ΔG_{solv}), and protonation (ΔG_{H^+}) by

$$\Delta G_S = \Delta G_{lat} - \Delta G_{solv} - \Delta G_{H^+}, \qquad (7.1)$$

in which the values of ΔG_{lat} and ΔG_{solv} for phthalocyanines of isocharge ions M^{2+}, M^{3+}, and M^{4+} remain practically constant, it may be maintained that ΔG_S varies correspondingly with ΔG_{H^+} and may serve as a sufficiently reliable quantitative characteristic of electron distribution at the ligand orbitals.

It has been established experimentally that the formation and strengthening of the σ bond, accumulation of a positive charge on the metal atom in MPc, and the steric ligand coplanarity disturbance factor give rise to a hypsochromic shift of the first absorption band in the spectrum of metal phthalocyanine. The formation and strengthening of the backward dative π bond and the increasing coplanarity of the ligand, on the contrary, cause a bathochromic shift of the first band with respect to its position in a σ-type standard complex (ZnPc). A similar manifestation of the coordination effects in spectra clearly follows from Fig. 4.7 and is mentioned in a number of works.[37,290,424]

It is interesting to follow the correlation between the amounts of variation of the dissolution energies ($\Delta\Delta G_S = \Delta G_S^{MPc} - \Delta G_S^{H2Pc}$) and the shift of the first absorption band of a phthalocyanine complex ($\Delta\nu' = \nu'_{H2Pc} - \nu'_{MPc}$) since both quantities are determined solely by coordination effects. The data are presented in Table 7.2 and Fig. 7.6. All complexes are grouped along three straight lines in the latter. The first line corresponds to Ga, Al, V, and Sn phthalocyanines which are characterized

Table 7.2 Shift of the first absorption band and changes in the free energy of dissolution of phthalocyanine in H_2SO_4 during coordination

Complex	$\Delta\nu'$ (cm^{-1})	$\Delta\Delta G_S$ (kcal mol^{-1})	Complex	$\Delta\nu'$ (cm^{-1})	$\Delta\Delta G_S$ (kcal mol^{-1})
H_2Pc	0	0	$(HSO_4)GaPc$	280	246
ZnPc	700	910	$(HSO_4)IrPc$	1130	−420
CoPc	650	580	$(HSO_4)AlPc$	330	1400
NiPc	850	−260	$(HSO_4)RhPc$	1010	−890
CuPc	600	0	$(HSO_4)RuPc$	680	−960
PdPc	980	−170	$(HSO_4)_2SnPc$	620	2640
PtPc	960	−200	$(HSO_4)_2VPc$	330	380
			$(HSO_4)_2OsPc$	1010	480

by the absence of π bonds and no distortion of the π ligand during coordination. The latter feature is due to the high intensity of light absorption in the first band of metal phthalocyanines (log $\varepsilon = 5.0$–5.3), which is indicative of ideal conditions for the π overlap in the macroring.[108] The correlation between the above values for this group of complexes is expressed as

$$\Delta\nu' = K_1(\Delta G_S^{MPc} - \Delta G_S^{H2Pc}), \tag{7.2}$$

where $K_1 = 0.135$. Here both $\Delta\nu'$ and $\Delta\Delta G_S$ decrease with the effective charge on the central atom in accordance with the σ and charge effects.

The second line corresponds to Zn, Co, and Cu phthalocyanines characterized by lack of distortion of the π ligand during coordination (log $\varepsilon > 5.0$) and by a greater capacity for formation of backward π bonds in a series from Zn to Cu (zero in the case of zinc). The correlation for this

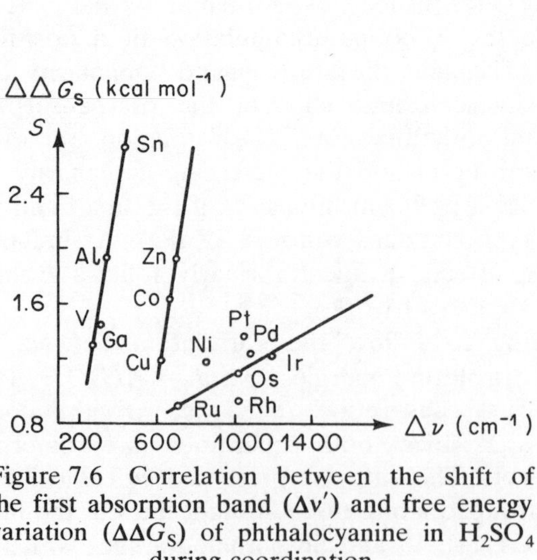

Figure 7.6 Correlation between the shift of the first absorption band ($\Delta\nu'$) and free energy variation ($\Delta\Delta G_S$) of phthalocyanine in H_2SO_4 during coordination

group is

$$\Delta\nu' = K_2(\Delta G_S^{MPc} - \Delta G_S^{H2Pc}), \tag{7.3}$$

where $K_2 = 0.115$. As the capacity for formation of π bonds increases, the values of both $\Delta\nu'$ and $\Delta\Delta G_S$ decrease as a result of the increasing π-electron effect (Co^{2+} and Cu^{2+}) being superimposed on the σ effect (Zn^{2+}).

The third line in Fig. 7.6 corresponds to phthalocyanines of metals of the platinum group and Ni. They are characterized by a wide dispersion of points. In this group of complexes, with the exception of Pt and Pd, a pronounced deviation of the bond lengths and valence angles of the phthalocyanine ligand from the standard values is expected without distortion of the planar configuration. This assumption follows from comparison of X-ray analysis data for NiPc, PtPc, and H_2Pc,[180–182] as well as low intensities ($\log \varepsilon = 4.6–4.0$) of the first absorption band. The latter feature is indicative of the worsening conjugation conditions in the chromophore of the macroring. Along with a decrease in the value of log ε, a distortion of the absorption spectra of Ir, Rh, Os, Ru, and Ni complexes is observed together with a considerable hypsochromic shift of the absorption bands. Therefore the decrease in the values of $\Delta\nu'$ and $\Delta\Delta G_S$ does not correspond either to the growing strength of the dative π bonds or the decreasing strength of the σ bonds but is determined entirely by steric effects.

The correlation between absorption spectra and chemical bonding type as well as basic properties of the donor atoms included in the conjugated system, considered in the context of phthalocyanine complexes, is without any doubt applicable to complexes of all types of tetrapyrrole molecules, including porphyrins and chlorins. Therefore the deductions as to the influence of coordination effects on the properties determined by the distribution of π and σ electrons at the ligand orbitals are equally applicable to metalloporphyrins. On the basis of the correlation between coordination effects and absorption spectra the range of applicability of the stability criteria of porphyrin and phthalocyanine complexes has been defined.

7.3 Spectral Criterion of Stability of Porphyrin and Phthalocyanine Complexes

The idea that the stability of a complex somehow affects its electronic absorption spectrum was proposed for the first time by Linstead *et al.* in a work[445] concerned with the absorption spectra of phthalocyanine and its complexes. It was pointed out that the position of the long-wave band (λ_I) of a complex correlates with the capacity of the metal for planar hybridization and suggested that the hypsochromic shift of this band may be related to the complex stability. However, no quantitative data on the stability of MPc that would support this idea were obtained.

Much later, in comparing the qualitative data on the kinetic stability of porphyrin complexes with Co, Ni, and Cu to their absorption spectra, Caughey and Corwin[406] used the maximum hypsochromism of the visible bands in the spectra of nickel porphyrins as a proof of their maximum stability. Williams[432] analysed the effect of the cation on the spectra of various ligands and arrived at a conclusion, on the basis of the data of Table 7.3, that wavelength λ_I decreases as the electron-acceptor properties of the cations which limit the conjugation in porphyrin become more pronounced. Still later Phillips[268] and Falk,[23] having no quantitative data on the stability of metalloporphyrins either, and proceeding entirely from the conclusions of the above-cited references, gave the following definition for the correlation between the spectra of metalloporphyrins and the nature of the metal: a pronounced hypsochromic shift of λ_I corresponds to a greater stability of the complex.

Hence, the following series of relative stability of porphyrin and phthalocyanine complexes with metals has been derived: $Pt^{2+} > Pd^{2+} > Ni^{2+} > Co^{2+} > Ag^{2+} > Cu^{2+} > Zn^{2+}$, which was used in this form in many fundamental works, including reference 8.

The spectral criterion was quantitatively verified, more carefully studied, and its applicability range examined for the first time in the laboratory of the present author, using a great variety of phthalocyanine complexes.[37,102,191,290,424,446] It was shown that the quantitative aspects of this problem are rather complicated. This complexity stems from a number of factors which affect the electronic absorption spectra and stability of complexes of the same type in different ways. At this point I should like to specify that the spectral criterion may be applied with a high degree of confidence only to similar complexes, that is compounds of the same stoicheiometric composition and geometric configuration.

For the analysis use was made of the position of the first absorption band, λ_I, because the energy levels of the ground and first excited states are more susceptible to those changes that involve the ligand in the course of complex formation. The higher excited π levels are associated with the molecular skeleton to a lesser extent and are less sensitive to the changes occurring in it. The correlation between the structure of the π ligand and

Table 7.3 Position of long-wave band (λ_I, nm) in absorption spectra of metalloporphyrins[432]

Metal ion	H_2MP	H_2PP	H_2HP	H_2TPP	H_2EP
Zn^{2+}	570	586	—	550	572
Cu^{2+}	561	578	565	540	560
Co^{2+}	552	573	560	530	—
Ni^{2+}	550	567	557	530	550

its complexes and the absorption band intensity is more complicated, its theoretical aspects are yet to be clarified,[447] and it cannot be applied at present to determine the stability of complexes in the general case.

Porphyrins, phthalocyanine, and similar molecules belong to π ligands, that is molecules featuring a system of interrelated π electrons (conjugated system), which can be acted upon directly by the central atom. The latter may interact with the ligand, involving formation of σ bonds $M \leftarrow L$, backward dative π bonds $M \rightarrow L$, forward dative π bonds $M \leftarrow L$, the effective positive charge of the central atom, and changes in the geometric structure of the ligand during coordination. If the complex formation process occurs in a solvating medium, the solvation sheath of the ligand may change during coordination.

In the case of phthalocyanine and porphyrins the latter factor may be ignored in view of the ligand rigidity. The formation of forward dative π bonds and the forward dative π effect are observed in the electronic spectra of complexes very seldom, only in the case of cations with empty *p* and *d* orbitals of π symmetry (XAlPc, XGaPc, X_2TiPc, X_2VPc, and respective metalloporphyrins). Hence, one should take into account the spectral manifestations of the σ, backward π, steric, and charge effects. As was shown in Chapters 4 and 5, the magnitude of these effects is determined by the acid–base properties and kinetic stability of porphyrin and phthalocyanine complexes.

In our earlier works[37,284,290] it was shown that the total shift of the first absorption band in the electronic spectra of MPc can be written as

$$\Delta\lambda_{\mathrm{I}} = \Delta\lambda_{\sigma} + \Delta\lambda_{\mathrm{ch}} + \Delta\lambda_{\mathrm{ster}} - \Delta\lambda_{\pi}, \tag{7.4}$$

where $\Delta\lambda_{\sigma}$ is the σ effect, $\Delta\lambda_{\mathrm{ch}}$ is the charge effect, $\Delta\lambda_{\mathrm{ster}}$ is the steric effect, and $\Delta\lambda_{\pi}$ is the backward π effect.

The value of $\Delta\lambda_{\mathrm{I}}$ has a positive sign for all complexes examined, which corresponds to a hypsochromic shift in going from the ligand to the complex. The hypsochromism of shift $\Delta\lambda_{\mathrm{I}}$ is due to the degeneration of the B_{3u} and B_{2u} states of the ligand during coordination (see Fig. 7.2).[8] However, this scheme does not explain the hypsochromism of shifts $\Delta\lambda_{\sigma}$, $\Delta\lambda_{\mathrm{ch}}$, and $\Delta\lambda_{\mathrm{ster}}$ and the bathochromism of the $\Delta\lambda_{\pi}$ shift. The directionality of these effects was experimentally corroborated by the present author and can be explained in terms of the general concepts concerning the filling pattern of the π orbitals of the ligand during coordination (see Fig. 4.7), as well as on the basis of the potential box model theory, that is changes in the potential of the molecular skeleton within which the π electrons move.[37,102,446]

In the phthalocyanine complex with Zn where the σ bond is heavily predominant the value of $\Delta\lambda_{\mathrm{I}}$ in a sulphuric acid solution is 470 Å, and it can be almost completely associated with $\Delta\lambda_{\sigma}$. It was shown that in the case of complexes XGaPc and O=VPc $\Delta\lambda_{\mathrm{I}}$ may be assumed approximately equal to $\Delta\lambda_{\sigma}$. In these complexes the shift equals 210 and 250 Å, respectively, and in the Ga complex it corresponds to a lower stability of the σ complex as compared to ZnPc. However, in the case of triply and quadruply charged metals one should also take into account the effect of the extra ligands on the spectra. In metal phthalocyanines this effect may be significant.[191] For this reason comparison of complexes of isocharge ions is more reliable.

Table 7.4 lists the values of $\log(1/k^{25}_{\mathrm{diss}})$ characterizing the kinetic stability of phthalocyanine complexes, the shift of the first absorption band of phthalocyanine during coordination ($\Delta\lambda_{\mathrm{I}}$) in H_2SO_4 solutions, and log ε of its intensity. All data relate to monoprotonated complexes of $X_n^-MPcH^+$ (where $X^- = HSO_4^-$). The values of true dissociation constants (k_{diss}) are taken from Table 5.3.

The stability of complexes of doubly charged metals does not correlate with the value of $\Delta\lambda_{\mathrm{I}}$. The considerable stabilization of complexes CoPc and CuPc is accompanied by a decrease in their shift as compared to ZnPc (Fig. 7.7). This can be explained only by formation and strengthening of the backward dative π bonds which stabilize the complex. Their formation is promoted by no distortions of the ligand taking place whatsoever (the values of log ε are extremely high). The formation of π bonds $M \rightarrow N$ reduces the positive charge of the chromophore skeleton as a result of the electron density shifting into the macroring. Consequently, the energy level of the ground state from which an electronic transition in the first band takes place increases.

The latter phenomenon leads to a bathochromic shift $\Delta\lambda_\pi$, that is a decrease in the hypsochromism of total shift $\Delta\lambda_{\mathrm{I}}$. It is quite obvious that the effect of the σ bond on the potential of the chromophore skeleton is opposite, therefore the strengthening of the σ bond will correspond to a greater hypsochromic shift $\Delta\lambda_\sigma$. As can be inferred from Fig. 7.7, the value of $\Delta\lambda_{\mathrm{I}}$ for X_nMPcH^+ complexes with Pt^{2+} and Pd^{2+} is much greater than for complexes with Cu^{2+} at a relatively small difference in their stability. Taking into consideration the great values of log ε for X_nPtPc and X_nPdPc,

Table 7.4 Kinetic stability, shift of the first absorption bands, and their intensity for phthalocyanine complexes $(HSO_4)_nMPcH^+$

Metal ion	$\log(1/k^{25}_{\mathrm{diss}})$	$\Delta\lambda_{\mathrm{I}}$ (Å)	log ε	Metal ion	$\log(1/k^{25}_{\mathrm{diss}})$	$\Delta\lambda_{\mathrm{I}}$ (Å)	log ε
Zn^{2+}	4.84	470	5.32	Ir^{3+}	5.70	730	3.90
Co^{2+}	6.11	450	5.12	Al^{3+}	5.85	—	—
Ni^{2+}	6.18	570	4.57	Rh^{3+}	7.50	—	—
Cu^{2+}	8.31	410	5.36	Ru^{3+}	8.11	—	—
Pt^{2+}	9.03	639	5.18	Sn^{4+}	3.50	430	4.89
Pd^{2+}	9.08	640	5.17	V^{4+}	5.62	250	5.23
Ga^{3+}	4.24	210	5.24	Os^{4+}	7.72	290	4.11
				H^+	—	0	5.03

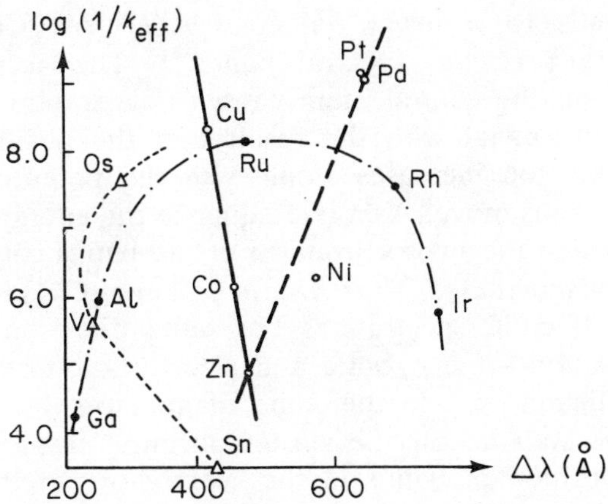

Figure 7.7 Correlation between the hypsochromic shift of the first absorption band (Δλ) and stability of metal phthalocyanines

which is indicative of no perceptible steric factor being involved in these complexes, one may assume that their high stability is determined primarily by a drastic strengthening of the σ bond and an insignificant increase in the strength of the π bond M → N. The same conclusions follow from analysis of free energy of dissolution $\Delta\Delta G_S$ (see Chapter 4).

Complex NiPc is much less stable than CuPc but its $\Delta\lambda_I$ shift is much greater. The only possible reason is the steric factor. According to X-ray analysis data[180–182] cation Ni^{2+} produces a tangible effect on the phthalocyanine ligand. By pulling together the four N atoms it reduces the size of the ligand's inner space, changing the valence angles of the C—N—C bonds and impairing the conditions for the π overlap in the macroring. Evaluation of the π-electron charge of the macroring (see Chapter 3) has shown that in NiPc it decreases by 2.7 units because of the less favourable conjugation conditions. As a result the *n* electrons are partially excluded from the conjugation, and the value of ΔG_S becomes lower than in H_2Pc or CuPc.

The pronounced hypsochromism of the λ_I band in NiPc corresponds to a steric factor which reduces the aromaticity of the chromophore, draws apart the π levels of the ground and excited states, and brings about hypsochromic shift $\Delta\lambda_{ster}$. In the case of NiPc the steric factor results in a considerably lower intensity of the first band (log ε = 4.57). Hence the dative π bonds in NiPc are much weaker than in CuPc.

In phthalocyanine complexes with triply charged metals the predominant role is played by the σ bonds and the steric factor. Analysis of spectra and free energies of dissolution ΔG_S of aluminium phthalocyanine has shown that, depending on the solvating capacity of the solvent (H_2SO_4 of different concentrations), the effective charge on the Al atom changes accordingly.

Its increase leads to a lower value of ΔG_S (see Chapter 4) and a hypsochromic shift of the spectral bands.[424] The accumulation of the positive charge on the central atom causes polarization of the π-electron cloud towards the metal with the result that the positive charge of the chromophore skeleton increases along with the potential of the field in which the π electrons move. As a consequence the energy of the bonding π levels decreases and the hypsochromism of the bands corresponding to the $\pi \rightarrow \pi^*$ transitions increases. Thus $\Delta\lambda_{ch}$ in equation (7.4) has a positive sign.

The charge effect in spectra has the same direction as the σ effect. Besides, both factors—the σ bond and electrostatic effect of the positive charge on the ligand—act in the same manner on the complex stability, increasing it, as well as on the proton affinity of the ligand, which is reduced. The position of bands in the spectra of purely ionic complexes (Li, Na, Be, Ca, Ba, Pb, Sn, Cd, Hg) is determined mainly by the charge effect, which increases with $\Delta\lambda_I$.

It was shown that in XGaPc the charge effect does not manifest itself, therefore, the increase in the $\Delta\lambda_I$ shift in the series Ga → V → Al corresponds to that in the strength of the σ bond.[290] The value of log ε diminishing greatly in the series of Ru^{3+}, Rh^{3+}, and Ir^{3+} phthalocyanines (from 4.60 to 3.90) is indicative of a severe distortion of the ligand during coordination. This distortion seems to be the same as in the case of NiPc. As the steric factor becomes more pronounced, the stability of XMPc (M = metal of the platinum group) not only corresponds to a significant hypsochromic shift [$(HSO_4)RuPc$] but also decreases in the series Ru → Rh → Ir with the aromaticity of the coordinated ligand. As a result the stability–$\Delta\lambda_I$ band shift curve passes through a maximum at ruthenium. The dative π effect does not seem to play a significant role in this group of complexes (except for Ru and Rh), and their stability is determined by a strong σ bond.

In the series of phthalocyanines of quadruply charged metals the charge effect was shown to be significant in $(HSO_4)_2SnPc$,[424] and $(HSO_4)_2VPc$ was shown to be immune to π and charge effects. The pronounced hypsochromic shift of the λ_I band in the Sn^{4+} complex corresponds to $\Delta\lambda_{ch}$ in its spectrum. In the case of the Os(IV) complex the spectral picture is considerably complicated by the structural factor and polymerization similar to that occurring in an extra complex.[110] Therefore, depending on the H_2SO_4 concentration, the $\Delta\lambda_I$ shift varies from 245 to 670 Å. In the case of $(HSO_4)_2OsPc$ the value of log ε is low.

Thus analysis of the spectra, stability, and other properties of phthalocyanine complexes has shown that the spectral stability criterion as defined at the beginning of this section is not applicable to them.

It cannot be applied to those complexes of the same type in which a major role is played by the backward dative π bonds and the steric factor due to distortion of the π ligand structure during coordination. This is why the spectral stability criterion is not applicable to phthalocyanine complexes, acetyl acetonates, and many other compounds. It is applicable, though, to metalloporphyrins because the role of the dative π bonds in them is not so important as in phthalocyanine (see Chapter 3). The steric factor of ligand distortion is also less pronounced than in phthalocyanine ligands because of the lower strength of the M—N bonds and lesser rigidity of porphyrin ligands.

Now that the range of applicability of the spectral stability criterion has been determined, it may be finally defined as follows: in the series of similar covalent complexes with π ligands the amount of shift of the first absorption band of the latter varies with the complex stability if the role of the backward dative π bonds is minor and the factor of ligand structure distortion can be ignored.

7.4 Spectral Data on the Cyclopentanone Ring of Chlorophyllic Acid and Its Complexes

The cyclopentanone ring (isocycle) is a distinguishing feature of porphyrin derivatives known under the generic name 'phorbins'. The presence of this ring in chlorophyll **(II)**, chlorophyllic acids *a* **(XXVII)** and *b* **(XXVIII)**, and their derivatives certainly influences the entire set of useful properties making chlorophyll the main pigment in the photosynthesis system. The functions performed by the cyclopentanone ring (CPR) have not been elucidated so far.

Rabinovich[2] regards it as a solvation centre, a source of bound water. The so-called Molisch phase test has long been known in conjunction with the CPR properties. In this test chlorophyll acted upon by a strong alkaline solution turns brown then becomes green again. The chemical mechanism of this phenomenon has been established.[448–450] Dunicz *et al.*[449] and Weller[450] obtained the electronic absorption spectra of intermediates which are essentially enolates of chlorophylls *a* and *b*. These unstable compounds are produced in ether or pyridine solutions after addition of a methanol solution of KOH. In the spectra of these compounds the Soret band disappears, the long-wave absorption is minimum, and significant absorption occurs in the region of 450–550 nm. In an alkaline medium the intermediate undergoes rapid decomposition with splitting of the CPR. The end-product of the decomposition is Mg chlorin e_6 **(XXXVII)** or Mg rhodine g_7 **(XXXV)**, which are green and have spectra similar to those of chlorophyll *a* and *b*.

In an earlier work[77] the present author published the first study of the kinetics of the reaction in which the intermediate of the phase test of chlorophyllic acid *a* turns to chlorin e_6; in a subsequent work[451] the effect of the solvent and alkali on this reaction was examined; and in yet another work[452] the role of the coordinated metal in the opening of the CPR was demonstrated. Of greatest interest were the following facts established in these works. It was found, for example, that the absorption spectrum of

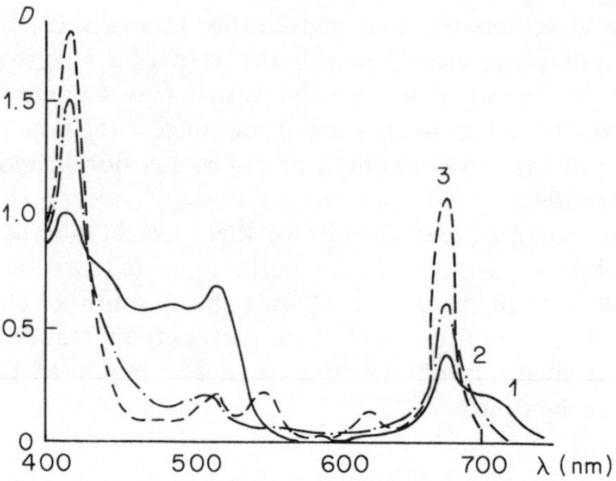

Figure 7.8 Absorption spectra of chlorophyllic acid *a* phase test intermediate in an ether–methanol solution of KOH, taken on a time scale. Time: curve 1, 0 min; curve 2, 135 min. The spectrum of phaeophytin *a* in ether (curve 3) is given for comparison

the intermediate of the phase test of H_2Chla, that is its enolate (Fig. 7.8), is in many respects similar to that of the chlorophyll intermediate. The only difference is that the shift of equilibrium keto-CPR ⇄ enol-CPR is incomplete because the Soret band does not disappear completely but only becomes much less intensive.

The intensity of the 668 nm band decreases sharply and a new band appears in the neighbourhood of 700 nm. This may be assumed to be the 668 nm band shifted bathochromically as a result of the effect of the cyclopentadiene group of enol on the π system of the macroring chromophore.

As was mentioned in Chapter 4, the 700 nm band in H_2Chl appears in ethanol solutions with large amounts of glycerin added, which suggests that glycerin shifts the keto–enol equilibrium (4.43) towards the enol. The enol is stabilized due to specific solvation by glycerin.[312] Under such conditions the enolic form is stable enough to air oxidation. In an alkaline medium the isocycle splitting rate obeys the following first-order equation:

$$-\mathrm{d}c_{\mathrm{int}}/\mathrm{d}t = k_{\mathrm{eff}}c_{\mathrm{int}}. \tag{7.5}$$

Here c_{int} is the intermediate (enol) concentration, k_{eff} is independent of the nature and concentration of the alkali and equals $9.0 \pm 0.6\ \mathrm{h}^{-1}$ on the average in an ether–methanol solution. It may be assumed that $k_{\mathrm{eff}} = k_{\nu}c_{O_2}$.

However, the phase test cannot be conducted in complete absence of oxygen in the solution, and the dependence of the ring splitting rate on the oxygen concentration has not yet been firmly established. It has only been shown that k_{eff} depends strongly on the nature of the solvent.[451]

The relationship between the enol isocycle splitting rate and the nature of the metal coordinated to H_2Chl is complex. Table 7.5 presents the values of k_{eff} in ethyl acetate in the presence of CsOH, the position of the first band in the spectrum, and the electronegativities of metals.[452]

The behaviour of the central atoms is substantially different: Mg and Cd promote the splitting, Zn and Co slow it down somewhat, Ni produces practically no effect on the splitting rate, Hg slows it down considerably, and CuChl fails to produce a phase test. The ring splitting rate in the case of Mg^{2+}, Cd^{2+}, Co^{2+}, and Hg^{2+} is linearly related to the electronegativity of the metals (see Table 7.5). Zn^{2+} and Ni^{2+} depart from the straight line.

As follows from the preceding section, the central atom acts upon the porphyrin molecule mainly through formation of σ bonds (σ effect) and through the charge effect in ionic complexes. Both effects manifest themselves in a hypsochromic shift of the λ_I band of the complexes with respect to the ligand. They reduce the σ-electron density in the isocycle, which prevents enolization. Therefore their spectra in an alkaline medium (or in glycerin–ethanol solutions) vary to a much lesser degree than in the case of H_2Chl.[452] The extent to which enolization is hindered by the metals does not always correlate with the value of λ_I. For example, NiChl ($\lambda_I = 649$ nm) is enolized, while in the case of CuChl ($\lambda_I = 650$ nm) the enolization is so weak that enol is not spectrally detectable.

As can be inferred from Table 7.5, coordinated Mg^{2+}, Cd^{2+}, and Ni^{2+} form enolates more active in the splitting reaction than in the case of H_2Chl*a*. This may also apply to enolates of Zn^{2+} and Co^{2+} complexes. Complexes with Hg^{2+} and Cu^{2+} behave in a different manner.

The substitution of the CH_3 group in H_2Chl*a* by a CHO group in H_2Chl*b* slightly increases the isocycle splitting rate. No definitive experimental proof has been obtained to draw any final conclusions as to

Table 7.5 Rates of splitting of the isocycle of enolic forms of chlorophyllic acid and its complexes in ethyl acetate

Complex	k_{eff}^{25} (h^{-1})	λ_I (nm) (in ethanol)	Electro-negativity
H_2Chl*a*	9.0 ± 0.6	668	—
MgChl*a*	32 ± 2	666	1.3
CdChl*a*	17 ± 1	665	1.7
HgChl*a*	1.7 ± 0.1	666	2.0
ZnChl*a*	6.7 ± 0.5	660	1.6
CoChl*a*	5.7 ± 0.7	644	1.9
NiChl*a*	11 ± 1	649	1.9
CuChl*a*	No reaction	650	1.9
H_2Chl*b*	15 ± 1	660	—
H_2PE	Fast	667	—

the mechanism of the reaction. I believe[77,452] that the opening of the isocycle in the enolic form involves formation of an O_2 π complex with a double bond of the isocycle and its breakage along C10–C9 with subsequent formation of chlorin e_6 (rhodine g_7). According to a different point of view,[453] the opening of the isocycle at low alkali concentrations involves an electrophilic attack of the C9 atom of the enolic form of the isocycle by OH^- rather than oxidative destruction by O_2. This point of view, however, is at variance with the present author's experimental data.[451,452]

Strong acids are known to favour the enolization of respective ketones as well. H_2Chl*a* is not enolized in H_2SO_4 because of the molecule being protonated at the nitrogen atoms. As a result the electron density at the C—O group of the isocycle is greatly reduced and it becomes incapable of enolization.[398] In MChl*a* complexes, as the metal becomes less likely to form a σ bond and the σ effect on the C=O group in the isocycle is less pronounced, the capacity for enolization increases, which manifests itself (see Chapter 5) in disappearance of the long-wave band in the spectrum as a result of H_2SO_4 addition. The sequence of transformations

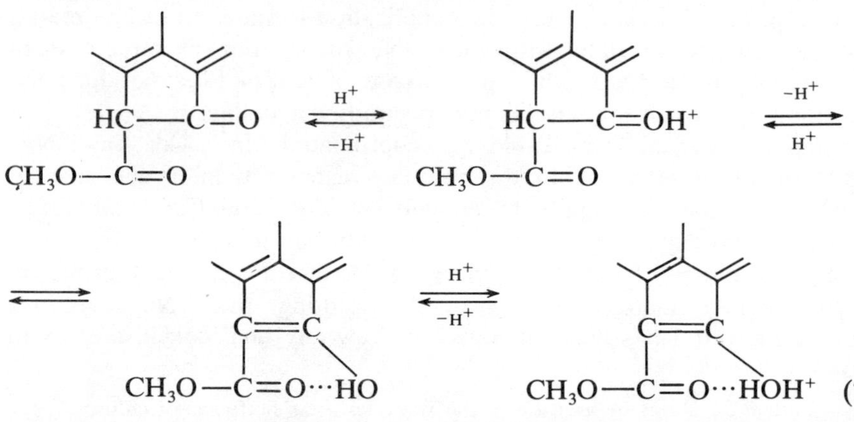

(7.6)

stabilizes the enol by way of protonation. The peripheral proton strongly polarizes the entire π system with the result that the low-intensity long-wave band of enol shifts bathochromically into the longer-wave region.

No data have been reported so far on other chemical properties of the cyclopentanone ring of phorbins.

7.5 Spectral Manifestations of Association of Porphyrins in Solution

Like most dyes porphyrins show a strong tendency to aggregation. Since chlorophyll performs its functions in green leaves in an aggregated state, a large body of literature is concerned with this state of chlorophyll. In reference 8 an entire chapter deals exclusively with this particular state.

Therefore I shall discuss only a few aspects and propose some new ideas concerning the association mechanism and structure of the associates.

Extensive studies into the spectral and other properties of films of chlorophyll and its structural analogues, conducted by Krasnovsky *et al.*,[454–457] have corroborated the conclusions[8] as the structural variety of associates and the absence of a common porphyrin association mechanism.

Most of the investigators' efforts are directed to elucidation of the aggregation of chlorophyll. The major role of water molecules in the formation of structural units of aggregates of chlorophyll and its analogues is emphasized.[454] The Mg atom and C═O group of the isocycle are mentioned as the most probable localization sites. According to one of the structural schemes of aggregation[8] the water molecule must favour parallel arrangement of the π systems in the associates of Mg porphyrins and their maximum π–π overlap.

This scheme, however, is difficult to reconcile with the existence of regions of π–π overlap of adjacent molecules, conclusively proved by NMR spectroscopy.[458,459] The geometry of superposition of these overlap regions indicates that the Mg atom coordinates to the C═O group of the isocycle. The coordination to the magnesium atom of such a strong electron donor as water certainly cannot be conducive to aggregation irrespective of the conditions. During such coordination the degree of protonation of the H—O bonds of water is not so great for the coordinated water to enter into an acid–base interaction of the Mg ← O—H···N type with the second Mg porphyrin molecule. The strength of such a hydrogen bond is negligibly small and the probability of its occurrence in a solution is low in the presence of water. The above scheme seems to be applicable only to solid Mg porphyrin samples.

Along with studies into processes involving molecule aggregates in solid films of metalloporphyrins, and the properties of aggregated molecules in native samples, of great interest and importance is investigation of aggregation in solution as one passes from molecularly disperse systems to dimers, trimers, and larger aggregates. Under such conditions of emergence of a new phase most properties of the molecule undergo significant changes which can be easily traced. During association the solvation sheath of molecules changes, the solvent molecules are gradually replaced in the close vicinity of molecule $M(solv)_i$ which transforms to dimer $(solv)_{i-m}M\cdots M(solv)_{i-m}$, trimer $(solv)_{i-m}M\cdots M(solv)_{i-p}\cdots M(solv)_{i-m}$, and so on.

Since the initial stages of aggregation unaccompanied by formation of the solid phase may be regarded as resolvation,[252–254] the spectral changes in dyes, including porphyrins, will depend on the difference between their association centres and respective solvent atoms. The association of large molecules such as porphyrins cannot involve a loss of the major part of the solvation sheath. In the solid phase the associate loses the entire solvation sheath, which is replaced by the force fields of analogous molecules. In the case of most large molecules in the category of dyes the impact of the solvent on the molecule is much greater than that of the adjacent molecules in the solid phase, whereby the dye molecules are polarized in the solution to a greater extent. Hence the hypsochromic shift of bands during aggregation of most dyes.[8,22,33]

The aggregation of water-soluble phthalocyanines is also accompanied by a hypsochromic shift of the first band.[287] In $CuPc(SO_2O^-)_4$ the shift amounts to 40 nm (from 672 to 632 nm). The bathochromic shift in the case of chlorophyll and its derivatives under conditions of insufficient solvation is associated with the formation of dimers and trimers.[8] There is no acceptable physical theory to account for the bathochromic shift of bands in the case of dimeric metalloporphyrins. From the physicochemical standpoint (solvation and resolvation) one can assume two cases of interaction between molecules in a dimeric associate composed of planar molecules of the porphyrin and phthalocyanine types, which may result in a bathochromic or hypsochromic shift of the first band in the spectrum. Both cases are postulated and substantiated in some works concerned with the association of such molecules.[8,454–459] In the first case chromophores are stacked like a deck of cards. Despite being corroborated by Closs *et al.*,[458] such an interaction is not considered to be typical of aromatic systems.[6] In my opinion, an interaction of this kind is possible in smaller associates. As a result of broadening of the region accommodating the π electrons of a chromophore a similar interaction must involve a bathochromic shift of band λ_I. Hence a single long-wave band must appear in the spectra of dimeric associates. As more molecules are accumulated in the associate, another bathochromic shift follows, and the number of bands increases with their subsequent broadening.

The bathochromic shift of the absorption bands of chlorophyllic acids *a* and *b* in pyridine and quinoline solutions with respect to their position in non-aromatic solvents may be explained by a π–π interaction of the above types.[305] The second case is 'plane-edge' association, in which the centre of the chromophore of the porphyrin molecule contacts the edge of a planar chromophore (Fig. 7.9). The energetic advantage of such an arrangement of planar aromatic molecules was shown by many investigators.[6] In such a dimer the first molecule is acted upon by the chromophore centre and loses most of its solvation sheath. The second molecule is acted upon by the first one at the edge of the planar chromophore and loses only an insignificant portion of the solvation sheath. The action on the centre of the porphyrin chromophore of molecule 1, involving hydrogen bonding, or the metalloporphyrin chromophore, accompanied by extra coordination,

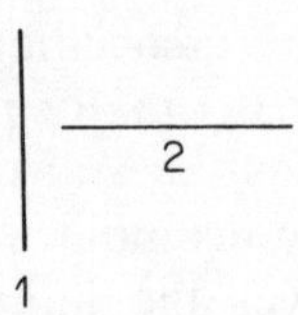

Figure 7.9 Dimeric 'plane-edge' associates

leads to slight shifts of the λ_I band, normally toward the short-wave side. The desolvation of molecule 1 promotes such a shift.

In contrast, the action upon molecule 2 at the edge exerts a strong influence on the spectra if the molecular interaction is sufficiently strong. This takes place in cases where the end portion of the chromophore contains an electron-donor atom (nitrogen atoms of phthalocyanines and azaporphyrins, O=C or CHO groups of chlorophyll). Such bathochromic shifts as a result of the plane-edge interaction were described in Chapters 4 and 5. The foregoing suggests that the association of chlorophyll as a result of the Mg···O=C interaction will be accompanied by a bathochromic shift. Enlargement of associates involves cooperative interactions and the situation becomes much more complicated.

As can be inferred from the above, of greatest interest from the theoretical point of view are studies into the association of porphyrin molecules of various structures, development of methods for determining the degree of association, mutual arrangement of the molecules, their optical properties, and reactivity. Accumulation of such data is necessary to develop the physical theory of association. At present only scant data are available. Most of them were presented in reference 8.

In a recent work[460] the spectral method was used to elucidate the effect of the nature and composition of binary solvents on the association of structurally different porphyrin molecules: chlorophyllic acid *a*, chlorin e_6, etioporphyrin, synthetic chlorin, and Cu^{2+} phaeophytin. It has been established that in the case of chlorin e_6 the first band shifts hypsochromically during association (Fig. 7.10), which corresponds to a strong influence of the carboxyl groups of one molecule on the N atoms of the coordination centre of the other (N···HOOC association).

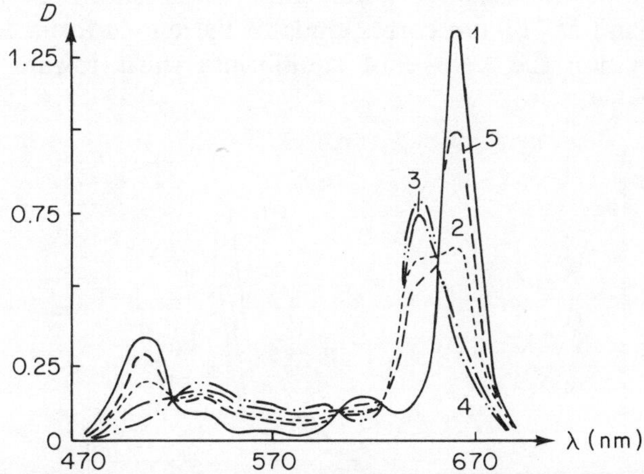

Figure 7.10 Absorption spectra of chlorin e_8 in a binary mixture. Ethanol–water ratio: curve 1, 10:0; curve 2, 8:2; curve 3, 2:8; curve 4, 1:9; curve 5, 5:5

As can be seen from Fig. 7.10, at an ethanol to water ratio of 3:7 a dimer is formed because the number of molecules acted upon by the centre of the chromophore is approximately equal to that of molecules acting upon the former by their COOH group. At an ethanol to water ratio of 1:9 a pentameric associate is formed, judging from the band intensity and shape. In this case the absorption curves naturally pass through the isobestic point since the states of the molecules in trimers, tetramers, and so on are practically the same as in dimers. Bearing this in mind, and knowing the interaction pattern, one can estimate the degree of association from the absorption spectra of a number of solutions.

Synthetic chlorin which lacks acid groups exhibits a weak bathochromic shift of the λ_I band during association. It cannot provide for association of the 'plane-edge' type, and the formation of associates is probably due to the π–π overlap.

An interesting kind of association has been revealed in H_2Chl*a* and CuChl*a* in a binary ethanol–water solvent. Figure 7.11 shows that when water is added to an ethanol solution of H_2Chl*a* so that the ethanol to water ratio becomes 6:4, the long-wave peak at 668 nm drops sharply in intensity and a clearly defined band appears at 700 nm. At an ethanol to water ratio of 3:7 the concentrations of both forms are equal, the concentration of the 700 nm form being lower than in the solution with the former ethanol to water ratio. As more water is added, the 700 nm form gradually diminishes and disappears completely at an ethanol to water ratio of 1:9. The spectrum displays only forms with a band at 674 nm. This phenomenon of disappearance of the band at 700 nm may be called 'hypsochromic band migration'.

In an earlier work[460] the appearance of the 700 nm band after addition of sizable amounts of water was attributed to stabilization of the enolic form of chlorophyllic acid by water for, as was shown in the previous section, the band at 700 nm corresponds to the enolic form. In the course of π–π association the keto–enol equilibrium shifts toward the ketone,

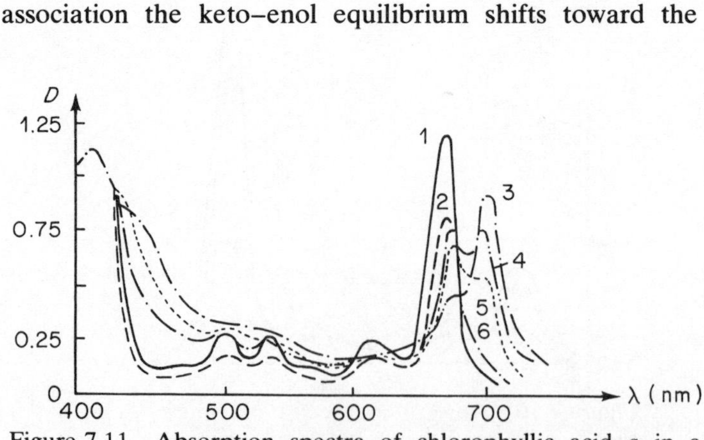

Figure 7.11 Absorption spectra of chlorophyllic acid *a* in a binary mixture. Ethanol–water ratio: curve 1, 10:0; curve 2, 8:2; curve 3, 6:4; curve 4, 4:6; curve 5, 3:7; curve 6, 1:9

which results in disappearance of the 700 nm band and appearance of a band at 674 nm, corresponding to the associate. Interestingly, similar spectral changes are observed in the ethanol–water system with CuChl*a*. Evidently, a 'plane-edge' association is out of the question here, and the only explanation of the strong bathochromic shift of the λ_I band (from 650 to 680 nm) occurring when 20–30% of water are added to ethanol is enolization. In the case of weak solvation of CuChl molecules, that is in the presence of 70% of water, favourable conditions are created for association, the 680 nm band disappears, and a single band corresponding to the associates appears at about 655 nm.

Thus the association of molecules with a cyclopentanone ring in an ethanol–water system is quite different at the initial stage. The final stage is absolutely the same as in synthetic chlorin. Characteristically, the effect of the solvent on the spectral manifestations of association of H_2Chl*a* and its derivatives is highly specific. For example, in an acetone–water solution migration of the 700 nm band is observed, while in dioxane and methanol it does not occur and the form corresponding to 700 nm persists even when an excess amount of water is present (90%).

The above facts are indicative of the complexity of the phenomenon of association, its diverse spectral manifestations, difficulties in experimentation, and theoretical interpretation of association.

7.6 Polymeric State of Phthalocyanine and Its Optical Properties

The incorporation of chromophore units into a polymer chain yields molecules which resemble the above-considered associates as far as some of their properties are concerned. Polymerization, just as association, provides for a greater spatial concentration of chromophores and their mutual orientation. Polymeric phthalocyanines and porphyrins arouse a great deal of interest as objects of investigation aimed at elaborating the theory of the polymeric state and its effect on the reactivity of molecules and their optical, semiconductor, catalytic, and other properties.

At present only polymeric phthalocyanines of various structures are known.[144–150,394–396]. True polymeric porphyrins have not yet been produced. Polymers of porphyrin complexes such as $(OSiP)_x$ are known in which the polymer chain develops due to extra coordination,[106] as well as molecules containing an ordinary C—C polymer chain with metalloporphyrin additives.[461]

This text will be restricted to manifestations of the polymerization of phthalocyanine and its complexes in electronic absorption spectra. The synthesis of numerous new groups of polymeric phthalocyanines and their complexes makes it imperative to study carefully the relationship between the degree of polymerization, structure, and properties of polymers. Because of the negligibly low solubility of phthalocyanines and their polymers most of the known methods for determining the degree of polymerization[462] are not applicable to them.

The analytical method based on comparison of the experimental data on percentage content of elements with values calculated for polymers of different structure, used by some investigators,[144,157] is too imprecise, the reasons being the poor accuracy of the analytical data for aromatic polymers, the multitude of various structures (ribbon, linear, branched, cross-linked, three-dimensional) of M^{n+} phthalocyanines, and the uncertainty of these structures (stemming from two molecules of different species normally taking part in the formation of the macroring, namely phthalic acid and pyromellitic acid derivatives, the latter yielding terminal groups). The method of terminal groups is more reliable but its area of application is limited. This makes it important to develop new procedures for estimating the degree of polymerization of such molecules.

One such procedure is the spectroscopic method proposed by Berezin and Shlyapova.[151,463] The polymers of metal phthalocyanines, based on pyromellitic acid, differ widely from MPc by the low intensity of the first band in the electronic absorption spectrum (Fig. 7.12). Characteristically, the position of bands remains practically the same (Table 7.6). This means that the state of the electronic energy levels of the phthalocyanine chromophore in a monomer and a polymer is the same, too. In the case of

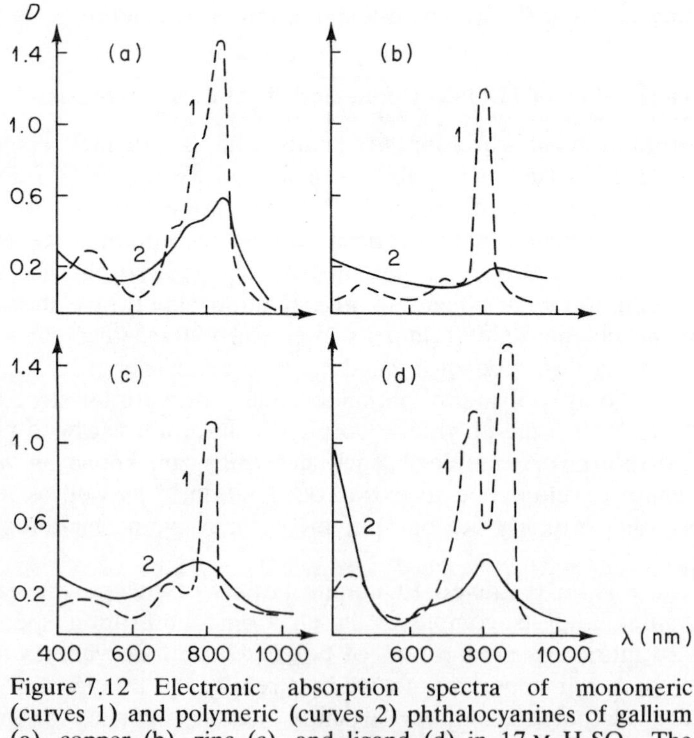

Figure 7.12 Electronic absorption spectra of monomeric (curves 1) and polymeric (curves 2) phthalocyanines of gallium (a), copper (b), zinc (c), and ligand (d) in 17 M H_2SO_4. The polymer sample is four to eight times larger than the monomer one

Table 7.6 Basic parameters of the first absorption bands of phthalocyanine complexes and their polymers with polymerization degree i

Complex	ν (cm^{-1})	δ (cm^{-1})	ε_{max}	i
CuPc	12 600	270	229 100	
$(CuPc)_i$	12 650	904	4 790	14
ZnPc	12 850	250	208 900	
$(ZnPc)_i$	12 800	655	13 270	6
$(HSO_4)GaPc$	12 280	324	173 800	
$[(HSO_4)GaPc]_i$	12 300	475	10 800	11
$(HSO_4)RuPc$	12 690	487	39 850	
$[(HSO_4)RuPc]_i$	12 900	1165	7 079	3
$(SO_4)OsPc^a$	—	—	12 880	
$[(SO_4)OsPc]_i{}^a$	—	—	2 795	5
O_2MoPc	11 960	520	45 710	
$(O_2MoPc)_i$	12 400	1650	2 203	7

[a]The values of δ are practically the same. The degree of polymerization has been calculated from ε_{max}.

polymeric phthalocyanine ligand $(H_2Pc)_i$ the two long-wave bands corresponding to H_2Pc and caused by the non-equivalence of the transitions along the coordinate axes in the chromophore plane are not degenerated. In polymer $(H_2Pc)_i$ only one symmetrical long-wave band appears. Its maximum coincides with the 'centre of gravity' of the two H_2Pc bands.

Such a behaviour of spectra in the transition MPc → $(MPc)_i$ suggests that the decrease in intensity of the long-wave band and the invariability of its position as the polymer chain grows are inversely proportional to the degree of polymerization and result from the impossibility of excitation of two or more chromophores of the conjugated polymer chain of phthalocyanine at a time. During excitation of one link the remaining $(i - 1)$ chromophores of $(MPc)_i$ are optically paralysed. The reason for such cooperation of the chromophores may be their spatial concentration without changes in their random distribution in the solution. The intensity of interaction between the chromophores (strong or weak) may be significant. For example, the association of dyes also leads to a lower intensity of the absorption bands, but the effect is much less pronounced than during polymerization of chromophores.

The foregoing permits definition of the degree of polymerization of $(MPc)_i$ as a ratio of the oscillator strengths in the first band of the monomer (P_0) and the polymer (P_i):

$$i = P_0/P_i = (\varepsilon_{max}\delta)_0/(\varepsilon_{max}\delta)_i. \tag{7.7}$$

The absorption spectra are expanded into Gaussian components.[464] Half-width δ and ε_{max} are taken from individual bands. Table 7.6 lists parameters obtained in this manner for six polymeric phthalocyanines and the degrees of their polymerization.

As can be seen from the table, the polymer synthesis methods described in Chapter 2 yield polymers with a low degree of polymerization (oligomers). The mean statistical value of i does not exceed 3–14 for various metal phthalocyanines. In the literature[144,157] a mean value of i approximately equal to six is given.

Thus the spectral method gives the actual value of i. In an earlier work[151] the results of determining the degree of polymerization of $[(HSO_4)RuPc]_i$ and $[(SO_4)OsPc]_i$ by the method of acid–base titration of side groups $(CO)_2NH$ in these polymers in a 1 N solution of KOH were presented. As was established from the values of the equivalent weights found in this manner, polymeric osmium phthalocyanine is a ribbon polymer (see Fig. 2.1) with $i \approx 7–8$ and contains 13 side groups $(CO)_2NH$, while polymeric ruthenium phthalocyanine with $i \approx 3$ has eight side groups. The values of i are in good agreement with the spectral data.

Combination of various methods for determining the degree of polymerization along with a study into the physicochemical properties of polymers brings to light the structural features of these complex molecules. Later[396,465,466] it was shown that the spectral method of determining the value of i is also applicable to bridged polymers of metal phthalocyanines, particularly to polyphthalocyaninesiloxane $(O\cdot SiPc)_i$. It was thus established that the method of vacuum dehydration of $(HO)_2SiPc$ produces an oligomer with $i \approx 3$, whereas heating of dihydroxysilicon phthalocyanine in nitrobenzene with $ZnCl_2$ at 175 °C yields a polymer with $i \approx 11$.

There is every reason to believe that bridged polymeric metalloporphyrins will be spectrally similar to respective metal polyphthalocyanines. As far as polymeric porphyrins of a different structure are concerned, one may expect a more complex correlation because of the poorer stability of the aromatic π system.

Chapter 8

Other properties of metalloporphyrins

The past decade has been marked by new significant advances in the studies of many physical and physicochemical properties of porphyrins and their complexes. The first encouraging results of infrared, EPR, and NMR spectral studies have been obtained. A wealth of valuable information on the relationship between the structure and properties of porphyrins and their complexes has been derived by the ESCA, polarized luminescence, and flash photolysis methods. I do not intend to dwell upon the results of these studies for they have already been summarized and analysed in the often cited reference 8.

New data have been obtained on the magnetic circular dichroism, Mössbauer spectra, and other properties, summarized in a specialized American publication.[467] Of particular interest are those properties which permit one to establish their relationship with the nature of the metal or porphyrin. They include photochemical stability, oxidation–reduction capacity, catalytic activity, and semiconductor properties.

8.1 Photochemical Stability of Chlorophyll and Its Metal Analogues in Solution

Complex formation involving metal salts and organic molecules acting as polydentate ligands often results in stabilization of the latter with respect to various chemical factors and in itself is an important factor of stabilization of dyes and pigments with respect to acid treatment (see Chapter 5). Such stabilization becomes particularly pronounced in all cases where the central metal atom forms, together with organic ligands, covalent complexes and does not perceptibly distort the geometric configuration of the ligand during coordination.

A typical example of such stabilization is phthalocyanine and its covalent complexes. In some cases, during dissociation of the complex, the ligand molecule may be released in an active form which is much more reactive than the uncoordinated ligand itself. Then the complex may be less chemically stable than the initial ligand. This is vividly illustrated by the decomposition of chlorophyllic acid and its covalent complexes in strong proton-donor media.[377,376]

The coordinated metal in the complex drastically changes the photochemical properties of organic ligands.[9,468–471] In this section I shall show the effect of the nature of the coordinated metal and the medium on the photochemical stability of chlorophyll and its metal analogues. Qualitative studies into the photochemical decomposition of several complexes of chlorophyllic acid (metal phaeophytins) have been conducted by Kukhtevich.[472] On the basis of these experiments it has been concluded that there is a complete parallelism between the thermodynamic and photochemical stability of metal phaeophytins and that complexes of chlorophyllic acid *b* are more stable to discoloration than those of chlorophyllic acid *a*.

An earlier study into the photochemical stability of H_2Chla and H_2Chlb, as well as their ionic and covalent complexes,[473] has led to the treatment of the findings of the above work as arbitrary. The photochemical decomposition of chlorophyllic acids and their complexes with metals was found to be accompanied only by a lower intensity of all absorption bands (Fig. 8.1). The shape of absorption spectra is not changed. The

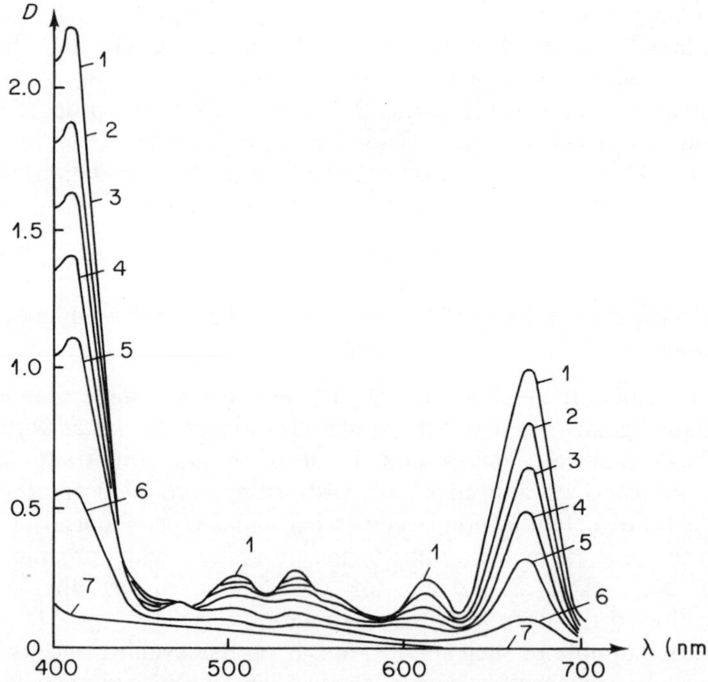

Figure 8.1 Photochemical decomposition of chlorophyllic acid *a* in glacial acetic acid at 20 °C, represented on a time scale. Time: curve 1, 0 h; curve 2, 10 h; curve 3, 20 h; curve 4, 35 h; curve 5, 55 h; curve 6, 125 h; curve 7, 150 h

end-products of photodecomposition are colourless. The rates of decomposition in ethanol and glacial acetic acid, as well as in their mixtures, are reliably characterized by the first-order rate constant equation

$$k_\nu^{h\nu} = (2.30/\tau)\log(c_{\mathrm{MChl}}/c_{\mathrm{MChl}}). \tag{8.1}$$

These constants remain invariable all the way to complete decomposition of H_2Chl and its complexes. Their values are taken from earlier work[473] and are listed in Table 8.1.

A wealth of data has been accumulated,[13,22,474] which indicate that the photochemical decomposition of complex organic pigments of the porphyrin type in solution is associated with their oxidation. The extremely low temperature coefficients of the reaction (see Table 8.1) are possibly due to the fact that the limiting step of the reaction is the interaction between the excited molecule and the oxygen present in the solution. Two paths of the elementary act of photodecomposition may be assumed:[475]

(1) Extra complexes $MChl \cdot O_2$ are subjected to photochemical

Table 8.1 Rate constants (k, h^{-1}) of photochemical decomposition of chlorophyll and its metal analogues

Compound	20 °C	40 °C	60 °C
Ethanol			
H_2Chl*a*	Very slow	Very slow	Very slow
MgChl*a*	0.400 ± 0.028	0.453 ± 0.040	0.588 ± 0.025
MgChl*b*	0.297 ± 0.008	0.402 ± 0.018	0.563 ± 0.010
ZnChl*a*	0.0350 ± 0.0010	0.0530 ± 0.0010	0.0815 ± 0.010
CdChl*a*	0.460 ± 0.044	0.466 ± 0.035	0.493 ± 0.055
CdChl*b*	0.744 ± 0.050	0.766 ± 0.070	—
HgChl*a*	0.0389 ± 0.001	0.0483 ± 0.003	0.0645 ± 0.0020
CoChl*a*[a]	—	—	0.0250 ± 0.0010
Glacial HAc			
H_2Chl*a*	0.0185 ± 0.0002	0.0388 ± 0.0004	0.0814 ± 0.0006
H_2Chl*b*	—	—	0.0191 ± 0.0006
ZnChl*a*	0.205 ± 0.020	0.304 ± 0.040	0.395 ± 0.035
ZnChl*b*	—	—	0.140 ± 0.010
ZnChl*a*[b]	0.063 ± 0.001	0.124 ± 0.002	0.193 ± 0.009
ZnChl*a*[c]	—	0.120 ± 0.003	—
CuChl*a*	0.0101 ± 0.0002	0.0147 ± 0.0009	0.0240 ± 0.0008
PdChl*a*	—	—	0.0133 ± 0.0005
NiChl*a*	—	—	0.00453 ± 0.0002
CoChl*a*[a]	—	—	0.0060 ± 0.0005
(Ac)FeChl*a*	—	—	0.0040 ± 0.0001

[a]The constant was determined with subtraction of the rate of the dark reaction in ethanol, $k_\nu(\text{dark}) = 0.021\ h^{-1}$, and in HAc, $k_\nu(\text{dark}) = 0.0083 \pm 0.0005\ h^{-1}$.

[b]Ethanol + HAc (1:1) mixture.

[c]Ethanol + HAc (1:1) mixture with a 10-fold excess of salt $ZnAc_2$.

decomposition. There are indications that the associates of molecule O_2 coexist with porphyrins and phthalocyanines.[476–478] The absorption of a light quantum brings $MChl \cdot O_2$ into one of the active electronic states: singlet S_{n,π^*}, S_{π,π^*} or triplet T_{n,π^*}, T_{π,π^*}.[479] This is followed by electron transfer from the conjugated system to O_2 with subsequent opening of the macroring.

(2) The O_2 molecule reacts with the MChl molecule brought, as a result of excitation, into a triplet state. The energy of the excited pigment is transmitted to the O_2 molecule which passes into an active metastable state:

$$\mathrm{MChl}[\uparrow][\uparrow] + \tilde{\mathrm{O}}_2[\downarrow\downarrow] \rightarrow \mathrm{MChl}[\uparrow\downarrow] + \tilde{\mathrm{O}}_2[\downarrow\uparrow]. \tag{8.2}$$

The $\tilde{O}_2$ molecule has a vacant quantum cell and acts as an oxidizing agent:

$$\mathrm{MChl} + \tilde{\mathrm{O}}_2 \rightarrow (\mathrm{O}_2^-)^{\cdot} + (\mathrm{MChl}^+)^{\cdot} \overset{\mathrm{HA}}{\rightarrow} \mathrm{HO}_2^{\cdot} + (\mathrm{MChl}^+)^{\cdot}. \tag{8.3}$$

The loss of a π electron by the conjugated molecule brings about a decrease in aromaticity and stability with the result that cation radical $(MChl^+)^{\cdot}$ rapidly dissociates into colourless products.

Wasser and Fuhrhop[485] propose a different scheme of photooxidative decomposition of the macroring of metalloporphyrins and metallochlorins, according to which a cyclic peroxide compound is formed first, then the macroring is opened as follows:

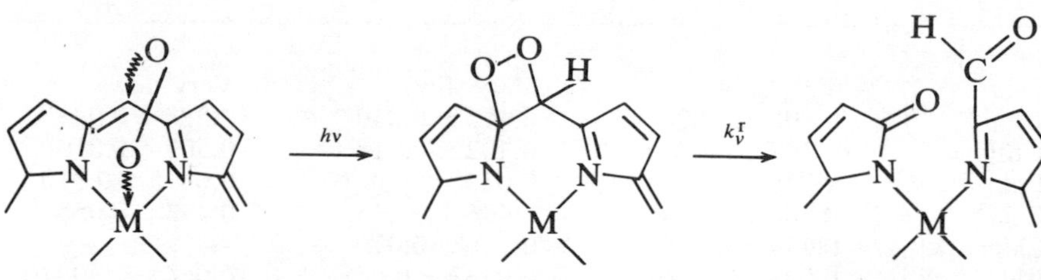

(8.4)

The subsequent photodecomposition of the open conjugated system is rapid and the limiting step must almost certainly be the ring opening stage (k_v^r). The transformations according to scheme (8.4) are possible through initial formation of cation radical of porphyrin. In the case of ionic and weakly covalent complexes both mechanisms may be assumed. This may account for the low photodecomposition rates of uncoordinated ligands and the sharp increase in these rates in coordinatively unsaturated complexes CdChl, MgChl, ZnChl, and HgChl.

In methanol chlorophyll *b* is photochemically more stable than chlorophyll *a*, whereas in the case of CdChl the photodecomposition rates are in an opposite ratio. The photodecomposition rate in ethanol does not depend on the covalence of the complexes. The highly covalent chlorophyll *a* and its ionic cadmium analogue are characterized by the same rates and temperature dependences. At the same time HgChl, which is one of the most labile analogues of chlorophyll, is almost as stable photochemically as the covalent ZnChl complex.

In contrast to ethanol the photodecomposition of the H_2Chl ligands themselves in glacial HAc is fast. This may be ascribed to the greater lability of protonated forms H_3Chl^+ and H_4Chl^{2+} which form in very small concentrations in glacial HAc. H_2Chl*b* is protonated to a lesser degree than H_2Chl*a* and its decomposition is slower. Of course, protonation is not the only responsible factor, although it brings the molecules closer to the oxidizing agents, having antibonding electron pairs. More important factors are the intramolecular processes of excitation energy transfer, properties of the singlet and triplet excited states, and the effect of substituents on these properties.[480–486] The zinc complex of chlorophyllic acid *b* is also more stable to photodiscoloration than ZnChl*a*. Addition of alcohol to its solutions in HAc slows down the discoloration, while addition of an excess amount of a metal salt (ZnAc) does not produce any effect.

The data of Table 8.1 permit tracing of the effect of the cation species and complex stability on the photodecomposition rates in HAc. They suggest that there is no strict correlation between the thermodynamic and kinetic stability (covalence) of the complexes on the one hand and their photochemical stability on the other. If the covalence (stability) series is AcFeChl, PdChl > NiChl > CuChl > CoChl > ZnChl, MgChl > CdChl $\gg$ HgChl, in glacial HAc the photodecomposition rates increase as follows: AcFeChl, NiChl < CoChl < PdChl < CuChl < ZnChl < HgChl < MgChl < CdChl. In these series Ni, Co, Pd, Cu, Hg, Mg, and Cd occupy different positions.

These data clearly suggest that the formation of ionic complexes (increase in electron density in the aromatic system) promotes photodecomposition of the chlorophyllic acid ligand. The formation of covalent compelxes (lower electron density in the macroring) practically always (except for Zn^{2+}) minimizes photodecomposition. Probably the degree of covalent bonding of the metal, the photodecomposition rate and capacity for luminescence are in a certain relationship with one another. The most covalent complexes of porphyrins with Fe^{2+}, Pd^{2+}, Ni^{2+}, Cu^{2+}, and Co^{2+} are not capable of luminescence, which is probably due to the rapid degradation of the excitation energy from the singlet levels, caused by radiationless transitions[482] and transitions to a triplet state (PdChl). Not only photodecomposition but also other photochemical properties of metalloporphyrins, for example capacity for photoreduction,[13] depend on the covalent bonding of the metal. Porphyrin complexes with Mg^{2+}, Cd^{2+}, and Zn^{2+} are capable of photoreduction, whereas the highly covalent complexes with Cu^{2+}, Ni^{2+}, and other metals do not have this capacity.

Despite the numerous studies into the photo-oxidative and photoreductive properties of chlorophyllic and other metalloporphyrin systems,[10–12,15,19] the available data are not sufficient for a better understanding of the relationship between the structural features of metal ions, porphyrin molecules, the medium, and the photochemical activity of metalloporphyrins in oxidation–reduction reactions. Extensive systematic research into the photochemical properties of metalloporphyrins with due account for structure are called for, along with spectroscopic studies into the intramolecular energetics of these molecules.

8.2 Oxidation–Reduction Reactions

The oxidation–reduction reactions, just as is photochemical decomposition, are among the most important reactions involving porphyrins and their complexes. This importance stems from the conditions under which metalloporphyrins perform their functions in biological systems (functions of energy storage, oxygen transport, oxidation, decomposition of H_2O_2).

The use of metalloporphyrins and metal phthalocyanines in commercial catalysis of many oxidation and reduction reactions, which will be discussed in the next section, is another factor making studies into the oxidation–reduction properties and transformations imperative.

Porphyrins and phthalocyanines exhibit low normal oxidation–reduction potentials (Table 8.2), which is why they are thermodynamically unstable systems in oxidizing media.[9–13,23] Their oxidation potential depends on the nature of the aromatic system (porphyrin, chlorin, azaporphyrin, benzoporphyrin), the functional substituents in pyrrole and meso positions, the central metal atom, and even the extra ligands coordinated along the z axis.

The thermodynamic data on redox potentials, which could elucidate the effect of the above structural factors, are scant. Table 8.2 suggests that the benzo substitution in porphyrins and the coordination to metals perceptibly destabilize the molecule with respect to the oxidizing agents. The increasing degree of π bonding of the metal (Cu > Ni > Pd, Pt) produces an opposite effect. The values of E_o are not in a simple relationship with

Table 8.2 Standard redox potentials (E_o) of phthalocyanine, porphyrins, and their complexes

Compound	E_o (mV)	pH	Compound	E_o (mV)	pH
H_2Pc	714 ± 5	7	$(Py)_2FeMP$	−63	9.6
H_2TAP	752	7	$(Py)_2FeEP$	−29	9.6
ZnPc	342 ± 4	7	$(Py)_2FeCP$	−36	9.6
CuPc	642 ± 7	7			
NiPc	668 ± 5	7	$(Py)_2FeHP$	4	9.6
PdPc	718 ± 3	7	$(Py)_2FePP$	15	9.6
PtPc	725 ± 5	7			
$(CN)_2FeMP$	−229	9.6			
$(L)_2FePP$[a]	−33	9.6	$(CN)_2FePP$	−183	9.6

[a] L = α-picoline.

the complex species and the nature of the coordinate bond, therefore they do not correlate as a rule either with the electronic absorption spectra, or the thermodynamic ionization and protonation constants or porphyrins, or the stability of the complexes, or their kinetic behaviour in oxidizing media.

As far as the relationship between E_o and the nature of the ligands entering into extra coordination to metalloporphyrins such as FeP (see Table 8.2) is concerned, one sees a strong tendency of E_o to decrease and the system to be oxidatively destabilized as the dative properties of the extra ligand become more pronounced ($CN^- > \alpha$-picoline $>$ Py).

The lack of systematic studies with measurements of the standard redox potentials of porphyrins and their complexes under comparable experimental conditions and in a broad range of structures, which is a consequence of enormous difficulties encountered in such experiments, compels us to outline the problem briefly from the kinetic viewpoint.

The early quantitative observations by Linstead have shown that the transition from azaporphrin to porphyrin molecules results in a sharp increase in their resistance to oxidizing agents. Linstead did not provide any explanation of this phenomenon. In later works[163,164] it was associated with the increase in the total π-electron charge in the macroring during aza substitution, while in other works[212,213] it was ascribed to the intimate mechanism of the oxidation reaction, particularly to the different localization of the oxidation site in phthalocyanines and porphyrins.

Quantitative measurements of the rates and other kinetic parameters of oxidation reactions involving phthalocyanine and its complexes,[212] chlorophyllic acid and chlorophyll analogues,[213] and polymeric phthalocyanine complexes were carried out for the first time in an earlier work.[149] Given in Table 8.3 are the effective oxidation rate constants of H_2Pc and its complexes oxidized by hydrogen peroxide in 17 M H_2SO_4 (the

Table 8.3 Rates of oxidative decomposition of phthalocyanine and its complexes by hydrogen peroxide in 17 M H_2SO_4

Complex	$c^0_{H_2O_2}$ (M)	k_{eff} (h^{-1})	Complex	$c^0_{H_2O_2}$ (M)	k_{eff} (h^{-1})
H_2Pc	0.1195	6.34	(Cl)AlPc	0.1525	2.17
MgPc	0.1662	7.56	(Cl)CrPc	0.1380	2.20
PbPc	0.1543	6.00	O—OsPc	0.1531	2.71
CaPc	0.1662	7.00	O—MnPc	0.1662	6.20
CdPc	0.1615	7.49	CoPc	0.0160	2.90
Ag_2Pc	0.1662	10.10	NiPc	0.0162	25.50
HgPc	0.1662	8.82	PdPc	0.0202	4.23
$(Cl_2)SnPc$	0.1644	1.36	PtPc	0.0199	15.05
CuPc	0.1624	0.402	(Cl)IrPc	0.0214	4.69
$CuPcCl_{15}$	0.1662	0.402	(Cl)RuPc	0.0214	37.10
ZnPc	0.1536	0.665	(Cl)RhPc	0.0214	21.60
(Cl)GaPc	0.1525	2.13	O—VPc	0.0199	30.40

phthalocyanine concentration was (3–9) × 10^{-4} M) at 25 °C. Labile metal phthalocyanines (Mg, Pb, Ca, Cd, Ag, Hg) feature the same stability as the ligand itself. Ions Ag^+ and Hg^{2+} catalyse oxidation.

Phthalocyanine complexes with a σ bond (Sn^{4+}, Zn^{2+}, Ga^{3+}, Al^{3+}) oxidize three to 10 times more slowly than H_2Pc. The greatest stability toward H_2O_2 is exhibited by CuPc. The chlorination of phthalocyanines in the benzene rings does not affect the oxidation rate. Co^{2+}, Ni^{2+}, and platinum metals drastically activate the oxidative decomposition of the phthalocyanine coordinated to them, which becomes five to 50 times faster as compared to H_2Pc. At H_2O_2 concentrations exceeding 0.05 M the rate of their oxidation cannot be measured at all. When H_2O_2 solutions are mixed with sulphuric acid solutions of phthalocyanine complexes with 'active' metals (Co, Ni, and platinum metals), the colour changes immediately from yellowish-green to deep dark green, then the solution becomes colourless after 1 min.

Active metals probably promote the formation of compounds similar to the hydroxyporphyrins obtained by Fisher by oxidizing porphyrins with hydrogen peroxide in concentrated H_2SO_4.[25] The latter compounds are essentially products of attachment of the hydrogen peroxide elements $(HO)_2$ to the pyrrole double bonds (in the case of free porphyrins) or, in the case of complexes, their incorporation at the meso and adjacent positions. In phthalocyanine complexes the attachment of H_2O_2 is possible at the aza position and the one next to it as the conjugated system passes from a cyclic to an open one.[212] At a low H_2O_2 concentration, or irrespective of it in the case of phthalocyanines of 'inactive' metals, no dihydroxy compounds are formed.

Oxidative decomposition proceeds[212,213] according to a general pattern through consecutive formation of intermediate complexes, then a charge-transfer complex, and cation radicals of phthalocyanine:

$$\mathrm{MPcH^+ + A{-}B \underset{}{\overset{k_1}{\rightleftharpoons}} MPcH^+ \cdots B{-}A \overset{k_2}{\rightleftharpoons} MPcH^+ \cdots B \overset{e}{\vdots} A}$$

$$\underset{\text{slow}}{\overset{k_3}{\rightleftharpoons}} \mathrm{(MPc^+)^{\bullet} \cdots BH + A^{\bullet}} \underset{\text{fast}}{\overset{k_4}{\longrightarrow}} \text{colourless products.} \qquad (8.5)$$

The rate of formation (k_1) of intermediate complex $MPcH^+ \cdots B{-}A$ is fast as a result of hydrogen bonding, because $MPcH^+$ is a much stronger acid than H_2SO_4[37] and all of the oxidizing agents examined (H_2O_2, HNO_3, $S_2O_8^{2-}$, $H_2Cr_2O_7$, $HBrO_3$) have antibonding electron pairs. This is immediately ($k_2 \gg k_1$) followed by formation of a charge-transfer complex with the charge being transferred from the phthalocyanine orbitals to the antibonding orbital of the oxidizing agent A—B. The next stage of formation of the phthalocyanine cation radical requires activation and is the slow (k_3) limiting step of the entire process. Solvated cation radical $(MPc^+)^{\bullet} \cdots BH$, in which BH is usually hydrated, is an extremely active unstable compound and rapidly disintegrates into phthalocyanine molecule fragments (phthalimide and its derivatives).

The above-described scheme is consistent with the experimentally established second order of the reaction of oxidation of MPc in a sulphuric acid solution with H_2O_2, $S_2O_8^{2-}$, and NO_3^- and with the absence of cation radicals and other partially oxidized forms. It may be assumed that the specific behaviour of phthalocyanine complexes with 'active' metals stems from the fact that the central atoms of Co, Ni, Pd, Pt, Ir, Ru, Rh, and V promote electron transfer towards the oxidizing agent as a result of the additional interaction with H_2O_2 involving its own *d* orbitals. However, it remains unclear why such active 'catalases' as (Cl)RuPc and O=OsPc fall into different groups. Os(IV) is found to be among repressors and Ru(III) among the strongest activators of oxidative decomposition of phthalocyanine. The group of 'active' central atoms does not include all metals apt to form strong dative π bonds (e.g. Os and Cu are not included). The 'active' metals include V(IV), which is incapable of coordinate π bonding of the M → Pc type, as well as Co, which forms only weak π bonds. Nevertheless, all 'active' metals are capable of changing their degree of oxidation in MPc. Therewith, judging from the values determined for metal octaethylporphyrins,[488] the half-wave potentials associated with transition CoOEP ± $\overset{+}{\text{Co}}$OEP ($E_{1/2}$ = +0.13) are low and much smaller than those for 'inactive' metals, that is $\overset{+}{\text{Cr}}$OEP → $\overset{2+}{\text{Cr}}$OEP ($E_{1/2}$ = +0.79). At the same time it should be pointed out that formation of cationic forms with delocalization of the positive charge over the entire macroring is observed in all porphyrin complexes, Mg^{2+} and Zn^{2+} complexes being no exception.[237] The cationic forms feature electronic spectra substantially different from those of the starting complexes. For example, the first band of $\overset{+}{\text{Mg}}$OEP ClO_4^- is bathochromically shifted by 100 nm, while the absorption spectrum of $\overset{2+}{\text{Mg}}$OEP $(ClO_4)_2$ features only the Soret band.

At present one can hardly provide an answer to the question of why some *d* metals activate porphyrin ligands in chemical and catalytic processes while others do not. These complexes have to be studied systematically using advanced physical methods such as magnetochemistry, EPR, ESCA, luminescence, and others.

A study[212] into the rates of oxidation of CuPc with hydrogen peroxide in the presence of the Fenton reagent and methyl methacrylate has shown that the main steps of reaction (8.5) are not chain processes involving free radicals. Free radicals $OH^{\bullet}$ emerging during oxidation at the limiting step are deactivated on the phthalocyanine decomposition products. It is also certain that the reactive site of oxidation is localized in the case of phthalocyanine at one of the peripheral nitrogen atoms, which readily enters into hydrogen bonding with the oxidizing agent after protonation in H_2SO_4, thereby creating a path for electron transfer. Table 8.4 lists the rates of oxidation of phthalocyanine and its complexes with persulphate and nitrate in H_2SO_4.

Comparison of the data presented in Tables 8.3 and 8.4 suggests that, in so far as their rates of oxidation with a persulphate and a nitrate are

Table 8.4 Rates of oxidation of phthalocyanine and its complexes with persulphate and nitrate in 17 M H_2SO_4

Complex	$c^0_{(NH_4)_2S_2O_8} \times 10^3$ (M)	k^{25}_{eff} (h^{-1})	Complex	$c^0_{KNO_3} \times 10^3$ (M)	k^{25}_{eff} (h^{-1})
H_2Pc	46.6	4.60	H_2Pc	0.665	Very fast
MgPc	60.0	2.63	ZnPc	7.77	1.10
(Cl)GaPc	60.0	1.53	CuPc	7.38	0.437
(Cl)AlPc	60.0	2.07	(Cl)GaPc	7.38	0.136
O=OsPc	60.0	1.29	O=OsPc	8.20	2.49
CoPc	5.82	7.88	Cl_2SnPc	7.38	0.072
NiPc	5.82	50.0	NiPc	7.26	14.8
PdPc	5.82	58.1	PdPc	0.665	24.2
(Cl)IrPc	5.82	13.3	(Cl)IrPc	0.665	13.5
			(Cl)RuPc	7.06	16.3

concerned, metal phthalocyanines can be divided into two groups: passive complexes with oxidation rates lower than in the case of the ligand and approaching the rates of oxidation with H_2O_2, and active complexes whose rates of oxidation with a persulphate are five to 50 times higher than with H_2O_2 and scores of times higher than those of passive MPc. The oxidation reaction is of first order with respect to phthalocyanine and first order with respect to the $S_2O_8^{2-}$ ion. Oxidation with a nitrate in H_2SO_4 is also characterized by being second order in both cases.[213] The oxidation rates are practically the same as with H_2O_2 and $S_2O_8^{2-}$. These facts indicate that the mechanism of oxidation with oxygen-containing inorganic oxidizing agents in a strongly acidic medium is the same and can be described by scheme (8.5). In all cases the limiting step of the process is the formation of practically the same cation radical. The oxidation site is always localized at the protonated peripheral nitrogen atom. The only exception is nitrate oxidation of H_2Pc. The extremely high rates of its oxidation, which could not be measured, show that in this case the central imino groups may perform the function of the reactive site.

Characteristically, the group of active MPc is constituted by the same metals regardless of the oxidizing agent. The impossibility of these MPcs forming extra complexes with NO_3^- or HNO_3 may serve as a convincing proof that the unusual state of the molecular orbitals of the MPc macroring is responsible for 'activation' of phthalocyanine by metals rather than the specific interaction between the oxidizing agent and the central atom.

The independence of the phthalocyanine oxidation rate of the H_2SO_4 concentration (14–17 M H_2SO_4) indicates that all of the phthalocyanine molecules in the H_2SO_4 solution are in the $MPcH^+$ state and that the protonation depth of MPc (see Chapter 4) does not affect the electron transfer rate. For $CuPcH^+$ to be oxidized with a nitrate the activation energy remains equal to 15.7 ± 0.3 kcal mol^{-1} and $\Delta S^{\ddagger} = -24$ e.u. Vulfson[487] has obtained a similar value of the activation energy

(16.1 kcal mol^{-1}) for oxidation of tetrabutyl-H_2Pc with benzoyl peroxide in nitrobenzene. He has established that the oxidation of H_2Pc, CuPc, and ZnPc is of first order with respect to both components and an overall third order in the case of NiPc, PdPc, and PtPc—first order with respect to the oxidizing agent (benzoyl peroxide and its benzene substituents) and second order with respect to the metal phthalocyanine.

The bimolecular reaction scheme proposed by Vulfson[487] is the same as (8.5). For a trimolecular scheme a second MPc molecule is introduced at the step of splitting of the charge-transfer complex. The rate constants of the bi- and trimolecular kinetic equation for H_2Pc, CuPc, NiPc, ZnPc, PdPc, and PtPc equal, respectively, 80, 1870, 250, 70 000, 43, and 21 h^{-1}.

The use of a different oxidizing medium drastically changes the interaction. If, for example, the rates of oxidation of H_2Pc in H_2SO_4 with H_2O_2 (true k_v = 53 h^{-1}) and in nitrobenzene with benzoyl peroxide (k_v = 80 h^{-1}) are almost the same, those of MPc and the order in which metals are arranged have nothing in common. This is a manifestation of the extremely important role of the nature of the solvent as the physicochemical medium (in which porphyrins react), which has been pointed out many times in this monograph.

In nitrobenzene the 'active' complexes include those with Cu^{2+} and Zn^{2+}, while the Pt^{2+} and Pd^{2+} complexes have been found to be passive. It is believed that this is due to the different localization of the oxidation site.

Listed in Table 8.5 are the oxidation rates[213] of chlorophyllic acid (chlorophyll behaves in a similar fashion) and metal analogues of chlorophyll with Cu^{2+}, Co^{2+}, Ni^{2+}, and Fe^{3+}, oxidized with potassium nitrate in glacial HAc.

Table 8.5 Rate constants of oxidation of chlorophyllic acid and its complexes with potassium nitrate in glacial acetic acid

Compound	$c^0_{NO_3^-} \times 10^3$ (M)	k_{eff} (h^{-1})		
		25 °C	35 °C	45 °C
H_2Chl	6.95	—	0.180	—
	6.83	0.067	0.160	0.387
	3.53	—	0.176	—
	1.78	—	0.171	—
CuChl	6.47	—	—	0.138[a]
CoChl	6.47	1.07	7.16	—
	3.17	—	3.67	—
NiChl	6.47	0.114	0.258	1.04
(Ac)FeChl	3.47	0.085	—	—

[a]For CuChl k_{eff} at 55 and 65 °C equals 0.271 and 0.627 h^{-1}, respectively.

The oxidation rates of H_2Chl are low and at the specified NO_3^- concentration are independent of the latter. This may imply that H_2Chl is entirely in the state of intermediate complex $H_2Chl \cdots NO_3^-$, which decomposes monomolecularly in accordance with (8.5). The formation of complexes with all of the above-mentioned metals, except for Cu^{2+}, speeds up the oxidation with the nitrate. In this case Co^{2+} is particularly active. The molecules of chlorophyllic acid (**XXVII**, **XXVIII**) and its complexes contain several centres exhibiting reductive properties: the isocyclic ring, the hydrogen atoms of the hydrogenated bond in the IV pyrrole ring, the vinyl group, and the macroring. However, the discoloration may be caused only by opening of the macroring. The oxidation of these molecules at the periphery may result only in a bathochromic shift of the absorption bands.[61]

This is why the optical density in the H_2Chl (MChl)–oxidizing agent system varies in a particular manner (Fig. 8.2). Due to oxidative splitting of the cyclopentanone ring it increases at first (stage I), then gradually diminishes (stage II). Stage II is determined by opening of the macroring. In complexes stage I is practically non-existent.

The oxidation of H_2Chl and its complexes proceeds at a measurable rate only at elevated temperatures. It is of first order with respect to H_2Chl as well as with respect to H_2O_2.[213] The rates of oxidation with a persulphate are twice as high. The oxidation rates of alkyl-substituted porphyrins seem to be much lower than those of phorbins. The oxidation of H_2EP with benzoyl peroxide is characterized by $k_v = 1.1\ h^{-1}$,[487] this rate being 80 times lower than in the case of H_2Pc and 500 times lower than in the case of H_2TBP. According to Vulfson[487] the oxidation of H_2TAP is the slowest ($k_v = 1.4 \times 10^{-2}\ h^{-1}$). No data are available on the oxidation rates of complexes with unsubstituted porphyrins.

In an earlier work[149] the oxidation rates of a polymeric phthalocyanine with a linear structure (see Fig. 2.2 between lines AA and BB) and its complexes with Cu^{2+}, Zn^{2+}, Ga^{3+}, and Os(IV) in 17 M H_2SO_4, and oxidized with hydrogen peroxide, a nitrate, and a persulphate, were measured. It was shown that the oxidation of each chain unit of the polymer is in

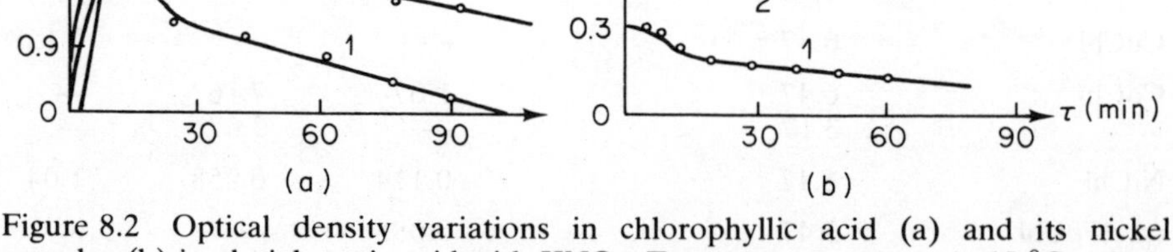

Figure 8.2 Optical density variations in chlorophyllic acid (a) and its nickel complex (b) in glacial acetic acid with KNO_3. Temperature: curves 1, 45 °C; curves 2, 35 °C; curves 3, 25 °C

accordance with (8.5) plus some specific behaviour typical of polymers. The first-step rate constants of oxidative decomposition of polymeric complexes are several times higher than those of monomeric complexes. The final-step rate constants practically coincide with those of monomers. Thus it has been established that the oxidation rate per chain unit increases with the degree of polymerization.

In another work[149] the mechanism of oxidative decomposition of polyphthalocyanines is presented along with interpretation of their specific behaviour.

In a recent work[489] the derivatographic method was used in studying the thermo-oxidative decomposition of phthalocyanine and its 12 complexes, as well as polymeric phthalocyanines of two different types: $(CoPc)_i$ and $H(OSiPc)_iOH$. The investigators arrived at a conclusion that the oxidation of phthalocyanines with air oxygen involves the central atom, which may be the most probable site of oxidation. Polymeric phthalocyanines have been found to be more stable to thermodecomposition.

In recent years a great deal of interest has been aroused in reactions of reduction of metal phthalocyanines and metalloporphyrins to negatively charged ionic forms which are partially or completely protonated. However, despite the enormous interest in this field of porphyrin chemistry, it goes well beyond the scope of the present monograph. The reader is referred to the recently published review by Sidorov and Maslov.[490]

8.3 Catalytic Activity

Metal phthalocyanines and metalloporphyrins are highly effective catalysts of many chemical reactions. FePc has long been known[491] to exhibit a high catalase activity as well as activity in the oxidation reactions involving certain carbocyclic compounds (benzaldehyde, tetralin, diphenyl methane, cyclohexene). As was established later,[492,493] NiPc is also catalytically active in the reactions of oxidation of saturated ketones, fatty acids, and alkyl aromatic hydrocarbons with oxygen, proceeding under more rigorous conditions (125 °C). Kropf spent several years in the study of the mechanism of oxidation of cumene with oxygen in phthalocyanine complexes,[494–496] many of which turned out to be effective catalysts.

Metal phthalocyanines were found to be catalytically active in the reactions of hydrogen exchange during the *n*-hydrogen → *o*-hydrogen conversion. A review of earlier works concerning the catalytic activity of phthalocyanines can be found in a book by Moser and Thomas.[32] Later Roginsky *et al.*[155,497] studied the catalase and oxidase properties of polymeric copper phthalocyanine, and the polymer was shown to be much more active than the corresponding monomeric compound. Japanese investigators[157,498] have established the selective catalytic activity of mixed complexes ($Cu^{2+} + Fe^{2+}$, $MoO_2^{2+} + Fe^{2+}$) of polymeric phthalocyanine in oxidation reactions, which have been found to exhibit synergism. The monometallic complexes of polymeric phthalocyanine and their mixtures were shown not to be active in the same reactions.

Tanaka *et al.*[499] examined the decomposition of HCOOH in MgPc and FePc in the temperature range 200–280 °C. The reaction involves both dehydrogenation to $CO_2 + H_2$ and dehydration to CO and H_2O. The course of the reaction depends on the metal species. It has been shown that in MgPc coordination $HCOO^- \rightarrow Mg$ takes place with simultaneous protonation of the extracyclic nitrogen atom. At the same time studies into the decomposition of gaseous formic acid to CO_2 and H_2 in Mg^{2+}, Co^{2+}, Ni^{2+}, Cu^{2+}, and Zn^{2+} phthalocyanines have revealed that the nature of the central atom does not produce any effect.[500]

It has been established[500] that the central atom exerts a strong influence during catalytic oxidation of cysteine and other H_2S derivatives. Only FePc and CoPc were shown to be catalytically active in these reactions, which are of first order with respect to the catalyst. The activation energies were found to equal 7.1 kcal mol^{-1} for CoPc and 4.2 kcal mol^{-1} for $CoPc(SO_3\overset{+}{N}a)_4$. Salt KCN was shown to be a catalyst poison.

In an earlier work[55] the present author studied the kinetics of H_2O_2 decomposition reactions on scores of phthalocyanine complexes and their polymers,[54,463,502] as well as the kinetics of oxidation of cumene in ruthenium and osmium phthalocyanines.[501] The catalase properties were found to be strongly dependent on the nature of the central metal atom and in some cases less dependent on the polymeric state.

The catalase activity of FePc on support $BaSO_4$ was studied by Balandin *et al.*[503] and that of the sulpho derivatives of FePc and CoPc by Waldmeir and Sigel.[504]

Many works have been concerned with the catalytic properties of ionic forms of Fe, Co, Ni, Mo, and Pt phthalocyanines with metallic sodium. These forms were shown[505] to catalyse such reactions as formation of NH_3 and H_2 and N_2 under mild conditions and formation of lower hydrocarbons from a mixture of CO and H_2. In this case the result of the reaction does not depend on the central atom. Other examples of such reactions can be found in a review by Sidorov and Maslov.[490]

In recent years much attention has been given to studies into the electrocatalytic properties of phthalocyanine and porphyrin complexes which are used to activate the processes of cathodic reduction of oxygen with hydrogen in fuel cells.[57,506–508] It should be noted that the catalytic properties of porphyrin complexes are practically unstudied because of the deficiency of these compounds and their poorer thermodynamic and kinetic stability as opposed to phthalocyanine complexes. It is believed that these complexes are highly promising in artificial enzyme systems.

In the present monograph there is no possibility of discussing this extremely important problem in its entirety. I shall only dwell on certain aspects of the mechanisms of reactions in phthalocyanine complexes and the effect of the central atom. The available data suggest that the role of the central atom is predominant. As was shown in an earlier work,[55] the catalase activity is exhibited only by complexes of Fe, Ru, and Os, that is metals which are at the beginning of each of the triads of Group VIII of the periodic table and have the same electronic structure of the valence shell in the free state. Their activity increases perceptibly as one goes from Fe down to Os within the group (Table 8.6). The activity of phthalocyanines of other metals, including those of the platinum group, is measured by the effective rate constant calculated from the integral equation of second order (with respect to H_2O_2)

$$k_{eff} = (1/\tau)[x/c_0(c_0 - x)], \tag{8.6}$$

where x is the amount of the decomposed peroxide.

Under experimental conditions this value is two orders of magnitude lower, on the average, than in the case of osmium and ruthenium phthalocyanines. The low activity may be the consequence not of the presence of these phthalocyanines proper but the possible presence of FePc traces as well as free radicals in MPc, resulting from bond breakage in the course of synthesis of phthalocyanines under rigorous conditions. These traces are easily recognizable in the EPR spectra of diamagnetic phthalocyanine complexes.[509]

The mechanism of the catalase reaction of iron, ruthenium, and osmium phthalocyanines may be similar to the action of a natural catalase.[29] At the first step the H_2O_2 molecule probably enters into a coordinative interaction with the central metal atom and becomes an extra ligand (see Chapter 6). In this case H_2O_2 is attached to the FePc and (HSO_4)RuPc complexes at the vacant fifth coordinate place, while in the case of $(HSO_4)_2$OsPc the HSO_4^- extra ligand must be substituted in the equilibrium process:

$$(HSO_4)_2OsPC + H_2O_2 \rightleftarrows HSO_4^- + (H_2O_2)OsPc(HSO_4). \tag{8.7}$$

Table 8.6 Catalase activity of various phthalocyanine complexes under heterogeneous conditions

Complex	$c^0_{MPc} \times 10^6$ (M)	k_{eff} (l mol^{-1} min^{-1})	$\tau_{1/2}$ H_2O_2 (min)
$(HSO_4)_2$OsPc	0.506	$(5.3 \pm 0.2) \times 10^{-2}$	45
(HSO_4)RuPc	0.625	$(4.5 \pm 0.6) \times 10^{-2}$	54
FePc	0.580	$(1.0 \pm 0.4) \times 10^{-2}$	240
(HSO_4)IrPc	0.550	$(6.5 \pm 0.9) \times 10^{-4}$	3 800
(HSO_4)RhPc	0.625	$(5.9 \pm 0.7) \times 10^{-4}$	4 100
$(HSO_4)_2$VPc	0.596	$(4.1 \pm 0.6) \times 10^{-4}$	6 100
PdPc	0.653	$(5.9 \pm 0.5) \times 10^{-4}$	4 100
CuPc	0.705	$(1.7 \pm 0.3) \times 10^{-4}$	14 700

Note: $c^0_{H_2O_2} = 0.412$ M; pH = 0.1; $t = 40$ °C.

The occurrence of such an equilibrium is supported by the catalase function of Fe, Ru, and Os phthalocyanines being inhibited by addition of catalyst poisons NH_3, HCN, and HF, as well as by the diminishing catalase activity in a homogeneous acetone solution stabilizing the extra complexes of Ru with HSO_4^-.[55] At the second step the peroxide extra complex reacts bimolecularly with the second H_2O_2 molecule:

$$(H_2O_2)MPc + H_2O_2 \rightarrow MPc + O_2 + 2H_2O. \quad (8.8)$$

Peroxide extra complex $(H_2O_2)MPc$ contains a hydrogen peroxide molecule which acquires pronounced oxidative properties following coordination to the metal atom. These properties become manifest as a result of the decreasing electron density of the O—O bond. The H_2O_2 may act as an oxidizing agent with respect to the second H_2O_2 molecule, which acts as a reducing agent and gives away an electron from the bonding σ orbital. The order of the O—O bond being reduced to 0.5 leads to its degradation.

The proposed mechanism is consistent with the experimentally established second order of the reaction with respect to MPc in an acetone solution[55] and almost first ($n \approx 1.3$) order with respect to MPc in a heterogeneous medium. The looseness of the crystal lattice of phthalocyanine complexes provides for free penetration of H_2O_2 and similar molecules into the crystal. Therefore, even in a heterogeneous medium, each phthalocyanine complex molecule is active.[25]

Thus the catalase activity is determined by the electronic structure of the central atom and its capacity for covalent σ bonding of H_2O_2. In this case the capacity of the central atom to enhance its valence state by losing a *d* electron is irrelevant, bearing in mind that most inactive elements (Cu, Cr, Mn, Co) are capable of such transformations. At the same time osmium, which forms the most active complex, is not capable of enhancing its valence state within the phthalocyanine complex. The sixth coordinate place in peroxide extra complex $(H_2O_2)OsPc(HSO_4)$ being occupied suggests that in the transition state of reaction (8.8) the second hydrogen peroxide molecule is probably out of contact with the metal, just as was assumed earlier.[55] The final solution of the problem requires investigation of alkoxy derivatives of Ru^{3+} phthalocyanine, in which the large R—O group completely occupies the fifth coordinate place. Characteristically, in a natural catalase such as the protein complex of protohaem of the (HO)FePP type the fifth coordinate place is also completely occupied by a covalently bonded HO group.

It can be easily seen that the mechanism of H_2O_2 activation in catalases is opposite to that of oxidation of porphyrins with hydrogen peroxide. In the latter case the H_2O_2 molecule is localized at the periphery, at those portions of the coordinated ligand which have a greater (and not lesser!) electron density.

An interesting aspect of the reactions of H_2O_2 decomposition in phthalocyanines is the dependence of the decomposition rate on the peroxide concentration.[497] When the H_2O_2 concentration grew from 0.1063 to 1.593 M in a reaction involving polymeric ruthenium phthalocyanine,[502] the effective rate constant dropped from 0.141 to 0.0032 h^{-1}, that is it became almost 50 times as low. A similar picture was observed with polymeric osmium phthalocyanine.[463] This phenomenon is similar to catalyst poisoning and may be associated with the formation of poorly active six-coordinated complexes $(H_2O_2)_2MPc$ in solutions with a high hydrogen peroxide concentration. At the same time the decomposition rate being independent of pH in the acidic region is a direct proof that the nitrogen atoms of the phthalocyanine do not participate in the H_2O_2 decomposition reaction.

The significant role of the central atom in the reactions of oxidation of cysteine, H_2S, and other sulphur-containing compounds is indicative of the predominance of the metal atom of MPc in the activation of the O_2 molecules or coordination to the sulphur atom. The equally selective participation of the metal atom in the reactions of electrochemical reduction of O_2 (porphyrin and phthalocyanine complexes with Co^{2+} and Fe^{2+} being particularly active) may suggest a coordination mechanism whereby the phthalocyanine reduced to anionic forms is π-coordinated to the O_2 molecule. The latter removes the excess electron from ion MPc^-:

$$\mathrm{MPc} + e \rightarrow \mathrm{MPc}^- + \mathrm{O_2} \rightarrow \mathrm{MPc}^-\cdots\mathrm{O_2} \rightarrow \mathrm{MPc}\cdots\mathrm{O_2^-} \xrightarrow{\mathrm{H^+}}$$

$$\mathrm{MPc}\cdots\mathrm{O_2H} + e \rightarrow \mathrm{MPc}^-\cdots\mathrm{O_2H} \xrightarrow{\mathrm{H^+}} \mathrm{MPc} + \mathrm{H_2O_2}. \qquad (8.9)$$

Structural changes in the ligand of a porphyrin or phthalocyanine complex affect its donor–acceptor properties and the electronic structure of the metal atom with the result that the catalytic activity changes, too. It has been established that the polymeric state of complex $(MPc)_i$ gives rise to a greater catalytic activity,[506] while the catalase activity may become greater, lower, or remain the same as a result of polymerization.[53]

Another large group of catalytic reactions involving phthalocyanine complexes depends on the nature of the metal, but to a lesser extent. However, these reactions are not catalysed by the phthalocyanine ligand without metal. One of the most important and extensively studied reactions of this group is catalytic liquid-phase oxidation of cumene to cumene hydroperoxide (CHP), yielding a number of valuable commercial products. Hock and Kropf[494–496] have shown that with complexes CuPc, ZnPc, and MgPc almost pure CHP is formed, while in the case of MnPc, FePc, CoPc, and NiPc it decomposes to acetophenone and dimethylphenylcarbinol. The induction period during which CHP is accumulated is followed by a period of initiation of a reaction in which cumene is oxidized as a result of CHP decomposition at $t > 95$ °C:

$$2\mathrm{ROOH} \rightarrow \mathrm{RO_2^{\bullet}} + \mathrm{RO^{\bullet}} + \mathrm{H_2O}.$$

The decomposition reaction is of second order with respect to CHP, first order with respect to CuPc, and its activation energy is 25.9 kcal mol^{-1}. Kropf ascribes the catalytic action of phthalocyanine complexes to the steric regulation of the process of collision of CHP molecules. At low temperatures ($t < 95$ °C) the cumene oxidation reaction is assumed[495,496] to be initiated by the chemisorptive activity of oxygen:

$$CuPc + O_2 \rightarrow (Cu^+Pc)^{\bullet} \cdots (O_2^-)^{\bullet} + RH \rightarrow CuPc + R^{\bullet} + HO_2^-,$$

$$(CuPc^+) \cdots (O_2^-)^{\bullet} + RH + RO_2H \rightarrow CuPc + RO_2^{\bullet} + R^{\bullet} + H_2O_2.$$

The activation energy of low-temperature oxidation is 10 kcal mol^{-1}, and the reaction is of second order with respect to the hydroperoxide.

According to Kropf[495] the proposed mechanism stems from the semiconductor properties of CuPc becoming more pronounced in the presence of O_2. The correlation between the semiconductor properties and heterogeneous catalytic activity of organic catalysts has become a well established fact and is explained theoretically elsewhere.[510] However, as we shall see later, the semiconductor properties of metal phthalocyanines change insignificantly and these changes are not always associated with the catalytic activities of MPcs in the above reaction, that can be arranged as follows:[501]

$$(HSO_4)RuPc > (HSO_4)_2OsPc > CoPc > NiPc > CuPc.$$

Figure 8.3 represents the oxidation rate of cumene in the temperature range 75.7–80 °C for these phthalocyanines. The ruthenium complex turned out to be the most catalytically active in the cumene oxidation reaction. The central atom in this complex has a single vacant coordinate place (the sixth). If this place is occupied by HSO_4, just as in osmium phthalocyanine, then, despite the coordinative saturation, the complex still

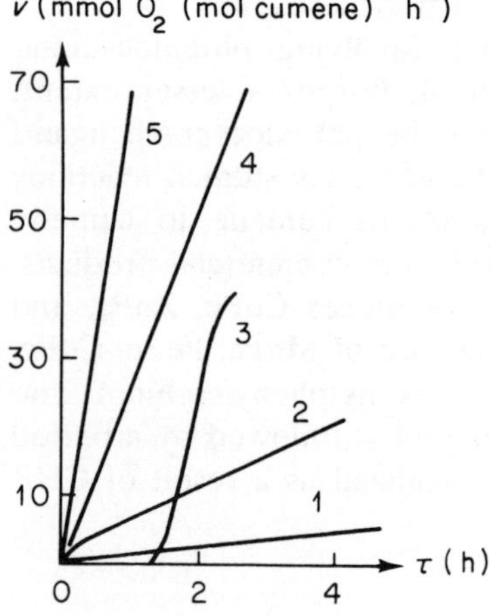

Figure 8.3 Rate of catalytic oxidation of cumene by oxygen in the air in the presence of metal phthalocyanines. Complex concentration: curve 1, 0.75 mmol CuPc (mol cumene)$^{-1}$ at 76.7 °C; curve 2, 0.5 mmol NiPc (mol cumene)$^{-1}$ at 75.7 °C; curve 3, 0.5 mmol CoPc (mol cumene)$^{-1}$ at 78.7 °C; curve 4, 0.15 mmol $(HSO_4)_2OsPc$ (mol cumene)$^{-1}$ at 80 °C; curve 5, 0.15 mmol $(HSO_4)RuPc$ (mol cumene)$^{-1}$ at 80 °C

exhibits a high catalytic activity which is much higher than in coordinatively unsaturated NiPc and CuPc. Unfortunately no data are available for comparison with the activity of other phthalocyanine complexes.

As can be inferred from the above results, the initiation of free-radical chains during cumene oxidation has nothing to do with the interaction between O_2 or the hydroperoxide and the central atom of the phthalocyanine since its possibility is ruled out in the case of $(HSO_4)_2OsPc$ in a cumene solution. This conclusion agrees with Minkov's data.[511] Minkov maintains that the reagents are most probably activated at the periphery of the phthalocyanine molecule. As far as the hydroperoxide is concerned, favourable conditions exist for formation of a molecular associate of the MPc···HOOR type, involving hydrogen bonding with a peripheral nitrogen atom. This associate, or charge-transfer complex, decomposes at the limiting step into cation radical $(MPc^+)^\bullet$, anion OH^-, and radical $RO^\bullet$. The cation radical of the phthalocyanine detaches a hydrogen atom from cumene and regenerates to phthalocyanine:

$$(MPc^+)^\bullet + HR + MPc + R^\bullet + H^+.$$

The subsequent transformations involve an oxygen molecule:

$$R^\bullet + O_2 \rightarrow RO_2^\bullet, \qquad RO_2^\bullet + RH \rightarrow RO_2H + R^\bullet, \qquad \text{and so on.}$$

The cation radical of MPc is stabilized by the metal cation and has no time to disintegrate into fragments as in the case of oxidation of H_2O_2 phthalocyanines in an acidic medium. The reasons why a phthalocyanine is inactive without the metal are not quite clear. Possibly cation radical $(H_2Pc^+)^\bullet$ is so unstable that it detaches a proton and is subject to rapid decomposition. Systematic studies are required to elucidate this and most of the other problems discussed in this section completely.

Of particular interest are data[501] on the oxidation rates of cumene on polymeric ruthenium and osmium phthalocyanines. These polymers have been found to exhibit a catalytic activity inferior to that of monomers. For example, at 80 °C (under strictly identical conditions) the oxygen absorption rates (in millimoles per mole of cumene) are: 20.90 for $(HSO_4 \cdot RuPc)_3$, 51.2 for $(HSO_4)RuPc$, 8.66 for $(HSO_4)_2OsPc_5$, and 33.22 for $(HSO_4)_2OsPc$.

By analogy with polymeric copper phthalocyanine one may expect polymeric ruthenium and osmium complexes to exhibit more pronounced semiconductor properties. However, their catalytic properties in the cumene oxidation reaction remain the same.

8.4 Semiconductor Properties of Phthalocyanine Complexes

The semiconductor properties of phthalocyanine and porphyrin complexes, along with their catalytic properties, are among the most valuable ones.

Hence the large volume of works concerning these properties, the results of which have been summarized in a monograph.[32]

Extensive research into the semiconductor properties of complexes of phthalocyanine, chlorophyll, and its analogues has been conducted by Vartanyan.[512–518] This, as well as other works carried out outside the USSR[519–524] indicate that substances with a conjugated system of double bonds exhibit pronounced intrinsic (rather than impurity) dark conductivity. This conductivity not only can be put to wide use in semiconductor engineering[510,529] but also may be responsible for energy migration in biological systems.[525,526] Due to the relatively easy migration of π electrons between molecules the dark conductivity of aromatic and conjugated systems in the condensed phase may reach a comparatively high value. The conductivity of most organic compounds at room temperature is 10^{-10}–10^{-20} Ω^{-1} cm^{-1}, which corresponds to application of a high potential difference (field strength of 10^4–10^5 V cm^{-1}).

The electrical conductivity of a dye layer increases exponentially with temperature:

$$\sigma = \sigma_0 \exp(-E_T/2kT). \tag{8.10}$$

Here σ is the sample conductivity at T kelvins; σ_0 is a constant (formally, conductivity at $T = \infty$, $k = 0.86 \times 10^{-4}$ eV K^{-1}); and E_T is the thermal conduction activation energy.

The most important characteristic of an organic semiconductor is E_T. It determines the minimum amount of energy required to produce a charge carrier (electron or hole) in the solid state, which moves in the electric field. The values of σ_0 and E_T may be strongly dependent on the sample preparation. Values of E_T (in electron volts) for H_2Pc obtained by different investigators using varous samples are presented in a handbook:[510] 2.4 for a fine crystalline unpressed powder, 1.7 for a single crystal, 1.7 for a vacuum-deposited layer, 1.5 for a layer deposited in air, and 0.9 for a layer rubbed on a quartz substrate. Such a spread of values is due to the various intermolecular and intercrystalline interactions. Therefore it is particularly important to prepare samples under identical conditions for analysis of σ_0 and E_T as a function of the structure of the conjugated system of bonds in the molecule.

If the samples are prepared under strictly identical conditions, the electrical conductivity method can be used in testing phthalocyanine complexes for purity.[527–529]. The dark conductivity method has made it possible to establish[512] an interaction between the O_2 molecule and the central atom of MPc, which increases the value of σ_0 and reduces that of E_T. Only after treatment in a deep vacuum and heat treatment of the sample can one determine the intrinsic conductivity of phthalocyanine complexes.

No systematic data have appeared in the literature so far as to the effect of the central metal atom on the conductivity of samples. Felmayer and

Wolf[530] found the values of E_T for CuPc and PtPc to be equal, respectively, to 1.79 and 1.52 eV. The difference between these values is attributed not to the dissimilarity of the aromatic systems of these complexes but to the manner in which the molecules are arranged in the crystal.[510] It is interesting to note that the complex (Cl_{15})CuPc has a slightly higher thermal activation energy (E_T = 1.86 eV) and a conductivity 10 times greater than that of CuPc.[530]

Studies of the dark conductivity of polymeric copper phthalocyanines based on pyromellitic acid (see Fig. 2.2 between lines AA and BB) have shown that the activation energy substantially decreases (down to 0.8–0.95 eV) in polymers. This finding is ascribed to the growing length of the conjugated system in polymers.[530] In interpreting such phenomena one should not forget that the conjugation in polymers develops via benzene rings which are essentially π systems practically independent of the macroring (see Chapter 3). Thus such polymers are conjugated systems only formally. The factors responsible for a lower value of E_T in polymeric phthalocyanine complexes are much more complicated than a mere elongation of the molecule and a formal increase in the number of π electrons whose different groups overlap only insignificantly.

Extremely interesting data can be found in a work by Danzig *et al.*[531] concerning the electrical conductivity of CuPc and two of its structural analogues: tetrapyridineporphyrazine and tetrapyrazineporphyrazine. Under comparable conditions, at 25 °C, the measured values of the activation energy were 1.3, 1.17, and 0.81 eV, respectively, and those of conductivity were 10^{-12}, 3×10^{-11}, $3 \times 10^{-8}\ \Omega^{-1}\ cm^{-1}$. The values of σ_0 were the same for all three compounds. It is now difficult to tell the variation in which property of phthalocyanine, occurring during aza substitution in the benzene rings, is responsible for the decrease in the activation energy because other properties of these compounds have not been examined so far. Possibly, more favourable conditions are created for intermolecular transfer of π electrons as a result of extra coordination of the adjacent molecules.

Koifman *et al.*[532] use extra complexes of osmium phthalocyanine X_2OsPc where X = Cl, Br, I, and $\frac{1}{2}SO_4$ as examples illustrating the strong influence of extra coordination on the semiconductor properties (Table 8.7). The authors attribute the decrease in the value of E_T in the series I, Br, Cl to the increasing electronegativity of these elements.

However, anion SO_4^{2-} does not fit in this series. In extra complexes of aluminium phthalocyanine with extra ligands Cl^-, F^-, OH^-, $C_6H_5O^-$, and O^{2-} [529] the increase in electronegativity from Cl^- to F^-, on the contrary, leads to a higher activation energy.

From the available data on the electrical conductivity of phthalocyanine complexes it clearly follows that the conductivity and thermal activation energy are determined by a large number of factors which are not properly understood at present. In this field, just as in most other areas of

Table 8.7 Semiconductor properties of extra complexes of phthalocyanine with osmium and aluminium

Complex	Conductivity at 25 °C (Ω^{-1} cm^{-1})	E_T (eV)	Complex	Conductivity at 25 °C (Ω^{-1} cm^{-1})	E_T (eV)
$(Cl)_2OsPc$	1.1×10^{-6}	0.44	(Cl)AlPc	—	1.49
$(Br)_2OsPc$	2.1×10^{-6}	0.61	$(C_6H_5O)AlPc$	—	1.56
$(I)_2OsPc$	4.5×10^{-5}	0.85	(HO)AlPc	—	1.59
$(SO_4)OsPc$	3.4×10^{-7}	1.09	(F)AlPc	—	1.60
			(KO)AlPc	—	1.67

physicochemical, and particularly physical, studies of porphyrin and phthalocyanine complexes, there is an acute lack of comprehensive and systematic data.

Conclusion

In this monograph I have discussed the coordination compounds of porphyrins and their synthetic structural analogue—phthalocyanine. The unusual molecular structure of such ligands puts them in a category radically different from most other inorganic and organic molecules capable of the formation of coordination compounds. The porphyrin complexes are so significantly different from most other groups of complexes that they must be examined separately, which is precisely what I have attempted to do in this volume.

In complexes with inorganic and simple organic ligands the central atom and the ligand are more often than not equal partners, the majority of the properties being determined by the nature of the central atom. In porphyrin and phthalocyanine complexes the ligand and the specific features of its geometric and electronic structures play a particular role.

It is the ligand that determines the formation of metalloporphyrins which, as was shown in Chapter 4, are less dependent on the metal species. The ligand provides for the extremely high kinetic stability of most complexes and their capacity for long life under the formidable conditions prevailing in human, animal, and plant organisms. The ligand is in fact solely responsible for the catalytic properties of complexes and their performing the function of enzyme systems. Here the predominant factor is the effect of the functional substituents and their electronic nature on the main coordination centre of porphyrin. It is the ligand and its structure that define the electronic absorption spectra of porphyrin and phthalocyanine complexes. This peculiarity is especially important in the case of chlorophyll and haemin.

The predominant role in transferring the electronic effects of the substituents to the central atom and back is played by the unique conjugated system of porphyrins and phthalocyanines. We are discussing an aromatic multicontour system with the main chromophore of the macroring, which is considerably disturbed by the ethylene double bonds of the pyrrole nuclei in porphyrins. Therefore all physicochemical properties of porphyrin are strongly affected by the substituents not only in the methine groups (meso positions) but also in the pyrrole nuclei.

The incorporation of a metal into the centre of the aromatic system of porphyrin produces a tangible effect on the electronic state of all atoms of the molecule. All properties of porphyrin complexes clearly exhibit the σ effect, dative π effect, backward dative π effect, central atom charge effect, and effect of stabilization or destabilization by the metal of the porphyrin ligand's planar system.

The close correlation between all properties of porphyrin and phthalocyanine complexes stems from the aromaticity of the ligand's conjugated system, which has prompted the so-called spectral criterion of complex stability. Studies into the relationship between the properties of a large group of porphyrin and phthalocyanine complexes have given a better insight into the physical factors of its validity and enabled the range of its applicability to be defined.

Of greatest scientific value are systematic structural studies of the physicochemical properties of complexes in a broad range of structures. Such studies into the ionization properties (Chapter 4) have shown that the degree of protonation of the N—H bond in porphyrins increases as the aromaticity of the system or σ- and π-acceptor properties of the substituents become more pronounced. The degree of protonation of the N—H bond manifests itself first of all in the constants of acid ionization of porphyrins (increasing with the degree of protonation of the N—H bond, particularly as one passes from porphyrin to phthalocyanine ligands), in the rates of formation of coordination compounds (which also increase), in the rates of dissociation of porphyrin complexes (which decrease), and in infrared spectra.

In the course of acid ionization of porphyrins the solvent takes an active part not only in the solvation of the proton but also in that of the emerging highly unstable anion. Hence the ionization of various porphyrins is accompanied by an overall thermodynamic compensation effect. The increasing degree of stabilization of the porphyrin anion by the solvent promotes ionization. The latter is characterized by very low constants (the total porphyrin ionization constant ranging from 10^{-30} to 10^{-34}), highly negative values of the entropies, and moderately positive enthalpies of ionization.

As far as protonation is concerned, systematic data have been obtained only for phthalocyanine and its numerous complexes. Investigators have clearly established the effect of the metal, coordinate bond type, extra coordination type, and so on. It has become imperative to obtain such data for various porphyrins. (They will permit developing the general theory of acid–base interactions of macrocyclic aromatic systems.)

The coordination properties of porphyrin and phthalocyanine (Chapter 4) have been systematically studied in a broad range of structures. Data are available on the rates and mechanisms of metalloporphyrin formation in many non-aqueous solvents for more than 20 porphyrins with systematically changing molecular structure. Porphyrins are protonated ligands with a rather weak protonation of the N—H bond. This is why the formation of complexes, or metalloporphyrins, is a relatively slow process (the formation rate is by three to seven orders of magnitude slower than in the case of non-cyclic unprotonated ligands). In coordination reactions the proton is in many ways similar to metal cations. It participates in the competitive interaction with the porphyrin ligand. Therefore the reactions of metal coordination and protonation in porphyrins are competitive. The proton concentration and activity determine not only the rate but also the possibility, in principle, of metalloporphyrin formation. In this case the porphyrin structure is a major factor. In protonating solvents (even in aqueous HCOOH) practically no formation of metalloporphyrins takes place. At the same time phthalocyanine forms stable complexes (Cu, Co, Ni, Pt) even when dissolved in concentrated H_2SO_4.

As regards the reactions yielding porphyrin complexes, a great number of interesting kinetic regularities have been established, which have made it possible to regard the reaction mechanism as a self-consistent process with associative–dissociative activation. The activation of the reagent salt involves a loss of two solvent molecules from the inner coordination sphere at a coordination number of the cation equal to six and more or deformation of the coordination sphere at a coordination number of four or less. Variations in the coordination number in the case of less coordinating solvents may result in superposition of the two activation paths.

The activation of a porphyrin molecule involves deformation of the N—H bonds, partial desolvation of the coordination centre, and additional solvation of the molecular periphery.

The structure of the aromatic part of the porphyrin ligand molecule strongly influences the stability of their covalent complexes in solution. This influence is non-existent as far as ionic complexes are concerned. A systematic investigation into the stability of metalloporphyrins and their models (metal phthalocyanines) was started two decades ago in the present author's laboratory and has yielded a number of interesting results. Stable MP and MPc complexes dissociate only in highly proton-donating media, that is the dissociation is typically protolytic. The effect of the solvent alone is not sufficient for dissociation. It is shown that a dissociation act involves simultaneous transfer of the proton and solvent molecule from the attacking reagent. As a result, the solvated proton is an effective reagent. The participation of acid molecules in dissociation is possible only in the case of ionic (labile) complexes.

The aza substitution in the macroring of porphyrins along with the benzo substitution produce a surprisingly strong effect on the stabilization of complexes. If stable MPs dissociate perceptibly in boiling acetic acid, metal phthalocyanines detach the ligand only in hot concentrated sulphuric acid.

In view of the fact that metalloporphyrins and similar complexes are effective catalysts, their stability is a matter of great practical importance. It has been found that the primary factors responsible for the stability of MP and MPc complexes include steric factors. In no other classes of complexes do these factors play such an important role as in cyclic tetrapyrrole ligands. Hence, in the series Zn, Cd, Hg, the stability of complexes drops catastrophically and the well-known stability series of Irving and Williams is disturbed because of the shifts in the stability of Fe, Ni, Co, and Cu complexes. Porphyrin complexes with Fe are the most stable. Co^{2+} forms much more stable complexes than Zn^{2+}. Porphyrin complexes with Ni^{2+} are more stable than those with Cu^{2+}.

One of the most important problems of organic physical chemistry and coordination chemistry concerns studies into the effect of the polymeric state of molecules on their reactivity. This problem is of paramount importance in a new, recently established branch of science—bioinorganic chemistry. The coordination phenomena in superpolydentate polyconformation molecules of such polymers as proteins, nucleic acids, and others are so specific and complex that their investigation requires that simpler models be created first. One such model is that of the polymeric phthalocyanines, which are considered briefly in this monograph (Chapter 4).

Despite the extremely high stability of porphyrin complexes the nature of the solvent substantially affects their kinetic and thermodynamic stability. The effect of the solvent is so significant that it may change the sequence in the stability series of complexes of the same porphyrin. Hence an important conclusion that even in the case of such stable complexes as metalloporphyrins one cannot derive universal stability series. Presumably, it is precisely this factor that is responsible for the great variety of enzyme catalytic reactions and determines their specificity and strong dependence on the environmental conditions.

Chlorophyll has been found to occupy a specific place in the series of its metal analogues. It is shown that the coordination of Mg to this porphyrin is optimal both for the stability of the system and for further coordination (extra coordination) of the molecules participating in photosynthesis. The metalloporphyrin dissociation mechanism is regarded as a multistep protonation–solvation process whose limiting step is a self-congruent trimolecular process involving associative–dissociative activation.

The adequacy of the mechanisms of formation and dissociation of metalloporphyrins has enabled calculation for the first time of the constants of instability of many metalloporphyrin complexes (Chapter 5). Analysis is made of the kinetic criterion of complex stability and the thermodynamic evaluation is shown to be necessary even for such stable complexes as metalloporphyrins. The basic factors of high stability of metalloporphyrins are indicated and methods for measuring their stability are outlined.

One of the most important phenomena in metalloporphyrins is extra coordination, that is continuing coordination of molecules at the central atom. Extra coordination is fully responsible for the capacity of metalloporphyrins to act as biocatalysts. A review of the literature has shown that in spite of the large number of works on extra coordination there are practically no systematic studies that would have established the general regularities of their formation and effect on other physicochemical properties of metalloporphyrins.

A most characteristic feature of porphyrins, phthalocyanine, and their complexes is their electronic absorption spectra. They represent the intramolecular energetics and clearly display all changes occurring in the molecule.

In dealing with the spectroscopy of porphyrins (Chapter 6) I have touched upon only those aspects that have not been adequately covered before.[8] In particular, detailed analysis is made of the spectral criterion of stability of complexes with π ligands, the state of the cyclopentanone ring of chlorophyllic acid and its complexes according to spectral data, specific features of association of porphyrins with the cyclopentanone ring, and the spectroscopy of polymeric states of phthalocyanines. All these problems are of great scientific importance and are in urgent need of experimental and theoretical solution. In all cases these subjects were treated from the standpoint of the effect of coordination to the metal on the above properties.

The coordination chemistry of porphyrins and phthalocyanines would not be complete without an at least brief consideration (Chapter 8) of such practically important properties of their complexes as the photochemical, oxidation–reduction, catalytic, and semiconductor ones. These properties are clearly manifest in metalloporphyrins, the predominant factors responsible for them being both the metal atom and the ligand structure. At the same time, this field of the coordination chemistry of porphyrins and phthalocyanines also calls for extensive systematic studies.

References

1. K. A. Timiryazev, *Nature*, **32**, 342 (1885); **109**, 414 (1895).
2. E. Rabinowitch, *Photosynthesis* [*Fotosintez*], IL Publishers, Moscow, Vol. 1, 1951; Vol. 2, 1953; Vol. 3, 1959.
3. T. N. Godnev, *Chlorophyll, Its Structure and Formation in Plants* [*Khlorofill, ego stroenie i obrazovanie v rastenii*] BSSR Academy of Sciences, Minsk, 1963.
4. *Mechanism of Photosynthesis* [*Mekhanizm fotosinteza*], Transactions of the 5th International Biological Congress, USSR Academy of Sciences, Moscow, 1961.
5. *1st All-Union Biochemical Congress* [*Pervy vsesoyuzny biokhimichesky syezd*], abstracts, Vol. II, Section of Photosynthesis Biochemistry, USSR Academy of Sciences, Leningrad, 1964, p. 204.
6. *Structure and Function of Photosynthesis Apparatus* [*Struktura i funktsiya fotosinteticheskogo apparata*], IL Publishers, Moscow, 1962.
7. A. N. Terenin, *Photochemistry of Chlorophyll and Photosynthesis* [*Fotokhimiya khlorofilla i fotosintez*], 6th Bakhov Readings, USSR Academy of Sciences, Moscow, 1951.
8. G. P. Gurinovich, A. N. Sevchenko, and K. N. Solovyev, *Spectroscopy of Chlorophyll and Allied Compounds* [*Spektroskopiya khlorofilla i rodstvennykh soedineniy*], Nauka i tekhnika, Minsk, 1968.
9. A. A. Krasnovsky, *Uspekhi khimii*, **29**, 736 (1960).
10. V. B. Yevstigneev, *Zh. fiz. khimii*, **32**, 968 (1958).
11. V. B. Yevstigneev, Unpublished doctoral dissertation, Institute of Biochemistry of the USSR Academy of Sciences, Moscow, 1956.
12. I. I. Dilung, Unpublished doctoral dissertation, Institute of Physical Chemistry of the UkSSR Academy of Sciences, Kiev, 1971.
13. *Elementary Photoprocesses in Molecules* [*Elementarnye protsessy v molekulakh*], Nauka, Moscow–Leningrad, 1966.
14. A. N. Terenin, in *Problems of Photosynthesis* [*Problemy fotosinteza*], USSR Academy of Sciences, Moscow, 1959, p. 9.
15. A. A. Krasnovsky, *Conversion of Luminous Energy in Photosynthesis. Molecular Mechanisms* [*Preobrazovanie energii sveta pri fotosinteze. Molekulyarnye mekhanizmy*], 29th Bakhov Readings, Nauka, Moscow, 1974.
16. A. A. Shlyk, in *Current Problems of Photosynthesis* [*Sovremennye problemy fotosinteza*], Moscow University Press, Moscow, 1973, p. 85.
17. *Chlorophyll* [*Khlorofill*], Nauka i tekhnika, Minsk, 1974.
18. B. D. Berezin and O. I. Koifman, *Uspekhi khimii*, **42**, 2007 (1973).
19. A. N. Sidorov, *Uspekhi khimii*, **35**, 366 (1966).
20. F. F. Litvin, in *Current Problems of Photosynthesis* [*Sovremennye problemy fotosinteza*], Moscow University Press, Moscow, 1973, p. 175.

21. V. M. Kutyurin, I. Yu, Artamkina, N. P. Melnikov, and I. N. Anisimov, in *Chlorophyll* [*Khlorofill*], Nauka i tekhnika, Minsk, 1974, p. 85.
22. A. N. Terenin, *Photonics of Molecules of Dyes and Allied Compounds* [*Fotonika molekul krasiteley i rodstvennykh soedineniy*], Nauka, Leningrad, 1967.
23. J. Falk, *Porphyrins and Metalloporphyrins*, Elsevier, Amsterdam, 1964.
24. K. B. Yatsimirsky, *Thermochemistry of Complex Compounds* [*Termokhimiya kompleksnykh soedineniy*], USSR Academy of Sciences, Moscow, 1951.
25. R. Lemberg and I. Legge, *Hematin Compounds and Bile Pigments*, Intern. Publ., New York, 1949.
26. L. A. Blyumenfeld, *Haemoglobin and Reversible Attachment of Oxygen* [*Gemoglobin i obratimoe prisoedinenie kisloroda*], Sovetskaya nauka, Moscow, 1957.
27. L. Pauling, *Nature of Chemical Bonding* [*Priroda khimicheskoy svyazi*], Goskhimizdat, Moscow–Leningrad, 1947.
28. *Structure and Bonding* [*Struktura i svyaz*], Mir, Moscow, 1969.
29. L. A. Nikolaev, *Biocatalysts and Their Models* [*Biokatalizatory i ikh modeli*], Vysshaya shkola, Moscow, 1964.
30. R. Linstead, *Berichte*, **72A**, 93 (1939).
31. A. Zander, *Chemie*, **55**, 33/34, 225 (1942).
32. F. Moser and A. Thomas, *Phthalocyanine Compounds*, Reinhold, New York, 1963.
33. B. I. Stepanov, *An Introduction into the Chemistry and Technology of Organic Dyes* [*Vvedenie v khimiyu i tekhologiyu organicheskikh krasiteley*], Khimiya, Moscow, 1971.
34. F. Gund, *J. Soc. Dyers Colour.*, **69**, 671 (1953).
35. O. M. Lifentsev and B. N. Melnikov, *Dyeing and Printing of Fabrics by Synthesis of Pigments on Fibre* [*Krashenie i pechatanie tkaney putem sinteza pigmentov na volokne*], Legkaya industriya, Moscow, 1973.
36. V. F. Borodkin, Unpublished doctoral dissertation, Ivanovo Chemical Technology Institute, 1966.
37. B. D. Berezin, Unpublished doctoral dissertation, Institute of Organic and Inorganic Chemistry of the UkSSR Academy of Sciences, Kiev, 1966.
38. F. Baumann, B. Bienert, G. Rösch, H. Vollmann, and W. Wolf, *Angew. Chem.*, **68**, 133 (1956).
39. *Organic Semiconductors* [*Organicheskie poluprovodniki*] (V. A. Kargin, ed.), Nauka, Moscow, 1968.
40. M. Wolf, *Electronics*, **36**, 35 (1963).
41. F. Haak and J. Nolta, *J. Chem. Phys.*, **38**, 2648 (1963).
42. A. A. Maslov and G. A. Yurova, *Zh. fiz. khim.*, **41**, 656 (1967).
43. Ye. F. Ishchenko and Yu. M. Klimkov, *Laser Generators* [*Opticheskie kvantovye generatory*], Sovetskoe radio, Moscow, 1968.
44. P. Sorokin, I. Lankard, E. Hammond, and V. Moruzzi, *J. Res. and Develop.*, **11**, 130 (1967).
45. Yu. M. Gryaznov, O. L. Lebedev, V. A. Serebryakov, A. D. Starikov, and Ye. M. Shvom, *Zh. prikl. spektrosk.*, **10**, 739 (1969).
46. Yu. M. Gryaznov, O. L. Lebedev, and L. A. Chastov, *Optika i spektrosk.*, **20**, 503 (1966).
47. V. N. Gavrilov, Yu. M. Gryaznov, O. L. Lebedev, and L. A. Chastov, *ZhETF*, **48**, 772 (1965).
48. I. V. Nizhegorodova, V. V. Fadeev, Ye. M. Shvom, and L. P. Shklover, *Vestn. MGU, seriya fiz.*, **4**, 103 (1966).
49. R. Sallé, I. Gallard, and P. Traynard, *C.r. hebd. Séanc. Acad. Sci., Paris*, **256**, 2588 (1963).

50. H. Inoue, S. Hayashi, and E. Imoto, *J. Chem. Soc. Japan,* **37**, 326 (1963).
51. E. Ochiai, *Tetrahedron*, **20**, 1819 (1964).
52. W. Hanke, *Z. Chem.*, **9**, 1 (1969).
53. A. N. Shlyapova and B. D. Berezin, *Transactions of the Ivanovo Chemical Technology Institute*, **15**, 89 (1973).
54. A. N. Shlyapova, Unpublished dissertation, Ivanovo Chemical Technology Institute, 1971.
55. B. D. Berezin and A. V. Loshchilova, *Kinetika i kataliz*, **8**, 592 (1967).
56. L. A. Nikolaev, *Khim. nauka i promyshl.*, **2**, 202 (1957).
57. R. Kh. Burshtein, K. A. Radyushkina, M. R. Tarasevich, B. D. Berezin, and S. D. Levina, *Elektrokhimiya*, **9**, 410 (1973).
58. H. Alt, H. Binder, and G. Sandstede, *J. Catalysis*, **28**, 8, (1973).
59. A. Lever, *Advances Inorg. Radiochem.*, **7**, 27 (1965).
60. P. Černy, *Chem. Zvesti*, **9**, 94 (1955).
61. H. Fischer and H. Orth, *Die Chemie des Pyrrols*, Vol. 1, Part 1, Akademische Verlag, Leipzig, 1937.
62. H. Fischer and A. Stern, *Die Chemie des Pyrrols*, Vol. 2, Part 2, Akademische Verlag, Leipzig, 1940.
63. S. Granick, *J. Biol. Chem.*, **175**, 333 (1948).
64. P. Wei, A. Corwin, and R. Arillano, *J. Org. Chem.*, **27**, 3344 (1962).
65. S. Baum and R. Plane, *J. Amer. Chem. Soc.*, **88**, 910 (1966).
66. S. Baum, B. Burnham, and R. Plane, *Proc. Nat. Acad. Sci. USA,* **52**, 1439 (1964).
67. R. Becker and I. Allison, *J. Phys. Chem*, **67**, 2669 (1963).
68. F. Zscheile and C. Comar, *Bot. Gaz.*, **102**, 463 (1941).
69. K. Winterstein and A. Stein, *Z. phys. Chem.*, **220**, 272 (1933).
70. M. Grinstein, *J. Biol. Chem.*, **167**, 515 (1947).
71. B. D. Berezin and A. N. Drobysheva, *Zh. fiz. khimii*, **42**, 2821 (1968).
72. I. Zalesky, *Z. phys. Chem.*, **43**, 11 (1904).
73. B. D. Berezin and A. N. Drobysheva, *Zh. fiz. khimii*, **44**, 2804 (1970).
74. A. N. Sidorov and A. N. Terenin, *Optika i spektrosk.*, **8**, 482 (1960).
75. E. Fleischer and I. Wang, *J. Amer. Chem. Soc.*, **82**, 3498 (1960).
76. B. D. Berezin and A. N. Drobysheva, *Zh. fiz. khimii*, **48**, 2742 (1974).
77. B. D. Berezin and O. B. Lapshina, *Zh. fiz. khimii*, **48**, 303 (1974).
78. J. Barnes and G. Donough, *J. Amer. Chem. Soc.*, **72**, 4045 (1950).
79. J. Erdman and A. Corwin, *J. Amer. Chem. Soc.*, **69**, 750 (1947).
80. I. Buchler, L. Puppe, K. Rohbock, and H. Schneehage, *Ann. New York Acad. Sci.*, **206**, 116 (1973).
81. J. Erdman, V. Ramsey, N. Kalenda and W. Hanson, *J. Amer. Chem. Soc.*, **78**, 5844 (1956).
82. J. Taylor, *J. Biol. Chem.*, **135**, 569 (1940).
83. A. Treibs, *Annalen*, **728**, 115 (1969).
84. A. Wolberg and I. Manassen, *J. Amer. Chem. Soc.*, **92**, 2982 (1970).
85. H. Thiele, I. Behlke, and W. Scheler, *Acta Biol. Med. German.*, **11**, 767 (1963).
86. I. Kida, *J. Chem. Soc. Japan*, **1967**, 23.
87. M. Tsutsui, R. Velapoldi, L. Hoffman, K. Suzuki, and A. Ferrari, *J. Amer. Chem. Soc.*, **91**, 3337 (1969).
88. B. D. Berezin, *Transactions of the Ivanovo Chemical Technology Institute*, **10**, 62 (1968).
89. R. Linstead, *J. Chem. Soc.*, **1934**, 1016.
90. C. Dent, R. Linstead, and A. Lowe, *J. Chem. Soc.*, **1934**, 1033.
91. G. Byrne, R. Linstead, and A. Lowe, *J. Chem. Soc.*, **1934**, 1017.
92. R. Linstead and A. Lowe, *J. Chem. Soc.*, **1934**, 1022.

93. C. Dent and R. Linstead, *J. Chem. Soc.*, **1934**, 1027.
94. P. Barrett, C. Dent, and R. Linstead, *J. Chem. Soc.*, **1936**, 1719.
95. P. Barrett, D. Frye, and R. Linstead, *J. Chem. Soc.*, **1938**, 1157.
96. V. F. Borodkin, *Zh. prikl. khimii*, **31**, 813 (1958).
97. C. Dent, *J. Chem. Soc.*, **1938**, 1.
98. A. Zander, *Angew. Chem.*, **55**, 255 (1942).
99. I. Owen and M. Kenney, *Inorg. Chem.*, **1**, 331 (1962).
100. H. Rutter and I. McQueen, *J. Inorg. Nuclear Chem.*, **12**, 361 (1960).
101. R. Taube, *Z. Chem.*, **3**, 194 (1963).
102. V. D. Berezin, *Izv. vuzov. Khimiya i khim. tekhnol.*, **7**, 982 (1964).
103. D. Colaitis, *C. r. hebd. Séanc. Acad. Sci., Paris,* **242**, 1026 (1956).
104. R. Hill, *Biochem. J.*, **19**, 341 (1925).
105. I. Elvidge and A. Lever, *Proc. Chem. Soc.*, **20**, 195 (1959).
106. R. Joyner and M. Kenney, *J. Amer. Chem. Soc.*, **82**, 5790 (1960).
107. I. Elvidge and A. Lever, *J. Chem. Soc.* **1961**, 1257.
108. B. D. Berezin and G. V. Sennikova, *DAN SSSR*, **159**, 117 (1964).
109. B. D. Berezin, *DAN SSSR*, **150**, 1039 (1963).
110. B. D. Berezin and N. I. Sosnikova, *DAN SSSR*, **146**, 604 (1962).
111. L. P. Shklover, V. Ye. Plyushchev, I. A. Rozdin, and N. A. Novikova, *ZhNKh,* **9**, 478 (1964).
112. V. Ye. Plyushchev, L. P. Shklover, and I. A. Rozdin, *ZhNKh*, **9**, 125 (1964).
113. B. D. Berezin, Yu. A. Zhukov, and O. I. Koifman, *DAN SSSR*, **167**, 1318 (1966).
114. B. D. Berezin, Yu. A. Zhukov, A. N. Shlyapova, and I. M. Sokolova, *Transactions of the Ivanovo Chemical Technology Institute*, **11**, 52 (1969).
115. M. Lowery, A. Starshake, J. Esposito, P. Krueger, and N. Kenney, *Inorg. Chem.*, **4**, 128 (1965).
116. A. S. Akopov, B. D. Berezin, N. V. Klyuev, and A. A. Solovyev, *ZhNKh*, **17**, 981 (1972).
117. P. Krueger and M. Kenney, *J. Inorg. Nuclear Chem.*, **25**, 303 (1963).
118. M. Cragh and W. Koski, *J. Amer. Chem. Soc.*, **87**, 2496 (1965).
119. M. G. Gurevich and K. N. Solovyev, *DAN BSSR*, **5**, 291 (1961).
120. L. P. Shklover and V. Ye. Plyushchev, *Transactions of IREA*, **30**, 89 (1967).
121. G. Watt and I. Davis, *J. Inorg. Nuclear Chem.*, **14**, 32 (1960).
122. R. Taube and H. Munke, *Angew. Chem.*, **75**, 639 (1963).
123. R. Taube, M. Zach, K. Stauske, and S. Heidrich, *Z. Chem.*, **3**, 392 (1963).
124. W. Kroenke and M. Kenney, *Inorg. Chem.*, **3**, 251 (1964).
125. P. Moskalev, I. Kirin, and G. Shapkin, *ZhNKh*, **18**, 3229 (1973).
126. I. Kirin, A. Kolyadin, and A. Lychev, *ZhNKh*, **18**, 2295 (1973).
127. P. Moskalev and I. Kirin, *Zh. obshch. khimii*, **38**, 1191 (1968).
128. B. D. Berezin, *Izv. vuzov. Khimiya i khim. tekhnol.*, **6**, 841 (1963).
129. B. D. Berezin, *ZhNKh*, **6**, 2672 (1961).
130. A. S. Akopov and B. D. Berezin, in *1st All-Union Conference on Porphyrin Chemistry and Geochemistry*, abstracts, Donish, Dushanbe, 1977, p. 19.
131. O. K. Krutoyarov, M. I. Alyanov *et al.*, *Izv. vuzov. Khimiya i khim. tekhnol.*, **14**, 1379, 1606, 1911 (1971).
132. Yu. I. Khoinov, M. I. Alyanov, V. F. Borodkin, and V. A. Bendersky, *Izv. vuzov. Khimiya i khim. tekhnol.*, **15**, 1595 (1972).
133. M. I. Alyanov, V. F. Borodkin, Yu. G. Kalugin, V. R. Larionov, Yu. L. Khoinov, Ya. I. Yashin, and R. P. Smirnov, *Izv. vuzov. Khimiya i khim. tekhnol.*, **16**, 1604 (1973).
134. J. Curry and R. Shaw, *J. Phys. Chem.*, **69**, 344 (1965).
135. V. N. Klyuev and L. S. Shiryaeva, *Izv. vuzov. Khimiya i khim. tekhnol.*, **12**, 774 (1969).

136. L. S. Shiryaeva and V. N. Klyuev, *Izv. vuzov. Khimiya i khim. tekhnol.*, **12**, 1386 (1969).
137. V. N. Klyuev, B. D. Berezin, and L. S. Shiryaeva, *Izv. vuzov. Khimiya i khim. tekhnol.*, **13**, 997 (1970).
138. V. F. Borodkin, Yu. G. Yerykalov, and K. V. Usacheva, *Zh. prikl. khimii*, **29**, 1606 (1956).
139. A. A. Spryskov and A. I. Kobenin, *Transactions of the Ivanovo Chemical Technology Institute*, **5**, 196 (1955).
140. K. Bernauer and S. Fallab, *Helv. Chim. Acta*, **44**, 1287 (1961).
141. E. Bradbrook and R. Linstead, *J. Chem. Soc.*, **1936**, 1744.
142. R. Linstead, E. Noble, and J. Wright, *J. Chem. Soc.*, **1937**, 911.
143. M. Danzig, C. Liang, and E. Passaglia, *J. Amer. Chem. Soc.*, **85**, 668 (1963).
144. C. Marvell and J. Rassweiler, *J. Amer. Chem. Soc.*, **80**, 1197 (1958).
145. B. D. Berezin and L. P. Shormanova, *Vysokomol. soed.*, **A-10**, 384 (1968).
146. L. P. Shormanova and B. D. Berezin, *Vysokomol. soed.*, **A-10**, 1154 (1968).
147. B. D. Berezin and L. P. Shormanova, *Vysokomol. soed.*, **B-10**, 784 (1968).
148. B. D. Berezin and L. P. Shormanova, *Vysokomol. soed.*, **A-11**, 1033 (1969).
149. L. P. Shormanova and B. D. Berezin, *Vysokomol. soed.*, **A-12**, 692 (1970).
150. L. P. Shormanova and B. D. Berezin, *Transactions of the Ivanovo Chemical Technology Institute*, **12**, 42 (1970).
151. B. D. Berezin and A. N. Shlyapova, *Vysokomol. soed.*, **A-15**, 1675 (1973).
152. L. P. Shormanova, O. I. Koifman, and B. D. Berezin, *Vysokomol. soed.*, **B-15**, 910 (1973).
153. W. Drinkard and L. Bailar, *J. Amer. Chem. Soc.*, **81**, 4795 (1959).
154. A. A. Berlin, L. G. Cherkashina, and Ye. M. Balabanov, *Vysokomol. soed.*, **4**, 376 (1962).
155. S. Z. Roginsky, A. A. Berlin, O. A. Golovina, Ye. S. Fokukina, M. M. Sakharov, and L. G. Cherkashina, *Kinetika i kataliz*, **4**, 431 (1963).
156. A. Epstein and B. Wild, *J. Chem. Phys.*, **32**, 324 (1960).
157. H. Inoue, I. Kida, and E. Imoto, *Bull. Chem. Soc. Japan*, **38**, 2214 (1965).
158. A. A. Berlin, L. G. Cherkashina, Ye. L. Frankevich, Ye. M. Balabanov, and Yu. G. Aseev, *Vysokomol. soed.*, **6**, 832 (1964).
159. A. A. Berlin and N. Matveeva, *Uspekhi khimii*, **29**, 277 (1960).
160. Ye. Frankevich, L. Busheva, Ye. M. Balabanov, and L. G. Cherkashina, *Vysokomol. soed.*, **6**, 1028 (1964).
161. B. D. Berezin and L. P. Shormanova, *Izv. vuzov. Khimiya i khim. tekhnol.*, **16**, 121 (1973).
162. B. D. Berezin, L. P. Shormanova, and N. I. Zakharova, *Izv. vuzov. Khimiya i khim. tekhnol.*, **16**, 442 (1973).
163. B. D. Berezin, *KhGS*, **6**, 939 (1965).
164. B. D. Berezin, *Zh. fiz. khimii*, **39**, 321 (1965).
165. R. Woodward, *Zhurnal VKhO im. D. I. Mendeleeva*, **7**, 384 (1962).
166. D. Peters, *J. Chem. Soc.*, **1960**, 1274.
167. K. Fukui, *Bull. Chem. Soc. Japan*, **33**, 1591 (1960).
168. H. Hueckel, *Z. Physik*, **70**, 204 (1931); **76**, 628 (1932).
169. Kh. S. Bagdasaryan, *Zh. fiz. khimii*, **28**, 1096 (1954).
170. K. Rudenberg and C. Scherer, *J. Chem. Phys.*, **21**, 1565 (1953).
171. L. A. Blyumenfeld and V. A. Bendersky, *Zh. strukturn. khimii*, **4**, 405 (1963).
172. V. A. Lifshits and L. A. Blyumenfeld, *Zh. strukturn. khimii*, **8**, 433 (1967).
173. B. Pullman and A. Pullman, *Quantum Biochemistry* [*Kvantovaya biokhimiya*], Mir, Moscow, 1965.
174. R. Taube and P. Meyer, *Angew. Chem.*, **78**, 1022 (1966).
175. R. Bonnett, D. Dolphin, W. Johnson, D. Oldfield, and G. F. Stephenson, *Proc. Chem. Soc. (London)*, **1964**, 371.

176. R. Bonnett, I. A. D. Gale, and G. F. Stephenson, *J. Chem. Soc. C*, **1966**, 1600.
177. W. Schlesinger, A. Corwin, and L. Sargent, *J. Amer. Chem. Soc.*, **72**, 2867 (1950).
178. G. Donough and J. Miller, *J. Amer. Chem. Soc.*, **74**, 6106 (1952).
179. K. N. Solovyev, On the molecular structure of porphin and its derivatives [O stroenii molekul porfina i ego proizvodnykh], preprint, Institute of Physics of the BSSR Academy of Sciences, Minsk, 1969.
180. J. Robertson, *J. Chem. Soc.*, **1935**, 615.
181. R. Linstead and J. Robertson, *J. Chem. Soc.*, **1936**, 1736.
182. J. Robertson and J. Woodward, *J. Chem. Soc.*, **1937**, 219.
183. B. D. Berezin, *Izv. vuzov. Khimiya i khim. tekhnol.*, **2**, 165 (1959).
184. J. Nowakowski, *Roczniki Chem.*, **35**, 257 (1961).
185. B. N. Stepanenko, *Uspekhi khimii*, **13**, 462 (1944).
186. M. Gouterman, *J. Chem. Phys.*, **30**, 1139 (1959).
187. A. S. Davydov, *Quantum Mechanics* [*Kvantovaya mekhanika*], Nauka, Moscow, 1973.
188. F. Matsen, *J. Chem. Phys.*, **24**, 60 (1956); **28**; 950 (1958).
189. V. B. Yevstigneev and A. A. Krasnovsky, *DAN SSSR*, **58**, 1399 (1947).
190. B. D. Berezin, *Izv. vuzov. Khimiya i khim. tekhnol.*, **4**, 45 (1961).
191. B. D. Berezin, *DAN SSSR*, **141**, 353 (1961).
192. B. D. Berezin, *ZhNKh*, **7**, 2507 (1962).
193. B. D. Berezin, *Zh. prikl. khimii*, **36**, 1181 (1963).
194. B. D. Berezin, *Izv. vuzov. Khimiya i khim. tekhnol.*, **6**, 1016 (1963).
195. V. P. Gachkovsky, *Fizichesky sbornik Lvovskogo universiteta*, **3**(8), 372 (1957).
196. G. V. Bykov, *Electronic Charges of Bonds in Organic Compounds* [*Elektronnye zaryady svyazei v organicheskikh soedineniyakh*], USSR Academy of Sciences, Moscow, 1960.
197. A. N. Terenin, *Photochemistry of Dyes* [*Fotokhimiya krasiteley*], USSR Academy of Sciences, Moscow, 1947.
198. E. Jacobs, A. Holm, and E. Rabinowitch, *J. Chem. Phys.*, **22**, 141 (1954).
199. W. Simpson, *J. Chem. Phys.*, **17**, 1218 (1949).
200. B. Rackow, *Z. Naturforsch.*, **15a**, 129 (1960).
201. C. Spanjaard and G. Berthier, *J. Chim. phys.*, **58**, 169 (1961).
202. H. Kobayashi, *J. Chem. Phys.*, **30**, 1362, 1373 (1959).
203. H. Higgins, C. Rector, and J. Platt, *J. Chem. Phys.*, **18**, 1174 (1950).
204. A. N. Sevchenko, *Izv. AN SSSR, Seriya fiz.*, **1962**, 53.
205. C. Rimington, S. Mason, and O. Kennard, *Spectrochim. Acta*, **12**, 65 (1958).
206. J. Hoard, *Ann. New York Acad. Sci.*, **206**, 18 (1973).
207. A. Tulinsky, *Ann. New York Acad. Sci.*, **206**, 47 (1973).
208. A. Treibs, *Ann. New York Acad. Sci.*, **206**, 97 (1973).
209. J. Erdman and A. Corwin, *J. Amer. Chem. Soc.*, **68**, 1885 (1945).
210. A. Jackson and G. Dearden, *Ann. New York Acad. Sci.*, **206**, 151 (1973).
211. W. McEwen, *J. Amer. Chem. Soc.*, **58**, 1124 (1936); **68**, 711 (1946).
212. B. D. Berezin and G. V. Sennikova, *Kinetika i kataliz*, **9**, 528 (1968).
213. B. D. Berezin and G. V. Sennikova, *Zh. fiz. khimii*, **43**, 2499 (1969).
214. P. Barrett, R. Linstead, and G. Tuey, *J. Chem. Soc.*, **1940**, 1079.
215. B. D. Berezin, *Izv. vuzov. Khimiya i khim. tekhnol.*, **7**, 111 (1964).
216. E. Fleischer, *J. Amer. Chem. Soc.*, **85**, 146, 1353 (1963).
217. L. Webb and E. Fleischer, *J. Amer. Chem. Soc.*, **87**, 667 (1965).
218. L. Webb and E. Fleischer, *J. Chem. Phys.*, **43**, 3100 (1965).
219. E. Fleischer, C. Miller, and L. Webb, *J. Amer. Chem. Soc.*, **86**, 2342 (1964).
220. J. Chen, *J. Mol. Spectrosc.*, **23**, 131 (1967).

221. J. Chen, *J. Mol. Spectrosc.*, **23**, 144 (1973).
222. V. N. Klyuev, B. D. Berezin, and F. P. Snegireva, *Izv. vuzov. Khimiya i khim. tekhnol.*, **13**, 209 (1970).
223. A. V. Shablya and A. N. Terenin, *Optika i spektrosk.*, **9**, 533 (1960).
224. V. Ye. Kholmogorov and A. V. Shablya, *Optika i spektrosk.*, **17**, 298 (1964).
225. V. Ye. Kholmogorov and A. N. Sidorov, *DAN SSSR*, **178**, 897 (1968).
226. A. N. Sidorov and V. Ye. Kholmogorov, *TEKh*, **7**, 332 (1971).
227. A. N. Sidorov, *Zh. strukturn. khimii*, **14**, 255 (1973).
228. V. B. Yevstigneev and V. A. Gavrilova, *DAN SSSR*, **174**, 476 (1967).
229. V. B. Yevstigneev and V. A. Gavrilova, *Molekul. biologiya*, **2**, 869 (1968).
230. V. B. Yevstigneev, N. A. Sadovnikova, and G. D. Olovyannishnikova, *DAN SSSR*, **187**, 1184 (1969).
231. V. B. Yevstigneev and V. A. Gavrilova, *DAN SSSR*, **188**, 219 (1969).
232. I. I. Dilung and V. Ye. Karpitskaya, *DAN SSSR*, **152**, 367 (1963).
233. I. N. Chernyuk and I. I. Dilung, *DAN SSSR*, **165**, 1350 (1965).
234. I. N. Ivnitskaya, V. A. Mishchenko, and I. I. Dilung, *Biofizika*, **13**, 329 (1968).
235. N. I. Barboy and I. I. Dilung, *Biofizika*, **14**, 980 (1969).
236. N. I. Barboy and I. I. Dilung, *Biofizika*, **15**, 608 (1970).
237. D. Dolphin, *Ann. New York Acad. Sci.*, **206**, 177 (1973).
238. J. Fuhrhop and D. Mauzerall, *J. Amer. Chem. Soc.*, **91**, 4174 (1969).
239. B. D. Berezin, *Zh. obshch. khimii*, **43**, 2738 (1973).
240. I. L. Kukhtevich, *Ukr. khim. zhurnal*, **28**, 791 (1962).
241. B. D. Berezin, O. I. Koifman, and V. G. Andrianov, *Zh. fiz. khimii*, **47**, 1466 (1973).
242. B. D. Berezin, O. I. Koifman, and V. G. Andrianov, in *Proceedings of the Scientific and Technical Conference* [*Doklady nauchno-tekhnicheskoy konferentsii*], ICTI Press, 1973, p. 10.
243. V. G. Andrianov, Unpublished dissertation, Ivanovo Chemical Technology Institute, 1974.
244. B. D. Berezin, *Ukr. khim. zhurnal*, **28**, 532 (1962).
245. B. D. Berezin and O. I. Koifman, *Zh. fiz. khimii*, **44**, 1219 (1970).
246. O. I. Koifman, Unpublished dissertation, Ivanovo Chemical Technology Institute, 1970.
247. A. N. Drobysheva and B. D. Berezin, *ZhNKh*, **17**, 2468 (1972).
248. E. Stern and C. Timmons, *Electronic Absorption Spectroscopy in Organic Chemistry* [*Elektronnaya absorbtsionnaya spektroskopiya v organicheskoy khimii*], Mir, Moscow, 1974.
249. G. A. Krestov, *Izv. vuzov. Khimiya i khim. tekhnol.*, **10**, 627 (1967).
250. G. A. Krestov, *Thermochemistry of Compounds of Rare-Earth and Actinide Elements* [*Termokhimiya soedineniy redkozemelnykh i aktinoidnykh elementov*], Atomizdat, Moscow, 1972, p. 221.
251. B. D. Berezin, O. I. Koifman, A. N. Drobysheva, O. A. Golubchikov, and N. G. Mineeva, *ZhNKh*, **18**, 1540 (1973).
252. G. A. Krestov and B. D. Berezin, *Izv. vuzov. Khimiya i khim. tekhnol.*, **16**, 1789 (1973).
253. G. A. Krestov and B. D. Berezin, *Izv. vuzov. Khimiya i khim. tekhnol.*, **16**, 1343 (1973).
254. G. A. Krestov and B. D. Berezin, in *Thermodynamics and Structure of Solutions* [*Termodinamika i stroenie rastvorov*], Vol. 1, ICTI Press, p. 5.
255. J. Schiller, K. Bernauer, and S. Fallab, *Experientia*, **17**, 540 (1961).
256. K. Bernauer and S. Fallab, *Helv. Chim. Acta*, **45**, 2487 (1962).
257. J. Schiller and K. Bernauer, *Helv. Chim. Acta*, **46**, 3002 (1963).
258. C. Vestling and J. Downing, *J. Amer. Chem. Soc.*, **61**, 3511 (1939).

259. J. Weige and R. Livingston, *J. Amer. Chem. Soc.*, **75**, 2175 (1953).
260. N. Fuson and M. Losien, *C. r. hebd. Séanc. Acad. Sci., Paris*, **232**, 833 (1951).
261. S. Mason, *J. Chem. Soc.*, **1958**, 976.
262. J. Falk and J. Willis, *Austral. J. Scient. Res.*, **4**, 579 (1951).
263. N. Naqui and Q. Fernando, *J. Org. Chem.*, **25**, 551 (1960).
264. B. D. Berezin, *ZhNKh*, **15**, 2093 (1970).
265. G. Niemann, *Hoppe-Seyle's J. Physiol. Chem.*, **146**, 181 (1925).
266. S. Aronoff and C. Weast, *J. Org. Chem.*, **6**, 550 (1941).
267. S. Aronoff, *Chem. Rev.*, **47**, 175 (1950).
268. J. Phillips, *Rev. Pure Appl. Chem. (Australia)*, **10**, 35 (1960).
269. J. Conant, B. Chew, and E. Dietz, *J. Amer. Chem. Soc.*, **56**, 2185 (1934).
270. N. Hall, *J. Amer. Chem. Soc.*, **52**, 5115 (1930).
271. S. Aronoff, *J. Phys. Chem.*, **62**, 428 (1958).
272. B. Dempsey, M. Lowe, and J. Phillips, *Haematin Enzymes*, Pergamon, London, 1961, p. 29.
273. J. Phillips, *Current Trends in Heterocyclic Chemistry*, Butterworths, London, 1958, p. 30.
274. N. A. Izmailov, *Electrochemistry of Solutions* [*Elektokhimiya rastvorov*], Khimiya, Moscow, 1966.
275. A. P. Kreshkov, L. N. Bykova, and N. A. Kazaryan, *Acid–Base Titration in Non-aqueous Solutions* [*Kislotno-osnovnoe titrovanie v nevodnykh rastvorakh*], Khimiya, Moscow, 1967.
276. B. D. Berezin, *Izv. vuzov. Khimiya i khim. tekhnol.*, **9**, 717 (1966).
277. T. I. Strelkova, G. P. Gurinovich, and G. N. Sinyakov, *Zh. prikl. spektrosk.*, **4**, 429 (1966).
278. A. I. Gelbshtein, G. G. Sheglova, and M. I. Temkin, *DAN SSSR*, **107**, 108 (1956).
279. V. A. Palm, *Zh. fiz. khimii*, **32**, 380 (1958).
280. B. D. Berezin, *Izv. vuzov. Khimiya i khim. tekhnol.*, **4**, 379 (1961).
281. B. D. Berezin, *ZhNKh*, **7**, 2501 (1962).
282. B. D. Berezin, *Zh. fiz. khimii*, **35**, 248 (1961).
283. A. S. Akopov, B. D. Berezin, V. N. Klyuev, and A. A. Solovyev, *ZhNKh*, **18**, 965 (1973).
284. B. D. Berezin, Unpublished doctoral dissertation, Institute of Organic and Inorganic Chemistry of the UkSSR Academy of Sciences, Kiev, 1966.
285. L. I. Savransky, *Zh. prikl. spektrosk.*, **13**, 1084 (1970).
286. S. Mataga and N. Mataga, *Bull. Chem. Soc. Japan*, **32**, 511 (1959).
287. U. Ahrens and H. Kuhn, *Z. phys. Chem.*, **37**, 1 (1963).
288. A. N. Sidorov and A. N. Terenin, *Optika i spektrosk.*, **11**, 325 (1961).
289. B. D. Berezin, *Izv. vuzov. Khimiya i khim. tekhnol.*, **11**, 537 (1968).
290. T. I. Strelkova and G. P. Gurinovich, *Zh. prikl. spektrosk.*, **7**, 882 (1967).
291. B. D. Berezin, *Izv. vuzov. Khimiya i khim. tekhnol.*, **2**, 10 (1959).
292. R. P. Smirnov and B. D. Berezin, *Zh. obshch. khimii*, **37**, 789 (1967).
293. R. P. Smirnov, B. D. Berezin, and L. A. Skvortsova, *Zh. fiz. khimii*, **41**, 1630 (1967).
294. R. P. Smirnov and B. D. Berezin, *Zh. fiz. khimii*, **41**, 673 (1967).
295. B. D. Berezin and N. I. Sosnikova, in *9th All-Union Conference on the Chemistry of Complex Compounds* [*IX Vesoyuznoe soveshchanie po khimii kompleksnykh soedineniy*], abstracts, Tashkent University Press, Tashkent, 1963, p. 56.
296. B. D. Berezin and N. I. Sosnikova, *Zh. fiz. khimii*, **38**, 1216 (1964).
297. B. D. Berezin and N. I. Sosnikova, *Zh. fiz. khimii*, **39**, 1348 (1965).
298. B. D. Berezin and N. I. Sosnikova, *Kinetika i kataliz*, **7**, 248 (1966).
299. B. D. Berezin and N. I. Sosnikova, *Izv. vuzov. Khimiya i khim. tekhnol.*, **10**, 1216 (1967).

300. N. I. Volkova and B. D. Berezin, *ZhNKh*, **15**, 961 (1970).
301. N. I. Volkova and B. D. Berezin, *ZhNKh*, **16**, 3037 (1971).
302. N. I. Volkova, Unpublished dissertation, Ivanovo Chemical Technology Institute, 1969.
303. B. D. Berezin, N. I. Volkova, and Ye. B. Karavaeva, *ZhNKh*, **19**, 3053 (1974).
304. B. D. Berezin, N. I. Volkova, and Ye. B. Karavaeva, in *1st All-Union Symposium on Bioinorganic Chemistry* [*I Vsesoyuzny simposium po bioneorganicheskoy khimii*], abstracts, Naukova dumka, Kiev, 1974, p. 6.
305. B. D. Berezin and G. I. Smirnova, *Izv. vuzov. Khimiya i khim. tekhnol.*, **9**, 749 (1966).
306. B. D. Berezin and G. I. Smirnova, *Zh. fiz. khimii,* **41**, 1323 (1967).
307. G. I. Smirnova, Unpublished dissertation, Ivanovo Chemical Technology Institute, 1968.
308. B. D. Berezin and G. I. Smirnova, *Zh. fiz. khimii*, **41**, 2008 (1967).
309. B. D. Berezin and G. I. Smirnova, *Zh. fiz. khimii*, **41**, 2359 (1967).
310. G. I. Smirnova and B. D. Berezin, in *All-Union Conference on Reactivity of Nitrogen-containing Heterocycles* [*Vsesoyuznaya konferentsiya po reaktsionnoy sposobnosti azotosoderzhashchikh geterotsiklov*], abstracts, Urals Polytechnical Institute Press, Sverdlovsk, 1967, p. 16.
311. B. D. Berezin and G. I. Smirnova, *Zh. fiz. khimii*, **42**, 1952 (1968).
312. G. I. Smirnova and B. D. Berezin, *Zh. fiz. khimii*, **43**, 659 (1969).
313. B. D. Berezin and L. V. Klopova, *Zh. fiz. khimii*, **45**, 1679 (1971).
314. B. D. Berezin and L. V. Klopova, *Zh. fiz. khimii*, **45**, 2193 (1971).
315. B. D. Berezin and L. V. Klopova, *Zh. fiz. khimii*, **46**, 2258 (1972).
316. B. D. Berezin and L V. Klopova, *Zh. fiz. khimii*, **48**, 312 (1974).
317. B. D. Berezin and L. V. Klopova, *Zh. fiz. khimii*, **48**, 2917 (1974).
318. B. D. Berezin, *Izv. vuzov. Khimiya i khim. tekhnol.*, **14**, 336 (1971).
319. B. D. Berezin, *Proceedings of the XIII ICCC, Cracow*, **2**, 293 (1970).
320. B. D. Berezin, *Proceedings of the XIV ICCC, Toronto*, 614 (1971).
321. B. D. Berezin, *TEKh*, **9**, 500 (1973).
322. B. D. Berezin, *Transactions of the Ivanovo Chemical Technology Institute*, **15**, 11 (1973).
323. B. D. Berezin, O. I. Koifman, N. I. Volkova, L. V. Klopova, O. A. Golubchikov, V. G. Andrianov, and O. B. Lapshina, *Proceedings of the XV ICCC, Moscow*, 394 (1973).
324. B. D. Berezin, in *Problems of Kinetics and Catalysis* [*Voprosy kinetiki i kataliza*], Vol. 2, ICTI Press, p. 26.
325. B. Baker and G. Hudson, *J. Phys. Chem.*, **65**, 1078 (1961).
326. I. B. Bersuker, *Structure and Properties of Coordination Compounds* [*Stroenie i svoistva koordinatsionnykh soedineniy*], Khimiya, Leningrad, 1971, p. 188.
327. A. N. Drobysheva and B. D. Berezin, *Zh. fiz. khimii*, **45**, 622 (1971).
328. K. B. Yatsimirsky and V. P. Vasilyev, *Instability Constants* [*Konstanty nestoikosti*], USSR Academy of Sciences, Moscow, 1959.
329. G. I. Smirnova, Unpublished dissertation, Ivanovo Chemical Technology Institute, 1968.
330. O. I. Koifman, N. L. Golubchikova, and B. D. Berezin, *Izv. vuzov. Khimiya i khim. tekhnol.*, **19**, 1853 (1976).
331. B. D. Berezin and L. V. Klopova, *Zh. fiz. khimii*, **51**, 9 (1977).
332. G. A. Krestov, *Thermodynamics of Ionic Processes in Solutions* [*Termodinamika ionnykh protsessov v rastvorakh*], Khimiya, Leningrad, 1973.
333. G. A. Krestov, *4th Mendeleyev Discussion of Solutions* [*IV Mendeleevskaya diskussiya po rastvoram*], abstracts, ICTI Press, 1975, p. 6.
334. B. D. Berezin and N. I. Volkova, *ZhNKh*, **18**, 1285 (1973).
335. B. D. Berezin and O. I. Koifman, *Zh. fiz. khimii*, **45**, 1451 (1971).

336. A. N. Drobysheva and B. D. Berezin, *ZhNKh*, **19**, 837 (1974).
337. O. A. Golubchikov, Unpublished dissertation, Ivanovo Chemical Technology Institute, 1974.
338. B. D. Berezin, O. I. Koifman, and O. A. Golubchikov, in *Problems of Kinetics and Catalysis* [*Voprosy kinetiki i kataliza*], Vol. 2, ICTI Press, 1974, p. 94.
339. L. P. Karmanova, A. N. Drobysheva, and B. D. Berezin, *Zh. fiz. khimii*, **51**, 1344 (1977).
340. R. P. Yevstigneeva, V. N. Luzgina, and Ye. I. Filippovich, in *3rd All-Union Symposium on Kinetics and Mechanisms* [*III Vsesoyuzny simposium po kinetike i mekhanizmam*], abstracts, ICTI Press, 1974, p. 107.
341. E. Choi and E. Fleischer, *Inorg. Chem.*, **2**, 94 (1963).
342. E. Fleischer, E. Choi, P. Hambright, and A. Stone, *Inorg. Chem.*, **3**, 1284 (1964).
343. F. Longo, E. Brown, D. Quimby, A. Adler, and M. Meotner, *Ann. New York Acad. Sci.*, **206**, 420 (1973).
344. W. Gutman, *Chemistry of Coordination Compounds in Non-aqueous Solutions* [*Khimiya koordinatsionnykh soedineniy v nevodnykh rastvorakh*], Mir, Moscow, 1971.
345. O. I. Koifman and B. D. Berezin, *Izv. vuzov. Khimiya i khim. tekhnol.*, **13**, 204 (1970).
346. B. D. Berezin, O. I. Koifman, O. A. Golubchikov, and V. G. Andrianov, *Izv. vuzov. Khimiya i khim. tekhnol.*, **13**, 513 (1970).
347. O. I. Koifman and B. D. Berezin, *Zh. obshch. khimii*, **40**, 2701 (1970).
348. B. D. Berezin and O. I. Koifman, *Zh. obshch. khimii*, **42**, 699 (1972).
349. B. D. Berezin and O. I. Koifman, *Zh. fiz. khimii* **46**, 42 (1972).
350. O. I. Koifman, O. A. Golubchikov, V. G. Andrianov, and B. D. Berezin, *Izv. vuzov. Khimiya i khim. tekhnol.*, **14**, 518 (1971).
351. A. Parker, *Uspekhi khimii*, **40**, 2203 (1971).
352. C. Langford and G. Grey, *Processes of Ligand Substitution* [*Protsessy zameshcheniya ligandov*], Mir, Moscow, 1969.
353. A. A. Grinberg, *An Introduction into the Chemistry of Complex Compounds* [*Vvedenie v khimiyu kompleksnykh soedineniy*], Khimiya, Moscow–Leningrad, 1966.
354. F. Basolo and R. Pearson, *Mechanisms of Inorganic Reactions*, New York, 1958.
355. K. P. Mishchenko and G. M. Poltoratsky, *Problems of Thermodynamics and Structure of Aqueous and Non-aqueous Electrolyte Solutions* [*Voprosy termodinamiki i stroeniya vodnykh i nevodnykh rastvorov elektrolitov*], Khimiya, Leningrad, 1968.
356. M. Lowe and J. Phillips, *Nature*, **190**, 262 (1961).
357. M. Lowe and J. Phillips, *Nature*, **194**, 1058 (1962).
358. N. I. Volkova, Ye. B. Karavaeva, and B. D. Berezin, *ZhNKh*, **20**, 1929 (1975).
359. S. G. Entelis and R. P. Tiger, *Kinetics of Reactions in the Liquid Phase* [*Kinetika reaktsiy v zhidkoy faze*], Khimiya, Moscow, 1973.
360. J. Weaver and P. Hambright, *Inorg. Chem.*, **8**, 167 (1969).
361. P. Hambright and E. Fleischer, *Inorg. Chem.*, **9**, 1757 (1970).
362. P. Hambright, *J. Inorg. Nuclear Chem.*, **32**, 2449 (1970).
363. P. Hambright, *Coord. Chem. Rev.*, **6**, 247 (1971).
364. B. Shah, B. Shears, and P. Hambright, *Inorg. Chem.*, **10**, 1828 (1971).
365. R. Khosopour and P. Hambright, *Chem. Comm.*, **1972**, 13.
366. B. Shears, B. Shah, and P. Hambright, *J. Amer. Chem. Soc.*, **93**, 776 (1971).
367. P. Hambright, *Ann. New York Acad. Sci.*, **206**, 443 (1973).

368. R. Longo, J. Finarelli, and E. Schmalzbach, *J. Phys. Chem.*, **74**, 3296 (1970).
369. B. D. Berezin, L. P. Shormanova, and R. I. Feldman, *ZhNKh*, **19**, 1833 (1974).
370. B. D. Berezin, L. P. Shormanova, V. N. Klyuev, and A. B. Korzhenevsky, *ZhNKh*, **19**, 2745 (1974).
371. J. Bjerrum, *Formation of Metal Amines in Aqueous Solution* [*Obrazovanie aminov metallov v vodnom rastvore*], IL Publishers, Moscow, 1961.
372. M. Joslyn and G. Mackinney, *J. Amer. Chem. Soc.*, **60**, 1132 (1938).
373. G. Mackinney and M. Joslyn, *J. Amer. Chem. Soc.*, **62**, 231 (1940).
374. G. Mackinney and M. Joslyn, *J. Amer. Chem. Soc.*, **63**, 2520 (1941).
375. B. D. Berezin, L. P. Karmanova, and A. N. Drobysheva, in *Proceedings of the Scientific and Technical Conference* [*Doklady nauchno-tekhnicheskoy konferentsii*], ICTI Press, 1973, p. 10.
376. B. D. Berezin and A. N. Drobysheva, *ZhNKh*, **13**, 2720 (1968).
377. B. D. Berezin and A. N. Drobysheva, *Zh. fiz. khimii*, **40**, 719 (1966).
378. B. D. Berezin and A. N. Drobysheva, *Zh. fiz. khimii*, **41**, 4021 (1967).
379. A. N. Drobysheva and B. D. Berezin, *ZhNKh*, **16**, 622 (1971).
380. L. P. Karmanova, A. N. Drobysheva, and B. D. Berezin, *Transactions of the Ivanovo Chemical Technology Institute*, **17**, 27 (1974).
381. A. N. Drobysheva and B. D. Berezin, *Zh. fiz. khimii*, **51**, 21 (1977).
382. L. P. Karmanova, A. N. Drobysheva, and B. D. Berezin, *Zh. fiz. khimii*, **51**, 629 (1977).
383. B. D. Berezin, A. N. Drobysheva, and Ye. A. Venediktov, *Koordin. khimiya*, **2**, 491 (1976).
384. Ye. A. Venediktov, A. N. Drobysheva, and B. D. Berezin, *Koordin. khimiya*, **2**, 346 (1976).
385. B. D. Berezin, *Zh. fiz. khimii*, **36**, 494 (1962).
386. B. D. Berezin, *Zh. fiz. khimii*, **37**, 2474 (1963).
387. B. D. Berezin, *Zh. fiz. khimii*, **38**, 2957 (1964).
388. B. D. Berezin, *Zh. fiz. khimii*, **38**, 850 (1964).
389. B. D. Berezin, *Zh. fiz. khimii*, **39**, 1082 (1965).
390. B. D. Berezin and A. S. Akopov, *Zh. obshch. khimii*, **44**, 1089 (1974).
391. A. S. Akopov and B. D. Berezin, *Transactions of the Ivanovo Chemical Technology Institute*, **19**, 85 (1975).
392. R. Shellgrove and R. Plane, *J. Amer. Chem. Soc.*, **90**, 3185 (1968).
393. B. D. Berezin and A. N. Shlyapova, *Vysokomol. soed.*, **A-15**, 1671 (1973).
394. B. D. Berezin, A. S. Akopov, and T. N. Kharina, *Izv. vuzov. Khimiya i khim. tekhnol.*, **18**, 932 (1975).
395. B. D. Berezin, A. S. Akopov, L. P. Shormanova, and T. N. Kharina, *Izv. vuzov. Khimiya i khim. tekhnol.*, **18**, 1446 (1975).
396. B. D. Berezin, A. S. Akopov, and O. B. Lapshina, *Vysokomol. soed.*, **A-16**, 450 (1974).
397. S. Schanderl, C. Chichester, and G. Marsh, *J. Org. Chem.*, **27**, 3865 (1962).
398. B. D. Berezin and A. N. Drobysheva, in *Problems of Kinetics and Catalysis* [*Voprosy kinetiki i kataliza*], Vol. 1, ICTI Press, 1973, p. 3.
399. L. P. Karmanova, Unpublished dissertation, Ivanovo Chemical Technology Institute, 1974.
400. Ya. K. Syrkin and M. Ye. Dyatkina, *Chemical Bonding and Molecular Structure* [*Khimicheskaya svyaz i stroenie molekul*], Goskhimizdat, Moscow–Leningrad, 1946.
401. A. N. Drobysheva, Unpublished dissertation, Ivanovo Chemical Technology Institute, 1970.
402. B. D. Berezin and G. V. Sennikova, *Izv. vuzov. Khimiya i khim. tekhnol.*, **10**, 563 (1967).

403. J. Phillips, *Rev. Pure Appl. Chem.*, **10**, 35 (1960).
404. J. Hoard, M. Hamor, T. Hamor, and W. Caughey, *J. Amer. Chem. Soc.*, **87**, 2312 (1965).
405. A. Corwin and J. Erdmann, *J. Amer. Chem. Soc.*, **68**, 2473 (1946).
406. W. Caughey and A. Corwin, *J. Amer. Chem. Soc.*, **77**, 1509 (1955).
407. B. D. Berezin, V. I. Shkrobysheva, and B. N. Melnikov, *Transactions of the Ivanovo Chemical Technology Institute*, **17**, 18 (1974).
408. B. D. Berezin, B. N. Melnikov, and V. I. Shkrobysheva, *Zh. fiz. khimii*, **48**, 2938 (1974).
409. V. I. Shkrobysheva, B. D. Berezin, and B. N. Melnikov, *Izv. vuzov. Khimiya i khim. tekhnol.*, **19**, 906 (1976).
410. S. Gaspard, M. Verdaguer, and R. Viovy, *C. r. hebd. Séanc. Acad. Sci., Paris, C*, **275**, 572 (1972).
411. R. Grahn, *Arkiv Fysik.*, **19**, 147 (1961).
412. I. S. Akimov and G. P. Korsunkovsky, *Optika i spektrosk.*, **8**, 427 (1960).
413. A. Albert, in *Physical Methods in the Chemistry of Heterocyclic Compounds* [*Fizicheskie metody v khimii geterotsiklicheskikh soedineniy*], Khimiya, Moscow–Leningrad, 1966, p. 15.
414. J. Anderson, E. Bradbrook, A. Cook, and R. Linstead, *J. Chem. Soc.*, **1938**, 1151.
415. L. Hammett, *Chem. Rev.*, **16**, 125 (1935).
416. D. Frank-Kamenetsky, *Uspekhi khimii*, **10**, 373 (1941).
417. B. Busch, *J. Chem. Educ.*, **33**, 498 (1956).
418. H. Taube, *Chem. Rev.*, **50**, 69 (1952).
419. A. A. Grinberg, *An Introduction into the Chemistry of Complex Compounds* [*Vvedenie v khimiyu kompleksnykh soedineniy*], Goskhimizdat, Moscow–Leningrad, 1951, pp. 42, 46, 296, 311, 355, 362.
420. Yu. N. Kukushkin, in *Problems of Kinetics and Catalysis* [*Voprosy kinetiki i kataliza*], Vol. 1, ICTI Press, 1973, p. 16.
421. B. D. Berezin and A. N. Drobysheva, in *10th All-Union Conference on the Chemistry of Complex Compounds* [*X Vsesoyuznoe soveshchanie po khimii kompleksnykh soedineniy*], abstracts, Naukova dumka, Kiev, 1969, p. 37.
422. B. Dempsey, M. Lowe, and J. Phillips, *Haematin Enzyme Symposium*, Pergamon, Canberra, 1959.
423. R. Bates, in *Electrochemistry of Metals in Non-aqueous Solutions* [*Elektrokhimiya metallov v nevodnykh rastvorakh*] (Ya. M. Kolotyrkin, ed.), Mir, Moscow, 1974, p. 305.
424. B. D. Berezin, *Izv. vuzov. Khimiya i khim. tekhnol.*, **4**, 380 (1961).
425. *Modern Chemistry of Coordination Compounds* [*Sovremennaya khimiya koordinatsionnykh soedineniy*] (D. Lewis and R. Wilkins, eds), IL Publishers, Moscow, 1963.
426. F. Cotton and D. Wilkinson, *Modern Inorganic Chemistry* [*Sovremennaya neorganicheskaya khimiya*], Vol. 1, Mir, Moscow, 1969.
427. D. Cowan and A. Baum, *J. Amer. Chem. Soc.*, **92**, 2151 (1970).
428. K. B. Yatsimirsky, *TEKh*, **1**, 41 (1965).
429. K. B. Yatsimirsky, *ZhNKh*, **11**, 2429 (1966).
430. S. Husain and J. Jones, *Inorg. Nuclear Chem. Lett.*, **10**, 105 (1974).
431. L. Boucher, *Coord. Chem. Rev.*, **7**, 289 (1972).
432. R. Williams, *Chem. Revs*, **56**, 299 (1956).
433. *Ann. New York Acad. Sci.*, **206**, 32, 247, 296, 397, 404, 453, 463, 579, 604, 614 (1973).
434. A. N. Sidorov, Unpublished doctoral dissertation, Institute of Physical Chemistry of the UkSSR Academy of Sciences, Kiev, 1969.

435. *Stability of Mixed Complexes in Solutions* [*Ustoichivost smeshannykh kompleksnykh soedineniy v rastvorakh*], Ilim, Frunze, 1971.
436. J. Gibson, D. Ingram, and D. Schonland, *Discuss. Faraday Soc.*, **26**, 72 (1958).
437. I. B. Bersuker and A. V. Ablov, *Chemical Bonding in Complex Compounds* [*Khimicheskaya svyaz v kompleksnykh soedineniyakh*], Shtiintsa, Kishinev, 1962, p. 31.
438. C. Storm, A. Corwin, R. Arellano, M. Martz, and R. Weintraub, *J. Amer. Chem. Soc.*, **88**, 2525 (1966).
439. J. Elvidge and A. Lever, *Proc. Chem. Soc.*, **20**, 123 (1959).
440. O. I. Koifman, T. A. Koroleva, and B. D. Berezin, *Transactions of the Ivanovo Chemical Technology Institute*, **20**, 78 (1976).
441. B. S. Neporent, *Zh. fiz. khimii*, **30**, 1048 (1956).
442. J. Nowakowski, *Roczn. chem.*, **49**, 141 (1966).
443. C. Weiss, H. Kobayashi, and M. Gouterman, *J. Mol. Spectroscopy*, **16**, 415 (1965).
444. W. Caughey, R. Deal, C. Weiss, and M. Gouterman, *J. Mol. Spectroscopy*, **16**, 432 (1965).
445. J. Anderson, E. Bradbrook, A. Cook, and R. Linstead, *J. Chem. Soc.*, **1938**, 1151.
446. B. D. Berezin, *Izv. vuzov. Khimiya i khim. tekhnol.*, **7**, 202 (1964).
447. K. B. Yatsimirsky, in *Spectroscopic Methods in the Chemistry of Complex Compounds* [*Spektroskopicheskie metody v khimii kompleksnykh soedineniy*], Khimiya, Moscow–Leningrad, 1964, p. 5.
448. J. Conant, *J. Amer. Chem. Soc.*, **53**, 1615 (1931).
449. B. Dunicz, T. Thomas, R. Marcel, and R. Livingston, *J. Amer. Chem. Soc.*, **73**, 3388 (1951).
450. A. Weller, *J. Amer. Chem. Soc.*, **76**, 5819 (1954).
451. B. D. Berezin and O. B. Lapshina, *Zh. fiz. khim.*, **50**, 2007 (1976).
452. B. D. Berezin and O. B. Lapshina, *Izv. vuzov. Khimiya i khim. tekhnol.*, **19**, 379 (1976).
453. P. Hynninen, *Acta Chem. Scand.*, **27**, 1771 (1973).
454. A. A. Krasnovsky and M. I. Bystrova, *DAN SSSR*, **174**, 480 (1967).
455. M. I. Bystrova and A. A. Krasnovsky, *Molek. biol.*, **1**, 362 (1967).
456. M. I. Bystrova and A. A. Krasnovsky, *Molek. biol.*, **2**, 847 (1968).
457. M. I. Bystrova and A. A. Krasnovsky, *Molek. biol.*, **5**, 291 (1971).
458. G. Closs, J. Katz, F. Pennington, M. Thomas, and H. Strain, *J. Amer. Chem. Soc.*, **85**, 3809 (1963).
459. J. Katz and T. Janson, *Ann. New York Acad. Sci.*, **206**, 579 (1973).
460. B. D. Berezin, O. I. Koifman, N. L. Golubchikova, and I. A. Lyzhnikova, in *Chemistry and Technology of Dyeing, Synthesis of Dyes and Polymer Materials* [*Khimiya i tekhnologiya krasheniya, sinteza krasiteley i polimernykh materialov*], ICTI Press, 1973, p. 81.
461. H. Kamogawa, *Polymer Letters*, **10**, 711 (1972).
462. A. A. Tager, *Physical Chemistry of Polymers* [*Fiziko-khimiya polimerov*], Khimiya, Moscow, 1968.
463. B. D. Berezin and A. N. Shlyapova, *Izv. vuzov. Khimiya i khim. tekhnol.*, **12**, 1641 (1969).
464. K. B. Yatsimirsky and T. V. Malkova, in *Spectroscopic Methods in the Chemistry of Complex Compounds* [*Spektroskopicheskie metody v khimii kompleksnykh soedineniy*], Khimiya, Moscow, 1964, p. 102.
465. A. S. Akopov, Unpublished dissertation, Ivanovo Chemical Technology Institute, 1972.

466. A. S. Akopov, T. N. Lomova, and B. D. Berezin, *Koord. khimiya,* **8**, 985 (1976).
467. *Ann. New York Acad. Sci.*, **206**, 7–720 (1973).
468. V. L. Yermolaev, in *Elementary Photoprocesses in Molecules* [*Elementarnye fotoprotsessy v molekulakh*], Nauka, Moscow–Leningrad, 1966, p. 147.
469. A. N. Sidorov, in *Elementary Photoprocesses in Molecules* [*Elementarnye fotoprotsessy v molekulakh*], Nauka, Moscow–Leningrad, 1966, p. 267.
470. K. N. Solovyev, S. F. Shkirman, and T. F. Kachura, *Izv. AN SSSR, seriya fiz.*, **1963**, 767.
471. A. N. Sevchenko, *Izv. AN SSSR, seriya fiz.*, **1962**, 53.
472. I. L. Kukhtevich, in *Problems of Photosynthesis* [*Problemy fotosinteza*], USSR Academy of Sciences, Moscow, 1959, p. 125.
473. B. D. Berezin and A. N. Drobysheva, *Zh. fiz. khimii*, **42**, 2092 (1968).
474. C. Kautsky, *Trans. Faraday Soc.*, **35**, 216 (1939).
475. B. Ya. Dain and M. S. Ashkinazi, in *Problems of Photosynthesis* [*Problemy fotosinteza*], USSR Academy of Sciences, Moscow, 1959, p. 57.
476. T. Moss, A. Bearden and W. Caughey, *J. Chem. Physics*, **51**, 2624 (1969).
477. A. V. Karyakin and A. V. Shablya, *Optika i spektrosk.*, **5**, 655 (1958).
478. A. A. Krasnovsky, in *Elementary Photoprocesses in Molecules* [*Elementarnye fotoprotsessy v molekulakh*], Nauka, Moscow–Leningrad, 1966, p. 228.
479. D. N. Shigorin, V. G. Plotnikov, and V. Potapov, *Zh. fiz. khimii*, **40**, 192, 295 (1966).
480. M. P. Tsvirko, K. N. Solovyev, and V. V. Sapunov, *Optika i spektrosk.*, **36**, 335 (1974).
481. M. P. Tsvirko, V. V. Sapunov, and K. N. Solovyev, *Optika i spektrosk.*, **34**, 1094 (1973).
482. T. F. Kachura, A. N. Sevchenko, K. N. Solovyev, and M. P. Tsvirko, *DAN SSSR*, **217**, 1121 (1974).
483. K. N. Solovyev, V. A. Mashenkov and T. F. Kachura, *Optika i spektrosk.*, **27**, 50 (1969).
484. K. N. Solovyev, M. P. Tsvirko, A. T. Gradyushko, and D. T. Kozhich, *Optika i spektrosk.*, **33**, 871 (1972).
485. P. Wasser and J. Fuhrhop, *Ann. New York Acad. Sci.*, **206**, 533 (1973).
486. A. A. Krasnovsky, *DAN SSSR*, **58**, 617 (1947).
487. S. V. Vulfson, Unpublished dissertation, NIOPIK, Moscow, 1974.
488. K. Kadish and D. Davis, *Ann. New York Acad. Sci.*, **206**, 409 (1973).
489. A. S. Akopov, B. D. Berezin, V. N. Klyuev, and G. G. Morozova, *ZhNKh*, **20**, 1264 (1975).
490. A. N. Sidorov and V. G. Maslov, *Uspekhi khimii*, **44**, 577 (1975).
491. A. H. Cook, *J. Chem. Soc.*, **1938**, 1761, 1772, 1774.
492. C. Paquot, *C. r. hebd. Séanc. Acad. Sci., Paris,* **209**, 171 (1939); **214**, 173 (1942); **226**, 258 (1948).
493. C. Paquot, *Bull. Soc. chim. (France)*, **8**, 965 (1941); **12**, 450 (1945).
494. H. Hock and H. Kropf, *Z. prakt. Chem.*, **9**, 173 (1959); **13**, 285 (1961).
495. H. Kropf, *Annalen*, **637**, 73, 93, 111 (1960); **739**, 95 (1970).
496. H. Kropf, *Tetrahedron Letters*, **13**, 577 (1962); **18**, 659 (1967).
497. S. Z. Roginsky, A. A. Berlin, N. G. Matveeva, L. N. Kutseva, R. M. Aseeva, L. G. Cherkashina, and A. I. Sherle, *DAN SSSR*, **148**, 118 (1963).
498. H. Inoue, I. Kida, and E. Imoto, *Bull. Chem. Soc. Japan*, **41**, 184, 684 (1968).
499. S. Tanaka, T. Onishi, and K. Tamaru, *Bull. Chem. Soc. Japan*, **41**, 2557 (1968).
500. N. N. Kundo and N. P. Keyer, *Kinetika i kataliz*, **8**, 796, 1325 (1967); **11**, 91 (1970).

501. B. D. Berezin and A. N. Shlyapova, *Izv. vuzov. Khimiya i khim. tekhnol.*, **14**, 1665 (1971).
502. A. N. Shlyapova and B. D. Berezin, *Izv. vuzov. Khimiya i khim. tekhnol.*, **14**, 1810 (1971).
503. A. A. Balandin, V. I. Spitsyn, A. P. Rudenko, and N. D. Dobroselskaya, *Kinetika i kataliz*, **9**, 635 (1968).
504. P. Waldmeir and H. Sigel, *Chimia*, **24**, 196 (1970).
505. M. Ishikawa, M. Sudo, M. Soma, T. Onishi, and K. Tamaru, *J. Amer. Chem. Soc.*, **91**, 1538 (1969).
506. K. A. Radyushkina, O. A. Levina, M. R. Tarasevich, R. Kh. Burshtein, B. D. Berezin, L. P. Shormanova, and O. I. Koifman, *Eletrokhimiya*, **11**, 989 (1975).
507. V. A. Yerin, S. N. Troshina, S. N. Pobedinsky, and K. N. Belonogov, in *Problems of Kinetics and Catalysis* [*Voprosy kinetiki i kataliza*], Vol. 2, ICTI Press, 1974, p. 99.
508. H. Alt, H. Binder, and G. Sandstede, *J. Catalysis*, **28**, 8 (1973).
509. V. Ye. Kholmogorov and D. N. Glebovsky, *Optika i spektrosk.*, **12**, 728 (1962).
510. *Organic Semiconductors* [*Organicheskie poluprovodniki*], Nauka, Moscow, 1968.
511. A. I. Minkov, Unpublished dissertation, Institute of Catalysis, Siberian Branch of the USSR Academy of Sciences, Novosibirsk, 1967.
512. A. T. Vartanyan, *Zh. fiz. khimii*, **22**, 769 (1948).
513. A. T. Vartanyan and I. A. Karpovich, *DAN SSSR*, **111**, 561 (1956).
514. A. T. Vartanyan and I. A. Karpovich, *Zh. fiz. khimii*, **32**, 178, 274 (1958).
515. A. T. Vartanyan, *Izv. AN SSSR, seriya fiz.*, **1956**, 1541.
516. A. T. Vartanyan, *DAN SSSR*, **143**, 1317 (1962).
517. A. T. Vartanyan, *DAN SSSR*, **149**, 812 (1963).
518. A. T. Vartanyan, *DAN SSSR*, **153**, 70 (1963).
519. D. Eley, *Nature*, **162**, 819 (1948).
520. D. Eley and G. Parfitt, *Trans. Faraday Soc*, **51**, 1529 (1955).
521. P. Filding and F. Gutman, *J. Chem. Phys.*, **26**, 411 (1957).
522. C. Hamann and M. Starke, *Phys. Status Solidi*, **26**, 411 (1957).
523. F. Gutman and A. Netschey, *J. Chem. Soc.*, **1962**, 2355.
524. S. Basu and W. Moore, *J. Phys. Chem.*, **67**, 1563 (1963).
525. A. N. Terenin, *Zhurnal VKhO im. D. I. Mendeleeva*, **5**, 498 (1960).
526. V. Arnold and E. McLeigh, in *Structure and Functions of Photosynthetic Apparatus* [*Struktura i funktsii fotosinteticheskogo apparata*], IL Publishers, Moscow, 1962, p. 9.
527. V. A. Shorin, G. N. Meshkova, A. T. Vartanyan, N. N. Pribytkova, M. I. Alyanov, and V. F. Borodkin, *Izv. vuzov. Khimiya i khim. tekhnol.*, **16**, 1904 (1973).
528. A. P. Martynenko, V. A. Shorin, V. F. Borodkin, and M. A. Alyanov, *Izv. vuzov. Khimiya i khim. tekhnol.*, **17**, 528 (1974).
529. V. A. Shorin, Unpublished dissertation, Ivanovo Chemical Technology Institute, 1974.
530. W. Felmayer and J. Wolf, *J. Electrochem. Soc.*, **105**, 141 (1958).
531. M. Danzig, C. Liang, and E. Passaglia, *J. Amer. Chem. Soc.*, **85**, 668 (1963).
532. O. I. Koifman, N. G. Zemlyanaya, M. I. Alyanov, and V. R. Larionov, *Izv. vuzov. Khimiya i khim. tekhnol.*, **18**, 1644 (1975).

Index